Origins of Psychopathology

Origins of Psychopathology

The Phylogenetic and Cultural Basis of Mental Illness

HORACIO FÁBREGA JR., M.D.

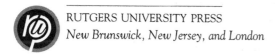

RUTGERS UNIVERSITY PRESS
New Brunswick, New Jersey, and London

Library of Congress Cataloging-in-Publication Data

Fabrega, Horacio.
    Origins of psychopathology: the phylogenetic and cultural basis of mental illness/ Horacio Fabrega Jr.
      p. cm.
    Includes bibliographical references and index.
    ISBN 0-8135-3023-7 (alk. paper)
    1. Mental illness—Etiology. 2. Genetic psychology. 3. Mental illness—Genetic aspects. 4. Mental illness—Etiology—Social aspects. I. Title

RC454.4 .F33 2002
616.89'071—dc21

                                        20001031783

British Cataloging-in-Publication information is available from the British Library.

Copyright © 2002 by Horacio Fábrega Jr.

All rights reserved
No part of this book may be reproduced or utilized in any form or by any means, electronic or mechanical, or by any information storage and retrieval system, without written permission from the publisher. Please contact Rutgers University Press, 100 Joyce Kilmer Avenue, Piscataway, NJ 08854-8099. The only exception to this prohibition is "fair use" as defined by U.S. copyright law.
Manufactured in the United States of America

En memoria de mi padre, Horacio Fábrega, con muchísimo respeto y cariño.

Contents

Preface ix
Acknowledgments xv

Part I Psychiatry and Evolutionary Biology

Chapter 1	Evolution and the Study of Psychopathology	3
Chapter 2	Evolutionary Theory Applied to Psychopathology and Psychiatry	26
Chapter 3	Clinical and Evolutionary Images of Psychopathology	40
Chapter 4	Accounting for the Universality of Psychopathology	76
Chapter 5	An Active Role for Psychopathology in Evolution	96
Chapter 6	On the Limits of an Evolutionary Conception of Psychopathology	112

Part II Psychopathology during Human Biological Evolution

Chapter 7	Searching for Psychopathology in Nonhuman Primates	143
Chapter 8	Responses to Psychopathology in Nonhuman Primates	189
Chapter 9	The Setting of Psychopathology during Evolution	213
Chapter 10	The Content of Psychopathology during Evolution	244
Chapter 11	The Impact of Meaning Systems on Psychopathology	273
Chapter 12	Dissociation, Psychopathology, and Evolution	294
Chapter 13	Psychopathology in Archaic Human Societies	313

Part III Recapitulation and Synthesis

Chapter 14   Phases of the Biological Evolution of Psychopathology   345

Afterword: Visualizing the Cultural Evolution of Psychopathology   368

References   375
Index   401

Preface

THE TITLE OF THIS BOOK may suggest a focus on the circumstances and events that take place during a person's lifetime that predispose to or even cause mental illness or emotional problems. Its subject is actually quite different. The human species, not individuals, are its focus. Millions of years of human evolution, and not just the lifetime of a person, are its temporal compass. Moreover, this book does not address why mental illness arises in individuals, but whether or not problems of social and psychological behavior analogous to those of contemporary psychiatric interest could have possibly existed during earlier phases of human evolution. It can safely be assumed that if a book of this length is written to address the question of the phylogenetic history of psychiatric disturbances, then my position on the question is in the affirmative.

The book has two complementary aims. First, it represents an effort to better understand the nature of the clinical material that psychiatrists deal with. Second, it seeks to accomplish this by drawing on an anthropological understanding of human evolution. This should produce benefits in two camps. For clinicians, the result should be a deeper, more comprehensive understanding of mental illness, rendering their work with patients more fruitful. For biologists, the result should be a better understanding of a class of behavioral problems usually all but neglected in evolutionary biology, creating opportunities for new academic emphases in research in relevant subfields.

From a more personal point of view, the subject matter of this book fills in the gaps and attempts to answer some of the questions that have bothered me as a practicing psychiatrist over the years. Questions about the origins and interpretation of psychiatric disorders, construed in a broad cultural and evolutionary frame of reference, are simply not a part of the knowledge base that psychiatrists are exposed to. Clinical psychiatry is presented to trainees as though the behavior problems that it deals with conform to clearly defined and easily recognized patterns. A behavior problem, and a psychiatric disorder, in particular, can be thought of as involving changes in overt actions and behaviors and changes in physiological functions (for example, energy, appetite, sleep, sexuality). In this sense, the problem can be said to have a tangible, objective existence and can be corroborated. It is this seeming externality that leads to the mistaken view that psychiatric disorders have a common identity across cultures and can be easily spotted. However, the alleged distinctiveness of these behavior problems soon raises a number of questions. A psychiatrist in training comes to learn that problems of behavior are not always brought to the clinic because they are not judged (either by the person showing them or by significant others) to be serious enough to require evaluation or treatment. Psychiatrists come to appreciate that behavior problems, including psychiatric disorders listed in the manual of diagnosis, differ in their manifestations and in their interpretation by laypeople. In other words, the variety of behavior problems, what they consist of, and how they are interpreted—how they affect the lives of persons exhibiting them and others—is much larger than what trainees are taught. Of those problems of behavior that are brought to the clinic, by far the largest proportion can be fitted into the descriptions of psychiatric disorders as outlined in the statistical manual of diagnosis. However, the price of fitting a person's personal behavior problems in the straightjacket of a diagnosis is that the subtlety and meaning of those problems is denied, and, hence, they elude our attempts at understanding them.

Psychiatric education and the actual practice of psychiatry are also burdened by a number of theoretical questions about the character of psychopathology. Among the most basic are why psychopathologies arise in the first place, where they come from, and how they become visible as social "entities" in a community. Of course, the behavioral sciences have standard responses to these questions, and the social sciences have others that are at times quite critical of the psychiatric profession. But answering the questions I have posed regarding the fundamental bases or origins—hence the significance—of psychopathology in the human species really requires taking hold of evolutionary theory and applying it to the domain of psychiatric practice. Here, in other words, one is required to incorporate knowledge not only of psychiatry but also, more important, knowledge about anthropology in all its comprehensive reach. This facet of psychopathology, which involves its origins viewed across the expanse of evolution, is totally neglected in psychiatric education, theory, and practice, and that deficiency is precisely what is addressed in this book.

In contemporary evolutionary biology, chimpanzees are regarded as the earliest persisting relatives of members of Homo sapiens. The last common ancestor of members of the human species and chimpanzees is assumed to have lived around 5 million years ago. Scientists have based these inferences on ethological analyses of social and psychological behavior, principles of evolutionary biology, and studies of molecular evolution. Hence, it is reasonable to assume that a tendency to show psychopathology, a universal possibility in Homo sapiens, arose some time during the line of evolution stretching from the last common ancestor to the emergence of modern humans. This is the orienting theme of this book and its intellectual terrain.

What is known about the normal and abnormal behavior of chimpanzees and the higher primates is the basic, foundational material that has to be reviewed and brought into the equation for explaining the origins of psychopathology in the line of human evolution. The anchoring point for an evolutionary study of origins is whether there is reason to believe that something like psychopathology is found in the higher apes, and, if so, how it is manifested. Following this point in evolution, the question of the origins of psychopathology requires us to consider the evolutionary trajectory of the hominids—that is, the changes that have transpired since the last common ancestor with respect to basic parameters of behavior such as social ecology, social organization, cognition, and culture. All of these issues can affect the character and meaning of behavior, and hence of psychopathology, in highly social groups like the higher apes, hominids, and especially humans. Explaining the origins of psychopathology entails clarifying what psychopathology consists of, how it can be understood from an evolutionary standpoint, and how it may have been configured and enacted in groups of hominids over the long process of evolution leading to Homo sapiens.

To deal with the question of the origins and significance of psychopathology in a satisfying way, I have reviewed and synthesized knowledge in paleoanthropology, archeology, social and behavioral ecology, human biology, evolutionary biology, comparative psychology, evolutionary psychology, cognitive science, cultural anthropology, and the behavioral sciences that bear on psychiatric problems. My purpose was to synthesize this material so as to relate phenomena pertaining to disturbances of social and psychological behavior glossed as psychopathology to the processes of human evolution.

The book is divided into three parts. The first six chapters address theoretical questions. Chapter 1 deals with general intellectual issues that are raised by simply contemplating the possibility that psychopathology has an evolutionary history. In other words, the problem is raised of crossing the animal-human divide by attributing something so human as psychopathology to animals. In addition, definitions of psychopathology that facilitate an evolutionary perspective with respect to its origins are reviewed. Chapter 2 looks at how researchers interested in human psychology, psychiatry in particular, have used evolutionary theory. These researchers generally refrain from entering into the behavioral space of hominids. This chapter explains the position adopted in this book, one that contemplates discussing possible characteristics of psychopathology during evolution. Chapter 3 discusses some of the differences between biologists' and clinicians' conceptions of psychopathology and suggests a way of reconciling these differences by presenting theoretical examples of psychopathology during different phases of evolution. Chapter 4 reviews genetic aspects of psychopathology that raise the question of its possible phylogenesis. Here examples are given of the kinds of behaviors that may have been selected for or against during human evolution. Chapter 5 extends the theme of genetic implications by reviewing the writings of biologists who ascribe an active role to behavior in the evolutionary process. This is done to raise the question of whether manifestations of psychopathology may have affected the process of human evolution, and if so, how. Finally, chapter 6 pushes to the limit an evolutionary conception of psychopathology. I extrapolate from the writings of comparative psychologists and anthropologists generalizations about the adaptive character of behaviors that constitute some forms of psychopathology, examining how they succeed or fail to provide a general evolutionary account of psychopathology.

The second section deals with descriptive material. Chapter 7 reviews writings in ethology and primatology about abnormalities in social behavior among natural, feral communities of nonhuman primates, while chapter 8 summarizes researchers' efforts to develop animal models of psychiatric disorders. Chapter 9 reviews the work of social and behavioral ecologists as it pertains to phases of human evolution. My purpose is to portray the hypothetical social terrain, or cultural and behavioral ecology, during phases of evolution that would have affected how psychopathologies were manifested and interpreted. Chapter 10 reviews the writings of cognitive psychologists, cognitive archeologists, and linguists who address the question of behavior during human evolution to provide realistic pictures of what manifestations of psychopathology might have looked like in prehistoric times. Chapter 11 concentrates on the question of culture, considered as a system of social systems and their meanings, and discusses the kinds of influences that culture in the symbolic sense might have played in the configuration and enactment of psychopathology. Chapter 12 gives special attention to altered states of consciousness, also termed *states of dissociation*, and discusses what these states consist of, their evolutionary history, and how they could have influenced the origins of psychopathology. Chapter 13 is an extended inquiry into the question of how culture, language, and conceptual structures and mechanisms, assumed to be three separate cognitive traits produced by the evolutionary process, may have affected characteristics of psychopathology during evolution. These three properties of cognition are obviously interrelated and mutually influential. Nevertheless, for heuristic purposes, I find it instructive to analyze their potentially different influences on behavior and psychopathology by concentrating on hypothetical groups of Archaic humans.

The final section consists of recapitulation and synthesis. Chapter 14 reviews and consolidates the information presented in earlier chapters by outlining phases of the biological evolution of psychopathology. Its two centers of interest are that of the individual, or how psychopathology was individually manifested, and that of the group or audience witnessing the psychopathology. An afterword marks the boundary or area of focus of this book, the biological evolution of psychopathology, from developments that follow the emergence of modern humans, namely, the social and cultural evolution of human societies, behavior, and psychopathology.

Acknowledgments

I WOULD LIKE TO express my appreciation to persons who kept me motivated and interested in this project and who provided me with help in its formulation and finalization as a document. Katerina Haka-Ikse unselfishly and without any flagging of energy provided me with her valuable time by reading and making suggestions about the very first drafts. Kent Bailey also read an early draft and made helpful suggestions. Conversations with Frans B. M. de Waal and Dario Maestripieri were helpful in formulating material in chapters 7 and 8. Merlin Donald provided expert advice and information in the early stage, and Gilbert Gottlieb furnished me with articles and then reviewed my formulations of his work. Only I am responsible for inferences that I have drawn from their ideas. Stephen Gaulin and John Marx, colleagues at the University of Pittsburgh, have consistently answered my questions and provided me with support and encouragement. Howard Boyer was particularly supportive and encouraging during early phases of my writing. My friend John Fong provided support and encouragement. Helen Hsu guided the project expertly, and her professionalism was a steadying influence. Monica Phillips' superb final editing is much appreciated. Alfonso Troisi read the entire manuscript and provided cogent and valuable suggestions during its final stages that sharpened the focus. Jane Flanders put up with the frustration of having to edit my heavy-handed writing and did so always with goodwill and patience. Kendall Stanley patiently entered and helped clean up awkward entries in the very many versions of the manuscript and helped me formalize the bibliography. David Kilinsky patiently and diligently provided his expert help in securing material from libraries I relied on. My daughter, Michele Marie Fábrega-Brown, read and offered suggestions regarding the summary statements at the end of the chapters. As always, my wife, Joan Rome Sporkin, created a structure of life for me that has allowed me to indulge and pursue my intellectual passions and obsessions.

Part I

Psychiatry and Evolutionary Biology

Chapter 1
Evolution and the Study of Psychopathology

I WILL USE THE TERM *psychopathology* to describe distressing conditions of an individual that consist of negative changes in social and psychological behavior and associated somatic manifestations. The latter involve, for example, bodily discomfort, negatively altered physiological functioning, and restrictions in achieving biological goals in a distinctive social and cultural setting. Psychopathology, then, is an ensemble of complex changes in an individual's general behavior, biology, and function that by definition causes distress and limits, constrains, and/or hinders adaptiveness or adjustment in the current social environment of the individual.

In clinical psychiatry, psychopathology is classified in terms of disorders, and the current diagnostic system in psychiatry identifies a large number of these. These include, for example, disorders featured by anxiety, depression, problems involving cognition or thinking, symptoms of body function that are not due to recognized general medical diseases, and disturbances in social functioning due to personality. A complex taxonomy and nosology, strategies of classification of disorders, and the supporting system of concepts and theories for achieving this, underlie the existing approach to psychopathology in clinical psychiatry (Jablensky 1988).

I will rely on the general term *psychopathology* in discussing aspects of its origins, for reasons that will become clear later. Distinctive preferences, orientations, and purposes underlie how psychopathology is classified. Modern psychiatry identifies a large number of disorders in its classification system. Examining each of these across the evolutionary terrain is simply too cumbersome and would also be inappropriate since a number of them are keyed to contemporary situations. In theory, one can examine psychopathology from the standpoint of how individuals showing it are hindered or disadvantaged in terms of achieving an agreed upon set of important goals and functions or with respect to salient emotional features. Psychiatrists influenced by evolutionary theory have actually pursued this. While it yields a classification of sorts, its categories are not clean and separate since biosocial goals and functions as well as emotions overlap. Furthermore, the problem of "co-morbidity" is one that plagues virtually all systems of classification in psychiatry.

Any one system of classification of psychopathology has advantages and disadvantages. I prefer to remain open and general about the best way in which this can be accomplished. However, in all instances it should be understood that my general psychopathology encompasses many different types of conditions featured by different types of manifestations, causes, and consequences. In order to emphasize the differences in the kinds of conditions that can fall under my general concept, I use the expression *varieties of psychopathology*. It should be understood that this expression refers not only to different types of conditions but also to different causes, manifestations, functional implications, and social, cultural consequences that the conditions can have in the evolutionary landscape. In discussions of origins of psychopathology I will attempt to provide examples whenever possible.

General Background

Conceptions of psychopathology and its treatment have undergone a major transformation in contemporary society. The behavioral problems that psychiatry addresses are increasingly defined and interpreted as caused by neurobiological disturbances. This reflects the prominent growth and importance of the neural sciences in the study of behavior. A brain-centered view of human behavior problems is complemented by an empirical approach to diagnosis and the formulation of treatment. Drugs now play a dominant role. Psychiatric theory and practice appear now more than ever to be securely anchored in the biomedical sciences and other medical disciplines.

This growing neurobiological perspective is paralleled by an increased awareness of the cultural biases inherent in psychiatric theory and practice (Mezzich et al. 1996). Many of the criteria and diagnostic indicators of the major psychiatric disorders, as well as their manifestations, have been shown to reflect cultural and social biases linked to the patient's ethnicity and social class. Claims about the descriptive specificity and theoretical soundness of psychiatric theory and practice have been challenged in a critical way by advances made by the social sciences, especially anthropology (Fábrega 1995; Kleinman 1988; Mezzich et al. 1996). One observes, then, that while a form of neurobiological reductionism dominates official psychiatry and seeks to anchor it in general medicine, there exists a challenging, critical psychiatry that on empirical and theoretical grounds questions the criteria for and general applicability of psychiatric knowledge. This critical view of psychiatry implies that psychopathology should be examined in relation to a range of social behavioral problems and the culture of the community.

The debate is over the nature and meaning of psychopathology. Central to the debate are such issues as how to deal with its manifestations, the roots and causes of the social problems seemingly integral to psychopathology, and what disciplines, institutions, and policy rationales should address it. (See, for example, Castel 1988; Fábrega 1974, 1975, 1990a, 1990b, 1990c, 1991, 1992a, 1992b, 1997; Goldstein 1987; Porter 1987; Scull 1993.) In such a climate it is appropriate to examine the origins and evolution of the kinds of problems that are addressed by psychiatric practice and its institutional rationale and purpose.

The topic of the origins of psychopathology (or psychiatric disorders) can be broken down into a number of related questions. What about the human (or hominoid) condition accounts for the presence of behavioral problems? In other words, what essential biological properties of human beings might explain the genesis and persistence of human behavioral problems? This raises questions about the possible genetic bases of psychopathology and about the basic character of adaptation or maladaptation that may have a bearing on psychopathology. Both the mechanisms underlying the unfolding of a behavior problem, which involve processes taking place within the individual (intrinsic causes or factors), and processes taking place in the ecological and social setting (for example, that operate as causes and that influence responses), also need to be considered, but in a more broad, general way.

Seeking the origins of psychopathology, then, naturally directs one back into the evolutionary past. Psychopathology has been a persistent, recurrent feature of human populations throughout recorded history and in all parts of the world. Could it have also played a role in human evolution? Given an evolutionary orientation, one is then compelled to inquire about the possible causes and manifestations of behavioral problems in groups living in very different ecologies and social settings; more specifically, one must consider the topic of what form and function psychopathology may have had during earlier phases of human evolution. To plausibly visualize and formulate such a scenario, one obviously needs to adopt the view of an outsider and have a useful set of criteria or guidelines by which to proceed. Furthermore, psychopathology does not just appear in a social group without warning. It has connections to social ecology and ongoing activities and it has social consequences. Thus, besides considering psychopathology from the outside and in the abstract, that is, from an etic standpoint, one can look at its manifestations and implications from inside the group's mode of operation or culture, that is, from an emic standpoint. In subsequent chapters, etic and emic aspects of psychopathology are both considered in relation to a group's social ecology and members' culture and cognition. Ultimately, the question of the origins of human psychopathology forces one to examine conditions of thinking, feeling, and living, and social ecological perturbations that have taken place during the course of human biological and cultural evolution.

Examining the roots of behavioral problems seen to be psychopathological entails concentrating on basic topics in the biological and social sciences. It requires knowledge from fields such as paleoanthropology, paleoecology, evolutionary biology, ethology, evolutionary psychology, and archeology—all of which concentrate on factors that shaped behavior and adaptation in the hominid line. Basically, one seeks to clarify the phylogenetic roots and bases of social behavior and its breakdowns. In the same way, one must consider how varieties of psychopathology affect the realization of key biological goals and functions such as survival, reproduction, and relationships in a group context. In short, how members of higher primate groups manifest behavioral disturbances and the effects these have on their adaptiveness, and, alternatively, how they respond to disturbances when they are manifested in group mates, provide clues about how evolving early hominid groups might have configured and dealt with psychopathology.

Human biological evolution was associated with distinctive changes in the social environment, cognition, language, and culture—all of which influenced the content, organization, and general interpretive framework of behavior (Deacon 1997; Donald 1991; Foley 1987, 1995a, 1995b). It was in the context of transformations affecting the very grounds of social behavior, and in the light of interferences in biological functions, that psychopathologies emerged during human biological evolution. Their form, content, and consequences were all modified during human biological evolution as a result of changes in hominid neuropsychological organization, biological needs, and social ecology.

Varieties of psychopathology may be likened to performances being enacted in the evolutionary landscape. In effect, both the stage and the play being acted out on it were subject to change. From what scholars have learned about the evolutionary biology of behavior and its aberrations, one may begin to imagine what kinds of behavioral pathologies might have appeared among hominids and Archaic human groups and how they may have been understood and dealt with—in other words, how the homologues and not just the analogues of psychiatric problems, and the social responses that they occasioned, might have been manifested during prehistory.

The Timeliness of an Evolutionary Perspective on Psychopathology

As ethology involving field studies of higher primates gathered momentum in the 1960s, theorists in the biological sciences and psychiatry were quick to point to some of the common themes shared by these and related disciplines (Bateson and Hinde 1976; Bowlby 1969, 1973, 1976; Goodall 1971, 1986; Hamburg 1975; Hinde 1974, 1987; Thorpe 1956; Tinbergen 1951; White 1974). This focus enhanced and deepened the quest for compelling models of human psychiatric disorders. The overlap and correspondence of interests, it is fair to say, was visualized purely in terms of biological and especially neurobiological underpinnings and implications. Few considered the possibility that by focusing on what rendered behaviors "pathological" in natural settings of primates (and, by extension, hominids and Archaic humans) one might achieve a deeper understanding of the social and biological sciences and psychiatry.

The discoveries of ethologists about behavior and its pathology in light of evolutionary biology may appropriately be used to paint the larger picture of human psychopathology—namely, psychiatric disturbances. Scholarly work paralleling recent developments in evolutionary biology and the neural aspects of behavioral pathologies brings into focus questions about the evolution and historicity of psychopathology in human populations (Cloninger 1986; Gambill 1981; Hamburg 1975; Lane and Luchins 1988; Marks and Nesse 1994; McGuire et al. 1981, 1992; Nesse 1984, 1990; Smith 1993; N. F. White 1974). A focus on the biology of psychiatric disturbances, if not an actual interest in evolutionary biology, is also manifest in cultural anthropology, psychology, sociology, economics, and the social history of medicine (Barkow et al. 1992; Berrios and Porter 1995; Brown 1991; Donald 1991; Dunbar 1988; Foley 1987, 1995a, 1995b; Gamble 1993; Gibson and Ingold 1993; Hinde 1972; Hodgson 1993; Mithen 1996a, 1996b; Porter 1987a, 1987b; Quiatt and Reynolds 1993; Renfrew and Zubrow 1994; Scull 1989, 1993). Social scientists are beginning to take the insights of Darwin and modern neoevolutionary theorists more seriously (Fisher 1958; Hamilton 1964; Mayr 1963, 1982, 1991; Williams 1966). This is reflected in the general interest given to the evolution of human behavior per se and the historical foundations of social institutions that control behavior (Durham 1991).

Anyone who approaches the study of the origins of psychopathology should begin with an open mind about what psychopathology consists of and carefully formulate its essential features. However, this is not possible for those with a preexisting perspective on what psychopathology is. Developing an essentialist conception is not easy, and philosophical dilemmas intrude. For example: Is psychopathology another phenomenon, like animals, plants, caves, or riverbeds, that is simply detected, witnessed, and catalogued? Is it not a human construction, its identity and nature deriving from functions arbitrarily assigned to it for specific reasons and/or with particular collective intentions? (Searle 1995).

The contemporary perspective in neuropsychiatry addresses psychopathology as though its essential characteristics constituted "real facts" or "real entities" in nature. Specifically, some psychiatrists reason that psychopathology is realized in at least some of the types of disorders defined by modern neuropsychiatry, that it has not only a genetic base or specificity but also a distinctive neurobiology and syndromic character consisting of well-defined manifestations that have been uncovered by the sciences undergirding clinical psychiatry. Implied in this position is that the disturbances themselves, the actual disorders described in the contemporary diagnostic manual of psychiatry, have a phylogenetic and historical basis.

Studying Origins of Psychopathology

Some examples of how contemporary biological psychiatrists handle the question of the origins of psychopathology will illustrate how these scholars depart from the perspective I have adopted in this book. Timothy Crow (1995, 1997) presents a formulation about the evolutionary roots of human psychosis, especially schizophrenia (clearly a fundamental form of psychopathology). Russell Gardner (1982) explores the evolutionary roots of mania and mood bipolarity more generally. Isaac Marks (1987) concentrates on anxiety disorders; Paul Gilbert (1992a) concentrates on depression, using his prior effort with respect to human nature and suffering (1992b); and Anthony Stevens and John Price (1996) present a comprehensive theoretical exposition about almost all psychiatric disorders, anchoring them in a broad-ranging theory about human evolution as well as knowledge pertaining to the behavioral sciences. The work of these and other theoreticians (such as Wenegrat 1984, 1990; D. R. Wilson 1992, 1993) regarding the nature of the evolutionary roots of psychopathology cannot be gone into further here.

To make my point about how other evolutionary approaches differ from mine, I will concentrate on Crow's ideas. Crow relies on cutting-edge knowledge about human handedness, cognition, genetics, psychosis, and schizophrenia and brings it to bear on the topic of why schizophrenia/psychosis is found in the human species, using an evolutionary perspective. To answer the question of the origins of schizophrenia, Crow begins his analysis knowing, or presuming to know, not only what schizophrenia *is* but also (according to his theory) finding or discovering it in nature arising from a series of impersonal, objective, highly abstract developments involving the punctate, steplike development of language, human speciation, the evolution of cerebral laterality, the process of sexual selection, and events involving molecular changes on the X and Y chromosomes.

The standard approach to origins of psychopathology is to start one's inquiry by looking for the functions of a particular and highly specific behavioral phenotype, such as salient features of schizophrenia, mania, depression, or anxiety, that is assumed to be a constant or universal in human societies, that is genetically based in some way, and that is more prevalent than can be accounted for by random mutation; in short, the genotype responsible for the variety of psychopathology appears to be serving some other positive function in human populations than producing psychosis. The task of the researcher is to conjure up a scenario that might have led to selection of the genotype/phenotype during earlier phases of evolution. This approach relies on the global properties of that phenotype—that is, salient behavior features or symptoms—yet it reflects a selective emphasis on distinctive types of psychopathology as outlined in contemporary systems of diagnostic classification such as schizophrenia, manic depressive disorder, depression.

In the standard approach, researchers are not interested in how their particular variety of psychopathology was configured and enacted in a distinctive behavioral ecological setting. Nor do they seek to trace how it evolved over time—for example, how landmarks taking place during evolution (for example, the emergence of mental attribution, linguistic competence, and culture) affected not only the prevalence or manifestations of the particular variety in question (such as "ancestral" forms of schizophrenia, mania) but also ways of handling it. If researchers show any interest in these topics, manifestations of the psychiatric disorder as observed in contemporary, usually Anglo-European, populations serve as models for analysis and interpretation of roughly analogous, but very general, postulates about behavior in prehuman populations.

As indicated earlier, the strategy adopted in this book is less concrete about what psychopathology *is*. I agree with van Praag et al. (1991) that it is best to go beyond diagnostic entities per se and to adopt a "functional approach to psychopathology," roughly, what it produces and how it can be modified via agents that affect brain function: "The basic units of classification in psychopathology are not syndromes or nosological entities, but psychological dysfunctions, such as disturbances in perception, cognition, memory, or information processing, among many others. *They* are the elementary constituents of which psychiatric syndromes are made" (302–303). I would complement an emphasis on neuropsychiatric parameters with the one that embraces biological functions and goals; in other words, how varieties of psychopathology undermine and are manifest in disturbances that hinder biological functions and the achievement of basic biological goals such as survival, reproduction, and social relations in group settings.

In short, psychopathology consists of fundamental properties of thinking, feeling, and bodily experience and how they are expressed in specifically social, psychological, behavior and physiological and general biological functioning. The question of how or at what level psychopathology should be defined is highly relevant but has no easy answer, creating a genuine theoretical, that is, epistemological, dilemma.

Second, in addition to the social constructionist argument about the cultural embeddedness of contemporary definitions, I consider the arguments of critical theorists and evolutionary biologists who assert that political, economic, and social conditions have greatly influenced both human adaptation (hence, maladaptation) and social and psychological behavior; further, I consider why selected forms of behavior are labeled as psychopathology in the first place (Fábrega 1994a, 1994b). Specific psychiatric disturbances reflect distinctive historical and cultural conditions and, moreover, have evolved and are still undergoing evolution. They highlight conditions that are obviously very different from those prevailing during earlier phases of human evolution.

My perspective differs somewhat from that adopted by Crow, Gardner, and others. I am interested in questions that are not part of the standard evolutionary account of psychopathology; specifically, I am interested in how it was played out during different phases of evolution and how it changed in relation to evolutionary landmarks. However, in seeking to formulate the character of psychopathology, I rely on material of primatologists, evolutionary biologists and psychologists, and comparative psychologists whose observations about behavior are necessarily beclouded by constructs and suppositions implicit in their culture. Furthermore, as a practicing clinical psychiatrist, I am obviously contaminated intellectually by the conventions inherent in the psychiatric enterprise. In many respects, this book constitutes my attempt to cut through the blinders of contemporary biological psychiatry to provide a better, more comprehensive approach by inquiring into origins.

I have two main reasons for using a general analytic category like psychopathology, instead of relying on specific types of disorders. First, it is easier to formulate the problem of origins by referring to one general category and handling it in a biological context rather than taking into account different types of disorders as currently conceptualized in psychiatry. Second, most properties and characteristics of psychiatric disorders, including such things as manifestations and even duration, are not fixed but rather shaped by physical environmental, social, and cultural conditions prevailing in contemporary societies. This is no less true today than it was during human biological evolution and earlier phases of cultural evolution. Since these conditions shape and contextualize how psychopathology manifests and then gets interpreted even today, namely, as the psychiatric disorders of a diagnostic system, it is only appropriate that a similar point of view be adopted with respect to other points in evolutionary time. Rather than assuming that I am dealing with ontologically specific "natural" kinds, I concentrate on an abstract formulation of social and psychological behavior problems, that is, psychopathology, and consider how it has manifested, changed, and evolved over vast spans of time.

The Need for a Comprehensive Definition of Psychopathology

Psychopathology brings into focus questions about an organism's well-being and fitness, notably its suffering. The relationship between psychopathology and subjective states will be described in later chapters. For present purposes, I will merely draw attention to the link between general biology and evolution, on the one hand, and animal well-being, fitness, and suffering, on the other.

Questions raised by the study of human psychopathology are related to questions involved in the study of animal stress and welfare as examined by veterinary specialists, general biologists, and zoologists. And the way this problem is formulated takes into account evolutionary concepts. Thus, in discussing motivation, fitness, and animal welfare, M. S. Dawkins (1980, 1990, 1993) explicitly links the key concept of animal suffering with natural selection: "Suffering occurs when unpleasant feelings are acute or continue for a long time *because the animal is unable to carry out the actions that would normally reduce risks to life and reproduction in those circumstances*" (M. S. Dawkins 1990, 2; emphasis in original).

Those who study animal welfare are understandably concerned with problems only peripherally related to human behavior pathologies. However, the logical tie between psychopathology and animal behavior and subjectivity, including suffering, is worth emphasizing to make apparent the relevant genealogy of the problem of the origins of psychopathology. Dawkins writes: "In assessing whether an animal is suffering, therefore, we must take into account not just the canonical costs of preserving its fitness but the 'perceived costs'—the costs as perceived by the animal; . . . [these] may or may not involve unpleasant subjective feelings. When they do have a subjective component, we call that state 'suffering.' We must accordingly consider both the physical and psychological health of animals in assessing animal suffering and well-being" (1990, 3).

A basic connection exists, then, between the topics of psychopathology, animal welfare and suffering, and human evolution. This epistemological amalgam necessarily includes the concept of adaptation (Brandon 1990, 1996; Dunbar 1988, 1996; Endler 1986; Gould and Lewontin 1979; Sober 1984; Williams 1966). In animal studies, the concept of adaptation has several meanings, and I shall later elaborate on its evolutionary significance with respect to psychopathology (see especially Harrison 1993). As used by veterinary specialists, *adaptation* refers to the waning across time of a physiological response to a particular situation or strain that an animal faces (Broom and Johnson 1993; M. S. Dawkins 1980, 1990, 1993). Evolutionary biologists use the concept to refer to any attribute of an animal, including its behavior, that better enables it to survive and reproduce compared to other members of the same species. This is the meaning that I will be using in this book. While adaptation refers to evolved or naturally designed traits that promoted survival and reproduction in the long run, *adaptiveness* refers to how the trait promotes adjustment in the short run and in the current environment. *Stress* refers to the effect of an environmental situation that militates against the fitness of the animal; that is, its ability to survive and pass on its genes. *Welfare* refers to the general state of the animal as it attempts to cope with its environment. Welfare is handled as a continuous, rather than a dichotomous, measure. "An animal's welfare is poor when it is having difficulty in coping or is failing to cope. Failure to cope implies fitness reduction and hence stress. However, there are many circumstances in which welfare is poor without there being any effect on biological fitness" (Broom and Johnson 1993, 75). Pain, considered as an aversive stimulation, and suffering, considered as an unpleasant subjective feeling, are important features of compromised welfare. Given the above definitions, pain and suffering can, but need not, imply stress or a change in the fitness of an animal.

The behavior of an animal enters into these considerations in two ways. Abnormal behavior, or behavior "which differs in pattern, frequency or context from that which is shown by most members of the species in conditions which allow a full range of behaviour" (Broom and Johnson 1993, 175), is considered a measure of poor welfare. Two others are the extent of behavioral aversion and to what extent normal behavior is suppressed. Animals can thus be characterized in terms of normal expectable behavior; and abnormalities and deviations of behavior have been linked by veterinary scientists to fundamental questions of ethology and evolutionary biology (Novak and Suomi 1988).

Psychopathology, in general terms, refers to abnormalities of social and psychological behavior. Examining the behavior of animals by extrapolating from considerations that pertain to human societies runs the obvious risk of misguided anthropomorphism (Kennedy 1992). However, I will consider primarily animals (monkeys and especially apes) in whom the question of awareness, intention, and social cognition is less problematic than in other species (but see Griffin 1992). I will concentrate, in other words, on behavior, mentation, and emotionality, as conjured up from what is known of the common ancestor of the pre-chimpanzee, pre-hominid clan, referred to as the last common ancestor (LCA), the hominids, archaic humans, and anatomically modern humans. In this way I hope to avoid misdirected anthropomorphism and metaphorical allusions to the social behavior of animals.

Ideally, one should employ a concept of psychopathology that encompasses the concerns of those interested in the scientific study of animal welfare, animal coping and stress, and animal fitness. One should also include the concerns of social scientists and psychiatrists interested in clinical questions of mental illness, as well as related questions pertaining to social deviance and stigma. Finally, one should address the concerns of neurobiologists and neuropharmacologists interested in general evolutionary aspects of psychiatry and animal models of psychiatric disorders. Ever since McKinney and Bunney (1969) put forward their criteria for evaluating animal models of human depression (similarity of inducing conditions, similarity of behavioral symptoms, common underlying neurobiology, and reversal of symptoms by clinically effective techniques), many have accepted the idea that psychopathology could have a genealogy or phylogeny (Geyer and Braff 1987; Harris 1989; Healy 1987; Kalin and Takahashi 1991; Iversen 1987; Öhman 1986, 1987). An evolutionary perspective on psychiatric disturbances has recently received more explicit recognition (Crow 1991a, 1991b, 1993a, 1993b, 1995, 1997; Marks 1987; McGuire et al. 1992; Nesse 1990; Nesse and Williams 1994; Stevens and Price 1996; Wenegrat 1984, 1990). Because this approach crosses many epistemological territories and boundaries, as well as questioning speciesism and ethnocentrism, finding a rigorous and comprehensive enough definition to accomplish all of this is difficult. Nevertheless, most of the symptoms of psychiatric disorders, and especially their social consequences, reflect suffering, distress, impaired ability to cope, stress in the transactions of living, and a potential impairment of fitness. Many of these disorders actually lead to increased mortality and can lower the fitness of individuals as postulated in general biology. Hence a concept as anthropomorphic and as species-bound and culture-bound as psychopathology, because of its epistemological roots in the scientific study of animal welfare, suffering, and fitness, as well as general evolutionary biology, is potentially useful despite its limitations.

Intellectual Quandaries

An interpretation of human psychopathology in the light of biological evolution reveals the value-laden, anthropomorphic, and ultimately ethnocentric character of the concept of pathology. Indeed, the question of how to understand a psychopathology in an evolutionary context is similar to the question of how to differentiate in a cross-cultural context normality from abnormality, a longstanding theme in cultural psychiatry that was most clearly articulated decades ago by George Devereux (1956, 1980).

In psychiatry, distinguishing between normality and abnormality highlights the crucial role played by cultural standards in the interpretation of behavior. What one cultural group views as normal or abnormal need not coincide with that of another; hence one reaches an intellectual impasse if one attempts to understand psychiatric problems or psychopathology in a panhuman or universal way (Fábrega 1977, 1987, 1994a, 1994b; but see Murphy 1976; Simons and Hughes 1985).

Instead of general categories such as normality versus abnormality, psychiatric disorder, or psychopathology, one could focus on human behavioral breakdowns (Fábrega 1993a). Human behavioral breakdowns refer to the ubiquitous, recurring malfunctions and dilapidations in the behavior of individuals that members of all human populations are likely to encounter. It refers to concrete instances of behavioral breakdowns that are undoubtedly cultural universals. In other words, members of all human groups or societies are exposed to behavioral breakdowns and develop a vocabulary and theory to explain them.

The concept of behavioral breakdown is more suitable for cross-cultural analyses in that it takes into account the variability of ideas and standards regarding the interpretation of human behavior. Because members of any particular social group or society share an interpretation of reality, their social behavior is judged in relation to this shared, culturally specific perspective. The baseline of reality and what is socially appropriate, or normal, behavior, against which an individual's actions are seen as a breakdown, is culturally variable and reflects beliefs and perceptions as well as ideas about social order, organization, and functionality, about marginality, dependence, and social malfunction. That is, an idea of behavioral breakdown takes into account cultural differences in how groups might construe or determine that an individual's behavior has broken down. The idea is similar to that of abnormal behavior as employed by general biologists and veterinary specialists (Broom and Johnson 1993; McFarland 1989, 1993). In discussing behavioral disturbances in relation to welfare, coping, stress, and fitness, veterinary specialists and ethologists invariably take into account deviations from their understandings of the normal behavior of members of the species.

Even the idea of behavioral breakdown, then, is still obviously burdened by the bias of the observer. It is the observer who stipulates that cognitive and intellectual competence or functioning as reflected in behavior constitutes a universal measure of the socially and culturally normative view of human behavior. In the same vein, it is veterinary specialists who draw their standards for abnormality. In a sense, relying on an idea of behavioral breakdown enables the analyst to sidestep the difficulty of specifying what exactly constitutes pathology, or abnormality compared to normality, in any particular group by relying on general criteria before the fact. This is so because the analyst uses culturally specific standards (or species-specific ones, in the case of animals). However, the analytic problem, especially in human studies, is more abstract, for it is transferred to the cognitive, intellectual, social plane or level of functioning and analysis. Such abstraction might be more suitable for the interpretation of psychopathology but reflects nevertheless an ethnocentric Western bias. This bias consists in stipulating that a particular cognitive and intellectual concept of personhood and social functioning constitutes the appropriate, universal basis for determining social impairment or behavioral breakdown.

In an evolutionary context, to attempt to understand psychopathology only underscores the same human bias of the observer regarding what is normal, healthy, or functional. The bias is again entangled in ideas pertaining to the centrality of culture (or the symbolic) for determining what constitutes normal, abnormal, or pathological behavior. But the bias is reflected in a somewhat different form. An evolutionary interpretation of psychopathology requires a frame of reference rooted in the phylogeny of human behavior. Such a frame of reference should allow for comparisons about behavioral normality, abnormality, and/or pathology across the divide that separates pongids, hominids, and Homo sapiens.

An evolutionary perspective on psychopathology also brings into question and challenges strict reliance on a concept of normality construed in purely statistical terms. Any trait needs to be viewed in terms of its functionality with respect to specific biological goals and in relation to distinctive social ecologies. Moreover, evolution implies trade-offs among different traits in relation to different functions and ecologies; for example, one of a set of biologically prepared behavioral routines or strategies may be adopted as a consequence of ecological clues about prevailing conditions. Taken together, these considerations imply that psychopathology analogized now to a biological trait of an individual creates tensions with the more static conception of psychopathology as statistical abnormality. In short, in an evolutionary context, the clinical psychopathology, namely, that which is not normal or healthy and functional in the short run, needs to be conceptualized in a dialectical, dynamic, and changing way.

Even presuming one could derive a suitable dynamic conceptualization of psychopathology, problems involving the continuity between human and animal behavior come to the fore. On the pongid and early hominid side, the idea of psychopathology has little basis, for the standards of the group regarding what is normal, healthy, or dysfunctional versus pathological obviously cannot be invoked because such standards may be unarticulated. They cannot even be easily inferred. Moreover, the concept of psychopathology is inextricably bound up with notions involving an individual's capacity to attach, bond, explore, feed, play, forage, compete with conspecifics, acquire status and rank in social hierarchies, court, mate, reproduce, nourish offspring, and ultimately survive. In other words, what an observer may term an archaic form of psychopathology is spread out in the ways members of a species or subspecies adapt to their social situation and physical habitat. This problem of the connectedness of behavior to social situation and to the arbitrary, value-laden, anthropomorphic act of labeling a nonhuman as showing signs of psychopathology cannot be avoided. (It is referred to as the problem of speciesism in M. S. Dawkins 1980, 1990, 1993.) Thus, an organism showing a form of behavior that an observer may call psychopathological may not be dealt with distinctively by group mates. On the other hand, the individual may be neglected, shunned, expelled, or perhaps even attacked and slain. In some instances, psychopathological behavior in an organism would likely cause its failure to thrive as a functioning member of the group.

In contemporary biological psychiatry, psychopathology is often equated with a brain-related physical-chemical "lesion" that produces abnormal mental processes and behavior. This has limitations from an evolutionary standpoint. For example, how could one determine that a putative psychiatric disorder exists in an organism? One must first diagnose the organism's failure to adapt or thrive in its group and habitat. However, in the absence of overt lesions or without a laboratory examination, an observer could not rule out general medical diseases and pathologies. Much more important is the fact that physical-chemical processes of an organism's brain are but mediators of its general ability to reach solutions to recurring biological problems and the latter are what is important. This emphasizes the importance of general criteria about behavior and function in relation to ecology. For this, special criteria need to be developed (see Erwin and Deni 1979 and Goosen 1981 for examples of animals in captivity). Only with (a) valid criteria pertaining to behavior and function in the species or subspecies and (b) knowledge about the environmental conditions could one infer that a particular variety of psychopathology is to blame for an organism's failure to thrive. Many of the relevant concepts introduced here, for example, pathology, health, disease, normal behavior, function, and goals or end states, are steeped in ethnocentric and anthropomorphic biases underscoring the value-laden character of the study of human psychopathology. These concepts would need to be redefined so that they apply to animals and then formulated in operational terms for purposes of reaching diagnoses of nonhuman varieties of psychopathology.

To understand psychopathology among hominids requires consideration of special distinctions that organisms make (or appear to make) about the behavior of group mates encompassing but not limited to the strict requirements of brute survival and competition. Because organisms are locked within a phylogenetically and ecologically rooted mode of adaptation reflected in spheres of behavior such as attachment, foraging, and competition, sheer survival and reproduction would appear to constitute basic criteria for health, normality, or success.

All organisms could probably be judged by their ability to forage, avoid predators, and gain social rank or power. In an evolutionary sense, "survival of the fittest" implies that organisms carry heritable material by which one might attempt to establish comparative functionality or adaptation. Consequently, notions about psychopathology would have to be reduced entirely to this type of currency unless one could establish that the organisms themselves relied on special criteria about group mates' social behavior.

Even in a group of free-ranging monkeys, individuals responding to alarm calls might be shown to reflect judgments about the efficacy, accuracy, or trustworthiness of another's vocalizations and behavior, for example, about the proximity of a predator (Cheney and Seyfarth 1990). A monkey that responds to a group mate's vocal communication in a special way, appearing to weigh it less heavily than another's, perhaps simply ignoring the call, could be said to be operating on some mental attribution about the social psychological competence of the group mate in question. This could conceivably be an elementary form of biased disqualification of the group mate. It could of course reflect a distinction based on rivalry, competition, and manipulation or deception, but this is a separate issue.

Even if one accepts the possibility of biased disqualification, a rhesus monkey's proto social judgments about the behavioral competence, normality, or pathology regarding the behavior of conspecifics (as reflected in how it responds to alarm calls), or more refined ones such as chimpanzees are capable of, given their more evolved capacity for social attribution, would be of little value for clarifying the evolutionary bases or aspects of human behavior and particularly psychopathology. This is the case because ideally one would wish to unambiguously establish criteria for how organisms judge and relate to group mates outside the dictates of strict competition and survival. Before we can conclude that early hominids had a notion of abnormal behavior analogous to our own, we need to know what special dispensations, accommodations, and forms of toleration and indulgence were available by which to judge the behavior of a particular member of a group. If members of a pongid, hominid, or human group killed all of its behaviorally compromised members who could not hold their own or constituted a burden—that is, who showed abnormality or breakdowns—one could still say that they held a concept of aberrant behavior analogous to psychopathology but simply expressed it in terms of a harsh evolutionary currency. Thus, one could say that they had a different moral economy or theory whereby they dealt with the abnormal. The environmental pressures that a group is exposed to, such as scarcity of food, the threat of predators, and the like, unquestionably influence whether a group (pongid, prehominid, hominid, or fully human) tolerates the behaviorally infirm and how it deals with them.

Evolutionary analysis of psychopathology brings to the fore a number of dilemmas in the study of behavior. Essentially, I am crossing the divide between animals and man in matters generally thought of as quintessentially human—the mental and moral. That is, I am choosing to enter hominid evolutionary spaces. Scholarly dilemmas include the relation between nature and culture, genetic versus environmental factors, innate versus learned capabilities, and biological forces versus social construction. In contemporary evolutionary studies, these dichotomies are viewed as historically based fallacies, given that behavioral traits are best seen as facultative and not obligate—in other words, as products of interactions between genes and the environment. Furthermore, what makes up human psychopathology or mental illness is highly complex. The question is much debated by social and biological scientists who study human populations, and it is no less problematic when applied to phases of human evolution. I attempt to deal with these dilemmas by using an abstract, general term—*psychopathology*—instead of classifying particular diseases such as schizophrenia, mania, and depression. However, I give attention to specific evolutionary contexts involving social ecology and the mental traits of hominids such as cognition and culture through which incidents of psychopathology would have been played out.

Clinical, Biological, and Social Science Conceptions of Psychopathology

A psychiatrist diagnoses psychopathology following a mental status examination and a review of pertinent biographical and laboratory information. Evidence of personal distress and of disturbances or deviations in physiological functions, and psychological and social behavior are essential criteria. However, although rendering a diagnosis entails merely fitting evidence to an appropriate standard, it nevertheless involves a value judgment. For one, the person is said to reveal or to have something negative—namely, a disorder or a disease. Moreover, to diagnose a psychiatric condition is to qualify that the person's mode of bodily experience, thinking, feeling, or acting is somehow faulty, since a diagnosis is based on deviance in these spheres of behavior. It also implies that the individual is somehow not functioning or not fitting in well according to social, civic, and moral criteria. Here, it is all too easy to translate a diagnosis of psychopathology as meaning abnormal, unacceptable, unproductive, bad, or simply wrong. Furthermore, since diagnosing a psychiatric condition is a form of social labeling, it has significant cultural implications in the society at large. The individual who is assigned a diagnosis is designated as in some way mentally ill, which has enormous negative social implications.

When a strict biologist affirms that an organism or individual is displaying maladaptive behavior, he or she brings into play a panoramic view of the evolutionary process and invokes the currency of fitness. Notwithstanding the scientific basis for their view, their diagnosis also has value implications. Putting aside the caveats discussed earlier, the claim of maladaptation negatively qualifies the individual in its group and ecology. When maladaptation is handled as a trait that has long-term negative consequences, it refers to future generations linked to the individual through a genetic chain. In either case, the claim says something negative about the individual and its kin and offspring. However, a biological diagnosis of maladaptation is inherently different from the clinical and especially the psychiatric diagnosis. It is made about hypothetical individuals long ago extinct. Furthermore, a diagnosis of maladaptation on the basis of fitness criteria could be said to apply to a trait and not to the whole individual, or to future generations that may in fact never materialize. It could be interpreted merely as a prediction of possible future events. The claim could also be made not about individuals at all but about demes, or populations, or species. But are scores of individuals thus being spared the stigmatizing label of negative fitness simply because an alleged natural type or taxon is being qualified? Regardless of how the biologist's affirmation of negative fitness is proffered, then, it is still a negative qualification. In some respects, then, both clinician and evolutionist are engaged in a form of negative attribution.

It could be said that what is different and what shields the biologist's value judgment, which is in some way a moral judgment about negative fitness from criticism is that it is made on the basis of scientific knowledge, and science is after all supposed to be neutral and value-free. But so is the knowledge base of biomedical and behavioral science that undergirds medicine and psychiatry. What this discussion implies is not that biologists' judgments do not apply to persons, but that they carry no unequivocal cash value in the community where the diagnosed individual resides and carries out the business of living. As far as one can tell, labeling individuals (or populations) as unfit does not damage their identity as individuals. After all, the biologist's diagnostic judgments are being made about animals, and we all know that (until recently, at least) animals do not have any rights or feelings.

Social scientists might appear to be free of the burden of negative labeling and values. This is so because they often focus precisely on the function of social labels, especially negative ones, in the operation of institutions and society at large and, most important, with respect to the social identity of persons. In fact, social scientists have been most penetrating in pointing out the arrogation of social power through expert but nonetheless negative labeling that characterizes the activities not only of physicians and the general medical profession but especially psychiatrists and the profession of psychiatry (Goffman 1959, 1963a, 1963b; Lemert 1951, 1967; Matza 1969). The labeling theory of deviance, of course, is a body of assumptions and generalizations about how society functions and operates; its claims apply not only to the effects of labeling by experts but also to how labels for psychopathology are generated in the first place. This emphasis in social science merges with social criticism not only of psychiatry but of science in general, as in the sociology of science. However, even labeling theorists in sociology acknowledge the reality of pathology and disease. This is so because they make a distinction between natural differences or primary deviance, or pathology, and secondary deviance, which is created by stigmatizing social labels. In other words, social scientists who study symbolic interaction and labeling would be forced to acknowledge that the claim of pathology is intrinsic to explaining how deviant human social behavior arises and gets promoted and sustained in social settings.

In conclusion, something like the concept of pathology and/or psychopathology is not only needed but may be logically necessary to make sense of the actions and behaviors of all types of organisms, including human beings. This is true regardless of whether the concept is to be applied to a human, animal, social science, or evolutionary biology frame of reference. It may seem that diagnostic labeling using the currency of fitness in evolutionary biology is scientific and neutral, and hence it spares the diagnostician committing the value-laden and socially arrogant sins of the clinician and especially the psychiatrist. And it may appear as if social scientists do not need to rely on the concept of pathology and of real deviance, thereby earning the social accolade of the scientific and social community by uncovering the arrogant, clinical diagnosing that takes place in medical disciplines. But this is illusory. The concept of psychopathology and the framework that provides its meaning is an indispensable and useful invention of science and medicine. The concept has been designed to set aside a class of phenomena so as to make better sense of them, so as to make the problems that they pose more amenable to cure.

We may assume that members of a group show variation in traits, states, conditions, and behaviors that may be called psychopathological. Psychopathology must be understood in light of the following questions, as summarized in four generalizations: First, what are the special characteristics that set psychopathology apart? This is the axiom of ontological specificity. Second, do group mates act as though they were aware of the condition in question? This is the axiom of recognizability. Third, if they do, do group mates support, understand, indulge, patronize, or avoid the individual in question? This is the axiom of social interpretation. Finally, does the individual showing psychopathology act and communicate with awareness of it? This is the axiom of personal insight.

Given these four generalizations about psychopathology, one can make six claims about its evolution, each of decreasing scope. These are the following:

1. Varieties of psychopathology are found in the naturally occurring social psychological behavior of our evolutionary ancestors—which encompass the last common ancestor (LCA), Australopithecine, Homo erectus or related Homo, Archaic humans like the Neanderthals, and/or Homo sapiens sapiens—that is designed to promote fitness, adaptability, well-being, and welfare. Relatedly, one also finds an adaptation that enables individuals to recognize psychopathology in conspecifics and to accommodate to the suffering or handicap caused by psychopathology. This can, depending on the content of the psychopathology, elicit compassion, assistance, and

toleration, or avoidance. As indicated earlier, varieties of psychopathology as conceptualized in the book really translate to varieties of conditions that differ with respect to causes, how they are classified, manifestation of signs and symptoms, and functional consequences. Finally, the individual who develops psychopathology is aware of it. This first proposition, then, is the boldest statement one could make about the existence of psychopathology during human evolution, and it incorporates all four of the generalizations.

2. Psychopathology is found among natural behavioral conditions that have a negative impact on fitness, and individuals can recognize, socially interpret, and react differently to psychopathology in others, depending on circumstances. This proposition incorporates the first three generalizations.
3. In any group of individuals, psychopathologies are real phenomena, and individuals can recognize psychopathology in others. This proposition incorporates the first two generalizations.
4. Psychopathology is a possible reality for any individual. This proposition derives from only the first generalization.
5. The claim that psychopathology exists in a group other than modern humans is ethnocentric and/or anthropomorphic. It is a projection based on one's cultural experience that is, if not logically unwarranted, at least questionable.
6. What are referred to as psychopathologies are found only in industrialized societies because so-called psychopathologies are politically and economically inspired reification that groups within these societies have promoted for reasons of self-interest. Thus, psychopathology is not a warranted concept and does not explain how natural communities operate and function.

Culture and Cognition in Relation to Evolution and Psychopathology

The emergence of culture and cognition are central to the study of human evolution. Episodes of psychopathology viewed as naturally occurring phenomena also have a cognitive and cultural dimension. Indeed, the commonsense view is that psychopathology is expressed as mental illness, emotional disorder, or aberrant behavior, all of which are conceivable only in persons capable of thinking and feeling, fully enculturated, and manifesting a preexisting pattern of interpersonal and institutional life in a society. An examination of the origins and evolution of psychopathology logically brings to the fore the topic of evolution of culture and cognition.

The study of culture, and particularly its evolution, involves complex issues that are at the heart of anthropological and sociological theory regarding the nature of human rationality and social organization. Since psychopathology by definition requires a consideration of culture and cognition, its evolution also brings into focus fundamental themes about humans and society that are hallmarks of Western intellectual history. Important figures of this tradition—indeed, the alleged founders of anthropology and sociology—began their inquiry into the nature of human society by comparing modern, scientifically influenced societies with primitive or savage groups and offering interpretations of the latter. I have in mind scholars such as Lévy-Bruhl, Spencer, Durkheim, Weber, Marx, Freud, Jung, Radcliffe-Brown, Evans-Pritchard, Lévi-Strauss, Needham, Leach, and Sahlins, whose ideas form a core text about the character and origins of human rationality and society. (See Hénaff 1998; Morris 1987.) In their classic works, these figures have wrestled with a series of interrelated intellectual quandaries relevant to culture and cognition as they have been studied by evolutionary biologists and psychologists. Their research involves (1) the nature of human rationality; (2) sacred as compared to profane ways of understanding the world; (3) differences between preliterate, mystical, or magical, as compared to logical, empirical, scientific ways of thinking and reasoning; (4) the nature, origins, and functions of the sacred, the numinous, and religion in general; (5) the origins of human mentation and classification; (6) the relationships between totemic, naturalistic symbols and classifications, on the one hand, and cosmological and social symbols and classifications, on the other; (7) the character and functions of social structures that make up a human cultural group, as compared to the character of the implicit, underlying, and/or unconscious symbolic structures and logical operations of myths and rituals, particularly in how these are composed and played out in narrative form and sacred practices; and (8) the role of the study of history and of historical transformations or evolution in the study of how primitive, savage societies operate and change and how systems of symbols operate and are transformed. While many of these founders of social science dealt explicitly with evolutionary themes, few reflected an informed and subtle understanding of the implications of Darwinian theory for human behavior and social organization. Further, none could draw on the modern synthesis of biological theory or the large body of information amassed by primatologists, comparative psychologists, paleoanthropologists, cognitive archeologists, and evolutionary psychologists.

Because psychopathology is necessarily linked to culture and cognition, an analysis of its origins must take into account central questions pertaining to the origins of human social organization, human rationality, and religious institutions as classically studied by social science theorists. I do not attempt to review and integrate this fundamental theoretical knowledge in this book. I make no concerted effort to integrate seminal ideas of key theoretical figures who have concentrated exclusively on the study of human cognition, culture, and religion—all staple themes in the study of human origins—even though all are likely to have affected how psychopathologies originated during evolution, or how during any one particular phase they emerged, were interpreted, and were dealt with as social phenomena.

While I do not attempt to integrate the ideas of the founding fathers of social science about the character of human mentation and society, especially culture, I will address broad-ranging theoretical issues. It is not possible to discuss a topic such as the origins and evolution of psychopathology, a category that encompasses culture and cognition, without broaching basic, seminal questions. However, I cover the relevant themes in restricted ways, using contemporary understandings. In particular, I interpret the concept of culture differently in different parts of the book. The concepts of culture and cognition during early phases of human evolution bear little resemblance to the collective representations of culture, with sacred, cosmological overtones, as conceived by the founders. As an example, when concentrating on the earliest hominids, I am guided by contemporary notions in primatology and comparative psychology about cognition and culture among the higher apes. Here, an ecological conception of culture and cognition dominates. As I discuss later phases of human evolution, I draw on ideas of evolutionary psychologists, linguists, and cognitive archeologists about culture and cognition. Here, ideas about earlier forms of reasoning, a protolanguage, and the evolution of modular forms of cognition come into play. None of these themes can be easily connected and integrated with ideas of the founders. Finally, only when I focus on modern humans will I draw on more broad, symbolically elaborated formulations of culture and cognition. In this context, I use the notion of culture and cognition as constituting an all-encompassing symbolic niche that defines the world and connects persons to the behavioral environment in a network of meanings, since a fully evolved form of language, culture, and symbolization may have entailed qualitatively different forms of adaptation. All of my ideas are drawn from contemporary evolutionary studies of human behavior, and while they do not connect directly with seminal ideas of the founding fathers, the themes covered are nevertheless clearly related to these broader intellectual themes.

As with the evolution of culture and cognition, to analyze the origins of psychopathology one must focus on central themes of evolutionary biology, evolutionary psychology, and the social sciences, especially cultural anthropology. Besides being a concept pertaining to Homo sapiens, who are enculturated cognizers par excellence, psychopathology refers to congeries of general and diverse symptoms and signs that are manifested in behaviors heavily influenced by culture and cognition. The impact of episodes of psychopathology on groups of individuals during human evolution would have depended on the mediating role of culture and cognition in interpreting such episodes. The intellectual territory that maps the evolution of culture and cognition thus overlaps with the study of the origins of psychopathology. Given the centrality of culture and cognition in the study of psychopathology, it follows logically that an analysis of origins of the latter will take into account key aspects about the evolution of the former. While the basic defining characteristics of human social organization and rationality are staple themes in Western intellectual and cultural history, themes that were central in the writings of the founding fathers of social science in particular, these issues are not examined directly in this book. Instead, ideas from the contemporary fields of evolutionary biology and psychology, contemporary research in the behavioral and social sciences, and theory pertaining to cultural and clinical psychiatry provide background material.

- Mental illness and psychiatric disorders refer to characteristics of human beings, and when used as concepts they carry a variety of meanings laden with biomedical concerns and directives as well as more general cultural values pertaining to persons and morals.
- There exist many types of psychiatric disorders and mental illnesses, but for the sake of economy of presentation and discussion, the general term *psychopathology* is used in this book to refer to them.
- Studying the evolutionary basis of psychopathology involves integrating material about primates with that about modern humans and all of the species along the evolutionary continuum in between. This crosses the fields of evolutionary biology and social science.
- Adopting a unitary conception of hominid behavior problems and disturbances, psychopathology, facilitates examining them in relation to changing characteristics of hominids in areas of cognition, language, and culture, including their social ecology.
- To claim that animal culture, wellness, and suffering have a biological basis is to provide a background and support for the validity that psychopathology originated during human evolution. It also blunts some of the philosophical criticisms of anthropomorphism.
- Normal and abnormal are key concepts in the cultural study of psychopathology in humans, and when one looks beyond the theoretical problems of speciesism, crossing from humans to animals, the concepts can be meaningfully applied to hominids.
- Actual descriptions and analyses of possible forms of psychopathology during phases of human evolution have not been the focus of evolu-

tionary psychologists and psychiatrists and, in general, the content of psychopathology among hominids has been neglected.
- To pose the question of origins of psychopathology is to contemplate basic evolutionary changes that led to modern humans; specifically, how language, culture, and cognition evolved and the role of emerging social institutions, including ritual, myth, and religion.

Chapter 2
Evolutionary Theory Applied to Psychopathology and Psychiatry

HUMAN BEHAVIOR, particularly social behavior, has traditionally been handled as a phenotypic trait of individuals which, along with anatomy and physiology, has been affected by the processes of natural selection and biological evolution. More recently, however, aspects of human psychology and psychiatric disturbances have come under the scrutiny of scholars influenced by evolutionary theory.

Evolutionary Theory and Psychology

An implication of Darwin's theory (1859) was that concepts like perception, cognition, emotion, motivation, and behavior should be viewed in terms of evolution. However, students of human behavior were little interested in this approach until the publication of E. O. Wilson's *Sociobiology* in 1975. In recent decades, this field of study has mushroomed to include many types of professionals, especially anthropologists, psychologists, and ethologists, who view themselves as Darwinians but who, to be sure, concentrate on somewhat different approaches to the study of human behavior in relation to ideas about evolution. (See Alexander 1979, 1987a, 1987b, 1989, 1990; Barkow 1989; Betzig 1997; Draper and Harpending 1988; MacDonald 1988a, 1988b; and chapters in Chagnon and Irons 1979.) Most recently, the study of human nature and behavior from the standpoint of evolutionary biology has assumed greater importance within the discipline of psychology; it is appropriately termed evolutionary psychology. (See Buss 1991, 1995a, 1995b, 1999; Crawford and Krebs 1998).

BASIC THEMES OF EVOLUTIONARY PSYCHOLOGY

Four core themes in cognitive, developmental, personality, social, and clinical psychology are (1) patterns of attachment between infants and mothers; (2) differences in temperament between infants and children; (3) the operation of psychological mechanisms, also termed adaptations or algorithms; and (4) factors underlying styles of adult personality function. These four ways of looking at behavior involve constructs and propositions about enduring characteristics inside the person that account for behavior throughout the life cycle. They will be referred to as dispositions that promote and shape behavior. All four emphasize heritable, genetic factors as well as environmental ones. Human behavior is explained as a consequence of the interaction between inherited factors and circumstances encountered in the environment.

First, the importance of attachment behaviors during infancy and their effects have received much attention since the pioneering studies of Bowlby (1969). Attachment refers to characteristics of bonding between infant and parent that develop early in life and influence how individuals cope with intimacy, frustrations that are emotionally arousing, and developmental challenges encountered later in life. The basic biology of attachment and bonding, including phylogenetic antecedents, hormonal and physiological aspects, and contribution to psychopathology, are an important focus in the sciences of behavior (see Atkinson and Zucker 1997; Carter, Lederhendler and Kirkpatrick 1997; Serban and Kling 1976). Second, temperament involves "aspects of behavior that reflect the intrinsic non-maturational, stylistic qualities that the individual brings to any particular situation. These intrinsic qualities will have a biological basis (including a genetic component) but . . . will also have been influenced by experiences" (Rutter and Rutter 1993, 185). Third, psychological mechanisms refer to ways of handling specific types of physical and environmental stimuli in relation to distinct types of tasks and challenges—for example, selection of a mate, achievement of social rank and status, color determination, and spatial reasoning and memory with respect to characteristics of the physical environment. The mechanisms are held to have been designed by natural selection. Finally, while not independent of psychological mechanisms and temperament, and influenced by attachment styles as well, personality refers to psychological structures or factors underlying social and psychological behavior that explain why individuals behave differently in different contexts and situations. For example, personality encompasses emotionality, gregariousness, inhibition, conscientiousness, and impulsiveness.

All of these behavior dispositions include subsidiary tendencies, traits, and factors, and they apply to different types of behavior. While the dispositions are analytically separate and their central focus or relevance is manifested at different times during the life cycle, they all influence enduring qualitative properties of social and psychological behavior; moreover, they influence each other in complex ways. In other words, the four dispositions are not mutually exclusive. As implied earlier, attachment behaviors may influence temperament, and vice versa; and an individual's prevailing psychological mechanisms, how they are combined and which ones dominate, contribute significantly to the individual's personality style. Furthermore, aspects of temperament clearly affect how personality traits and factors become manifest. Together, the four dispositions point to more or less intrinsic characteristics of individuals that explain patterns and styles of behavior, including adaptive and maladaptive behavior.

Many scholars have examined these four dispositions in relation to human evolution and have explained them as responses to ecological conditions and problems of adaptation in early ancestral environments. This book undertakes to show how the four central themes in evolutionary psychology—attachment, temperament, psychological mechanisms, and personality—interpreted from the standpoint of their origins, rationale, and biological goals and functions, also provide a useful framework for examining the origins of psychopathology. (See Belsky, Steinberg, and Draper 1991; Belsky 1997; Bowlby 1969; Buss 1991, 1995a, 1995b, 1999; Draper and Belsky 1990; MacDonald 1991, 1995; McGuire and Troisi 1998; Stevens and Price 1996.)

PSYCHOLOGICAL DISPOSITIONS AND EVOLUTIONARY THEORY

The relevance of these four dispositions to the question of the origins of psychopathology rests on how they operated and functioned in the environment of evolutionary adaptedness (EEA). They consist of mechanisms and processes that regulated specific social and psychological behaviors to fulfill important biological functions. These include adaptive behaviors such as looking for food (foraging, scavenging, hunting), avoiding predators, competition within the species and outside it, status seeking and competition, and mutual and reciprocal behavior. Sex differences in reproductive behavior, mating and parenting, have received the most direct attention, especially with respect to attachment, temperament, and the basic affective and motivational factors that are believed to underlie personality and that repeatedly turn up in empirical studies (such as Geary 1998; MacDonald 1991, 1995). Even sex differences in perception and cognition have been related to differences in reproductive strategies when examined in light of how males and females obtain food and rear their young (Buss 1999; Geary 1998). The four dispositions are keyed to and influenced by one or several brain systems that regulate such factors as behavioral avoidance, behavior inhibition, behavior surgency, emotion and motivation, extroversion in interpersonal relations, and intensity of affect.

Human behavior is crucially influenced by genetic factors. Individuals inherit genetically based material that contributes to the realization of behavior programs that in conjunction with influences from the social environment, such as maternal sensitivity and ecological stability, shape their behavior toward the characteristics described by the four dispositions. Infants reveal considerable behavioral plasticity, and, in relation to perceptions evoked by distinct types of stimuli, may develop any number of behavioral strategies extremely influential in shaping subsequent behavior. Perceptual and cognitive tendencies and strategies of social behavior are realized by the various evolved structures of the brain, but their salience, emphasis, and manifestations can vary among individuals and across populations, depending on genetic and environmental factors and developmental experiences.

Strategies of behavior, reflected in patterned forms of activity and not necessarily conscious, that develop as a result of developmental experiences are keyed to the achievement of biological goals and functions and are described by the four dispositions. As attachment proclivities and temperamental differences unfold, as they are modulated and channeled as the more or less universal psychological mechanisms and algorithms become operative during development, and as they are eventually joined to the evolved motive dispositions that underlie personality functioning, individuals are able to call on a mixture of inherited strategies to help them reproduce and survive.

The adaptive problems that selected for the four types of dispositions are manifested as behavior strategies carried out as a result of entrainment during sensitive periods of development. These are influenced later in development by the social and behavioral environment. The strategies operate as evolutionary residuals geared toward meeting life history schedules, reproductive strategies, interpersonal needs, and survival, more generally.

While the four behavior dispositions are generally viewed as having been selected for during the process of environmental adaptation, and hence they depend on inherited psychological structures and mechanisms, they do not involve invariant, universally realized, and/or heavily canalized traits. Rather, the four behavior dispositions are assumed to rest on polyvalent genetic systems and to respond to environmental conditions. While their components refer to qualitative and often, as in the case of personality, value-laden aspects of behavior, they differ in salience and emphasis between individuals. Most of the individual differences in behavior, between individuals and within individuals over time, that are observed in psychological research can be explained as a result of differences in population genetic systems, cultural and environmental influences, family rearing patterns, and characteristics of the social environment, such as group demographics, access to resources, and tendencies within the group and the family.

Moreover, manifestations of the four behavior dispositions differed during various phases of human evolution because of differences in the evolving character of language, cognition, and culture.

Differences in the characteristics and salience of the four dispositions in contemporary human populations, and arguably during earlier phases of human evolution, are perhaps best illustrated in the five-factor model of personality. Different levels of the five factors help explain differences in how individuals relate interpersonally and in relation to varying objectives and goals. However, characteristics of each of the four behavior dispositions should be viewed as functions of the relationship between inherited tendencies and environmental contingencies.

BEHAVIOR DISPOSITIONS AND PSYCHOPATHOLOGY

Understanding attachment modes, temperament differences, evolved psychological mechanisms, and personality characteristics in light of evolutionary theory entails viewing the four constructs as adaptive and keyed to important biological goals and functions. Few would dispute this idea. For example, to say that infants develop a secure attachment to the mother, that children show an even and warm temperament, that adults acquire "natural" cognitive reasoning power with which to spot social cheaters, and that an adult's personality is extroverted or conscientious, and that these four behavior dispositions are somehow based on inherited tendencies if not on evolutionarily designed mechanisms and processes is to affirm their adaptive character, since they can be assumed to have contributed to realizing any number of biologically important functions during human evolution—and, in fact, in most social environments. However, when applied to the question of psychopathology, these inherited and selected traits or dispositions create intellectual tension. For example, it seems unreasonable to claim that an infant's failure to develop a strong attachment, a child's predisposition to anxiety or avoidance, a cognitive trait that makes an individual suspicious of unhostile strangers, and an adult personality marked by anger and undue competitiveness should be viewed as expressions of adaptive behavior acquired and partly selected for in the EEA. In this instance, behavioral proclivities that are qualitatively negative and appear to thwart the realization of biological functions and goals ordinarily linked to social relations in group settings are cited as examples of adaptive tendencies that were selected for in the evolutionary environment.

This tension may be resolved in several ways. For example, psychopathology can be explained as resulting from pure hardware defects in the mechanisms serving the four dispositions (traceable to disease, injury, and/or pathology). An excess of a particular trait that ordinarily fosters adaptation can give rise to psychopathology; for example, one can imagine maladaptive extremes of the five-factor model of personality (see Costa and Widiger 1994a, 1994b). The theory of environmental mismatches is also relevant: psychopathology can be explained by EEA-based mechanisms and processes that are poorly matched to environmental conditions. In other words, social stimuli may today trigger behavior that was adaptive in the EEA but which in the contemporary setting is problematic or frankly harmful to survival and reproduction. (See, for example, Belsky 1997; Draper and Harpending 1982.) Environmental adaptation took many forms, since change and variety characterized the behavioral ecology of hominids during evolution. Different strategies and different motive systems underlying behavior could have proved adaptive and advantageous in various environments. This explains the existence of alternative strategies that were forged out of each of the four sets of behavior programs or dispositions. Aspects of attachment, temperament, psychological mechanisms, and personality that contribute to psychopathology in vulnerable individuals today are thus explained as the result of the entrainment of behavioral strategies that later prove maladaptive and/or because of extremes of the distribution of traits that in the normal range are adaptive.

The strategies of behavior generated by the four dispositions can be viewed as evolutionary residuals that underlie the behavior of Homo sapiens and hominid populations. The term *evolutionary residuals*, then, denotes the naturally designed and, hence, ultimately caused mechanisms and algorithms responsible for carrying out adaptive behavior associated with survival, reproduction, and associated social exchanges with kin and non-kin. While sometimes thought of as psychological in nature, they need to be seen as including and/or bringing into play integrated behavior modules relating to functional, biological, and motivational goals.

Some varieties of psychopathology can result from unusual combinations or extreme amounts of the aforementioned evolutionary residuals that function as behavior dispositions and imperatives. This would bring into play behavior strategies, once adaptive in certain niches of the EEA, that in contemporary social environments prove highly maladaptive. This account ignores how specifics of ecology may have operated in relation to specifics of evolved culture and cognition to give rise to specific scenarios of psychopathology during specific phases of human evolution. This book considers only the more descriptive details of social and psychological behavior and psychopathology, seen in relation to evolving characteristics making up culture and cognition. Later chapters attempt to use the EEA not just as a general abstract backdrop for understanding general evolutionary determinants of psychopathology but rather to extrapolate from what is known about the evolution of culture and cognition, and about the presumed characteristics of ancestral social environments, to speculate about varieties of psychopathology that may have emerged during different phases of evolution.

Evolutionary Theory as a Framework for a Clinical Psychiatry

Most uses of evolutionary theory in the study of the origins of psychopathology concentrate on specific contemporary psychiatric disorders—that is, those recorded in the *Diagnostic and Statistical Manual*, the DSM-IV—and seek to explain how these, or striking features of them, may have been selected for in ancestral environments. Many of the disorders recognized in psychiatry today are traced to conditions assumed to have prevailed in the EEA, where certain features of them may have proved adaptive. To the extent that this is true, scholars presume the existence of a process of natural selection, which implies that some features of these disorders may have once played an important biological role. Evolutionary theory, then, is sometimes used to explain both the characteristics of selected psychiatric conditions and their relatively high prevalence and universality. A basic theme is that there have been environmental mismatches between genetically designed mechanisms and contemporary aspects of the social and physical environment. This "theory" plays a role in Crow's argument with respect to schizophrenia, Marks on anxiety and fear-related conditions, Gardner (1982) on mania, and Gilbert (1992a, 1992b) on depression.

A more comprehensive application of evolutionary theory to psychiatry is to show how virtually all of the conditions recognized in the DSM system, their important features, such as causes, key manifestations, and, more generally, problems of behavioral maladaptation reflect the activation, or at least the mediation, of behavioral mechanisms that were functional during evolution. Indeed, this latter emphasis is part of a general orientation whereby most aspects of human behavior are seen in relation to tenets of evolutionary theory. This is the approach adopted by Wenegrat (1984, 1990), Stevens and Price (1996), and McGuire and Troisi (1998), who apply principles derived from evolutionary biology to explain the origins of problematic social and psychological behavior, as well as specific psychiatric conditions. In a sense, they seek to demonstrate that a functional approach to behavior in line with evolutionary theory can be used as a basic science by which to understand how individuals cope with their social environment and how failures to do so lead to diagnosable psychiatric disorders or symptoms of such disorders.

Evolutionary Psychiatry: A New Beginning (Stevens and Price 1996) integrates basic tenets of evolutionary thought regarding behavior and applies them to the analysis of psychiatric disorders. Rather than concentrating on specific disorders, Stevens and Price aim to provide a general theory about behavior, both normal and abnormal, and apply it to an interpretation about the causes and the actual manifestations of virtually all psychiatric conditions. They supplement the work of scholars who have examined human behavior with an eye to its evolution with their own analysis and interpretation of evolutionary biology and clinical psychopathology. Both authors have made seminal contributions to evolutionary psychology and psychiatry.

After reviewing prior efforts to link evolution and human behavior and summarizing the evolutionary ideas that are relevant to its understanding, with special emphasis on psychiatric disorders, Stevens and Price offer a basic theory about the dynamics of behavior, including those that have relevance for psychiatric practice, seen in relation to evolutionary imperatives. As an example, they conjoin the contemporary idea of stress (that is, strain on an organism that perturbs its function) with ideas about archetypes and the phylogenetic psyche of Carl Jung, who attempted to root psychoanalytic thinking in general speculations about human evolution. Stevens and Price conclude, "The probability is that the greater the gap between archetypal needs and environmental fulfillment of those needs, the greater the stress and the more incapacitating the illness" (1996, 37). They propose five laws of psychodynamics: (1) universal attributes of behavior are expressions of innate propensities or archetypes; (2) the goal of archetypes is actualization in behavior and psyche; (3) the fulfillment of archetypal goals results in mental health; (4) frustration of archetypal goals results in psychopathology; and (5) manifestations of psychiatric disorders are persistent exaggerations of adaptive psychophysiological responses.

Stevens and Price make a special effort to explain the etiology of psychiatric disorders, relying on rather general tenets about behavior drawn from evolutionary theory. As an example, they postulate that adaptive strategies such as fight, flight, and dominance are normally distributed in human populations and that "exaggerated or inadequate endowment with these strategies might account for psychopathology among individuals placed towards the tails of the distribution of adaptive traits" (1996, 39). Similarly, interactions between specific genotypically based adaptive strategies in different habitats result in different adjustment patterns and possible manifestations of psychopathology. Psychological traits that reflect adaptive strategies can be activated or inhibited, giving rise to distinctive types of symptoms. For example, a panic attack may constitute the activation of an adaptive strategy whose function was to protect humans in open spaces. In a similar way, Stevens and Price draw on social homeostasis theory, inclusive fitness theory, and ontogenetic theory as a basis for general frameworks that can be used to understand psychiatric symptoms and disorders.

More compelling is their claim that "the immediate cause of a large number of psychopathological conditions is a subjective prediction of probable failure in competing for two highly valued social resources: *attachment* and *rank*" (1996, 43; emphasis in original). In elaborating on this theme, Stevens and Price draw on basic tenets of ethological studies of social behavior to explain further the causes of psychiatric conditions. For example, they elaborate on various aspects of attachment theory and parenting modes, ethological notions about ritualized behavior regarding social status and rank, and the ideas of comparative psychologists about styles of competition in social groups. They postulate the existence of two fundamental archetypal systems, attachment/altruism and rank/territoriality, and two different modes of competition, dominance and submission.

Stevens and Price emphasize how in human groups attraction and affiliation as a form of competition arose to complement and in many instances to supplant competition by pure intimidation and frank opposition. Using ideas initially developed by Chance and Omark (1988a, 1988b), they describe two distinct types of social functioning, the agonic and hedonic modes. Agonic modes are associated with hierarchically stratified societies where the concern is to dissipate and socially equilibrate aggressive tensions, whereas hedonic modes are found in more egalitarian societies that emphasize attracting, affiliating with, and disarming rivals and seeking group approbation for activities that command respect and prestige. Stevens and Price use this theoretical apparatus to understand not only social and psychological behavior but also almost all psychiatric disorders, which they classify in terms of their causes and manifestations in light of their synthesis of evolutionary theory. As an example, using ideas derived from ethological and interpersonal theory, such as dominance, submission, social isolation, social integration, allows Stevens and Price to give emphasis to insiders and outsiders relative to the "in group." This serves as their basis for classifying psychiatric disorders: "If an 'insider' develops a psychiatric disorder, it will tend to be a *disorder of attachment and rank*, whereas an 'outsider' will tend to develop a *spacing disorder*. . . . Individuals who are uncertain as to their allegiance and who hover uneasily on the cusp between 'insider' and 'outsider' status will, if they develop a psychiatric disorder, tend to present with a *borderline state*" (1996, 54; emphasis in original). The rest of the book is devoted to this classification scheme, along with more in-depth analyses of clinical material and suggestions for treatment.

A similar blend of evolutionary theory and clinical psychiatry is illustrated in *Darwinian Psychiatry* (McGuire and Troisi 1998), a textbook for the practice of psychiatry. The authors aim to teach would-be clinicians how to approach, understand, explain, and deal with psychiatric conditions (their term for phenomena of psychiatric relevance, as compared to Stevens and Price's *disorders*), using fundamental principles of evolutionary biology such as natural selection, inclusive fitness, and reciprocal altruism. Much like Stevens and Price, who concentrate on a different mix of principles and tenets, McGuire and Troisi present a comprehensive and theoretically unified picture of how knowledge of evolutionary biology—particularly, aspects of human physiology, psychology, behavior, and emotion—can be used to explain how individuals behave, how psychiatric conditions come about, and why they are configured as they are. They employ, in other words, a functional approach to behavior that is derived on the one hand from a consideration of ultimate causes associated with coping and adaptation in ancestral environments (EEA), and on the other considerations related to proximal circumstances that pose challenges and obstacles to persons in their contemporary lives.

McGuire and Troisi's textbook, like that of Stevens and Price, draws on the literature about experimental analyses of primate behavior, as well as on research findings in clinical, epidemiological, and psychopharmacological aspects of psychiatric conditions, which is highly consistent with standard textbooks of clinical psychiatry. The basic science aspects of their approach is illustrated as follows. They list four behavior systems, those involving survival, reproduction, kin assistance, and reciprocation, that help to make sense of how individuals cope with their circumstances in line with principles derived from evolutionary biology. Each behavior system is associated with four "ultimately caused" infrastructural systems of the organism or person. The latter come into play in response to information derived from the environment or from internal biological mechanisms and processes. The infrastructural systems include biological motivations and goals, automatic systems that filter, select, and prioritize information and algorithms of the individual's perceptual and conceptual apparatus that filter information in adaptive ways so as to realize functional capacities that involve the actual execution of behavior. McGuire and Troisi summarize their approach as follows: "Hence, the view that we have staked out: Observe behavior and its function; introduce ultimate cause concepts; use comparative (cross-species) data as well as laboratory and clinical research findings to assess infrastructural functionality; infer the effects of genetic information, physiological changes, and environmental variables on infrastructural function and behavior; and develop and test hypotheses dealing with interactions between putative condition-contributing states, traits, or events and conditions" (1998, 92).

Virtually all of the clinically important aspects of behavior that a clinician might obtain from interviewing a patient (or from others who have knowledge of the patient) about the unfolding of a clinical problem or condition are formulated in terms of the logic of behavior derived from evolutionary biology. This includes, for example, the social meaning, maintenance, and dynamics of behavior and of the self, considered as a social and biological entity. McGuire and Troisi illustrate their system by concentrating on a selected sample of clinical conditions, as, for example, schizophrenia, depression, anorexia nervosa, and personality disorders, providing case illustrations of patients afflicted by them and listing intervention strategies that rely on the logic of evolutionary biology, such as the possible functions and/or causes of a patient's condition formulated in terms of ultimately based but proximally contextualized considerations. They outline an evolutionary framework for psychiatric treatment consisting of a set of specific principles that can be used to gather information, analyze causal factors, and set up options and strategies for treatment that are rooted in their theory of behavior, which, as illustrated above, is derived from evolutionary biology.

COMMENT

The general concept of psychopathology is adopted in this book as a shorthand, economical device for referencing different varieties of conditions. It includes disturbances, dilapidations, and dysfunctions of social and psychological behavior and associated physiological perturbations that are the concern of clinical disciplines. In clinical practice, these are described as psychiatric disorders. The history of psychiatry teaches how these disorders were initially identified, later sharpened as to clinical characteristics, and eventually codified in terms of criteria of diagnosis and modes of treatment (Berrios and Porter 1995). Psychiatric disorders have been examined from the standpoint of evolutionary theory (McGuire and Troisi 1998; Stevens and Price 1996; Wenegrat 1984, 1990). In this viewpoint emphasis is also placed on symptoms, signs, personal distress, and concerns pertaining to social life with a view toward treatment. The fundamental contribution of an evolutionary slant on psychopathology is the role attributed to "ultimate causes," natural selection influences in the environment of evolutionary adaptation (EEA), which brought into play the major behavior systems and their underlying structures and mechanisms. In the evolutionary viewpoint, psychopathology including, in particular, psychiatric disorders and their manifestations, is examined in terms of proximal causation—namely, as the outcome of perturbations of the ultimately caused behavior systems that actually originate in the person's current environment and produce impairments and deficiencies in areas of biological function and motivation. An ideal classification of psychopathology for evolutionary purposes would seek to merge valid and reliable indices of disorders as found in the existing diagnostic system of psychiatry with equally useful indices of impairment of functional capacities of the major behavior systems involved in the pursuit of biological goals. This would yield a classification of behavior disturbances or breakdowns that blends insights of psychiatric nosology with biological knowledge of the ways in which social organisms function and fail to function in the economy of fitness. The latter, which includes survival, reproduction, and associated social relationships, realizes the power of the evolutionary viewpoint about behavior, including psychopathology. A suitable classification of psychopathology that draws on evolutionary principles would create the conditions for the application of evolutionary insights about behavior for the purpose of treatment. These would complement treatments now employed in standard, establishment psychiatry.

What about Psychopathology during Evolution?

Scholars like McGuire and Troisi, Wenegrat and Gardner, and Stevens and Price, all of whom seek to advance the practice of clinical psychiatry using tenets of evolutionary biology, indirectly and by implication present an interpretation of psychopathology during phases of human evolution. In other words, in arguing for a universal interpretation not only of the prevalence—underpinnings in the human genome—but also of the character—causes, manifestations, and meanings—of psychiatric disorders and conditions, all who use evolutionary theory to explain aspects of psychiatry are assuming that psychiatric entities, or selected features of them, have been an integral component of hominid groups throughout human evolution. Adopting an evolutionary biology perspective on psychiatric diagnosis and clinical practice implies that phenomena that pertain to psychiatry were somehow present in species that antedated Homo sapiens, or at least came into being during distinctive phases of human evolution, since the theory in question implies a phylogeny and continuity of behavior across species ancestral to man.

Thus, to claim that in a particular contemporary social situation a psychiatric condition, or selected features of it, that constitutes a universal in human populations, regardless of ethnicity and culture, and moreover, is caused by a malfunction of archetypal strategies—modes of behavior based on rank seeking and competition, and/or infrastructural systems that are integral to the organism's social relations and/or pursuit of basic biological functions and goals—is to imply that in ancestral humanoid populations the same scenario could have prevailed because contemporary human populations trace their origins to earlier hominids. This basic claim about a panhuman as well as panhominid distribution and function of a psychiatric condition is also implicit in Wenegrat (1984, 1990).

Characteristically and for more or less transparent reasons, such as, for example, parsimony, emphasis on empiricism, desire for theoretical restraint, and a need to maintain philosophical purity, scholars like those cited above who pursue an evolutionary or sociobiological approach to psychiatry desist from actually speculating about what psychiatric conditions or disorders might have consisted of during human evolution. However, one might offer a tentative extrapolation of what accounts based on strict evolutionary tenets would look like when applied to earlier phases of human evolution.

The theories developed by evolutionary psychiatrists to explain disorders are necessarily abstract, general, and relatively unconcerned with the complex personal experiences and social situations of contemporary individuals. Using their theories to extrapolate accounts of psychiatric phenomena in ancestral populations is necessarily an even more abstract undertaking. First, in depicting how a particular psychiatric condition or disorder makes sense in light of generalizations about contemporary human behavior viewed in light of evolutionary biology, researchers have to reduce much of the complexity of psychiatric phenomena. In other words, compared to nonhuman primates and early hominids, human beings think and communicate in special ways, by means of symbols and language, and their activities and situations make sense in light of cultural and historical factors—the human symbolic niche—as well as adaptations derived from ancestral or archetypal conditions. Aspects of human cognition, communication, social behavior, and culture diverge sharply from related qualities (even if homologous and not merely analogous) of earlier humanoid groups. The evolutionary psychiatrist's casting of human experience and activity in the logic of adaptive strategies, archetypes of the human psyche, types of behavior systems and/or infrastructural systems, and interpreting how contemporary psychiatric phenomena make sense in the light of precisely those same general, that is, panhuman and interspecific, formulations about behavior suggests rather too easily and without proof that causes and manifestations of disorders identified in contemporary psychiatry can be unproblematically equated with related activities and problems of ancestral species. Left out of this equation is a fine-grained account of differences between contemporary and ancestral culture and cognition, as well as psychopathology. In other words, researchers discussed earlier provide a relatively abstract and decontextualized account of the possible origins of psychopathology.

To be sure, all of these researchers point out that human circumstances and situations today differ from those in ancestral environments. However, the importance of such a mismatch for explaining psychopathology differs among them. To Stevens and Price environmental mismatch is all important, whereas to McGuire and Troisi it is the way behavior systems are played out in relation to biological functions and goals during each and every phase of human biological and cultural evolution that is central. However, all researchers acknowledge enormous differences between a society with high demographic density, a sedentary life style, and highly stratified and urbanized social systems, as compared to social relations in evolutionary environments, small kin-related groups that are egalitarian and nomadic. When carrying out standard clinical tasks of diagnosis and treatment, evolutionary psychiatrists apply general evolutionary principles pertaining to fundamental biosocial goals. They take for granted cultural variation. When speculating about earlier phases of human evolution, they are also more concerned with general issues; for example, as compared to how psychiatric phenomena might have been configured and enacted during earlier phases of human evolution.

This book attempts to enter into the evolutionary space occupied by ancestral human groups and to use knowledge drawn from paleoanthropology and comparative psychology so as to offer an informed account about the character of their psychological experience, along with their social and ecological circumstances. This is the space where psychopathology emerged. To repeat, the objective is to provide a more or less realistic picture of the origins of psychopathology, including an analysis of its implications in light of contemporary theory in the social and biological sciences, as well as psychiatry.

- General tenets of evolutionary psychology, the study of how evolution has shaped human behavior, are reviewed and discussed in terms of four interrelated parameters: attachment, temperament, psychological adaptations, and personality.
- Changes taking place during evolution shape these parameters into evolutionary residuals of behavior that have to do with basic biological functions pertaining to survival, social competition, and reproduction as well as allocations of and trade-offs among resources.
- Because evolutionary residuals underlying human behavior were shaped in ancestral environments and conditions, they can cause problems when played out in social and ecological environments that do not match those of the evolutionary past.
- Evolutionary psychiatry is the field that seeks to understand and to treat psychiatric disorders using insights about evolutionary biology and psychology.
- Some of the major concepts and principles of evolutionary psychiatry were reviewed.
- Evolutionary psychology and evolutionary psychiatry emphasize ultimate, ancestral, EEA-related questions about origins of behavior as compared to proximal, causal ones tied to the effect of the contemporary environment on brain/behavior relations.
- In contrast to the more contemporary focus on psychopathology of evolutionary clinicians, this book enters the social and psychological behavior spaces of hominids and visualizes how psychopathology was configured and played out in them.

Chapter 3
Clinical and Evolutionary Images of Psychopathology

AN EVOLUTIONARY CONCEPTION of psychopathology implies that varieties of behavior disorders were manifest and could have in theory been diagnosed in selected members of animal species that preceded Homo sapiens. In other words, regardless of whether one relies on a reasonable, abstract formulation of psychiatric disorders—for example, drawn from selected categories of the diagnostic system of psychiatry—or on a diagnostic system based on an ecological and functional analysis of behavior, as illustrated in McGuire and Troisi (1998), one must hold that varieties of psychopathology constituted realities in ancient hominid societies. Yet despite the naturalness of this conclusion, it is a fact that neither evolutionary psychologists nor psychiatrists have grappled with the philosophical quandary of conceptualizing psychopathology as relevant to animals or the theoretical task of specifying what the configuration and pattern of enactment—for example, as per manifestations and consequences—of varieties of psychopathology may have consisted of during earlier phases of human evolution.

Researchers have tended to show a conceptual inertia or reluctance to cross the "species barrier" as it pertains to psychopathology and its many varieties. There are two interrelated factors that I believe have created such a barrier. One is a general bias to associate psychopathology with things human, namely, aspects of the mental, the moral, and the cultural. At its most extreme, this factor seems to imply that among animals one is not likely to find mental illnesses, and why should one care about this anyway. The other factor, very much related to the preceding one, is a contrast in value and sentiment said to prevail in a social, cultural, as compared to an evolutionary, context as these are depicted by representatives of the respective disciplines. This factor asserts that helping individuals to cope with an affliction or disablement, such as by healing, helping, and prolonging life, constitutes a culturally evolved concern associated with medicine, a quintessential cultural institution (but see Fábrega 1974, 1975, 1997), whereas in nature, where survival of the fittest is the rule, "life of man is solitary, poor, nasty, brutish, and short," hardly a place in which to find humanitarian instincts that seem at the core of the moral, cultural, and, indeed, clinical approach to disease, generally, and psychopathology, in particular. These two factors are complementary, and together, I believe, have played a role in keeping researchers away from formulating scenarios about psychopathology during earlier, precultural, phases of human biological evolution.

I hasten to add that any thoughtful interpretation of evolutionary theory, contemporary social science, and medical practice quickly convinces that the alleged difference and opposition between the two points of view is incorrect. The material discussed previously, in particular, makes abundantly clear that evolutionary influenced clinicians are not only strongly committed to general human, cultural viewpoints but are also seeking quintessentially compassionate, humane goals and ends (Gilbert 1998; McGuire and Troisi 1998; Nesse 1998; Stevens and Price 1996; Wenegrat 1984, 1990). Similarly, a reading of evolutionary theory and cultural anthropology makes abundantly clear, for example, that culture, cognition, and morality and the fact of altruism, mutuality, and social reciprocation, constitute key areas of theoretical concern (Alexander 1979, 1987a, 1987b, 1990; Brown 1991; Buss 1999; Chisholm 1996; M. S. Dawkins 1980, 1990, 1993; Donald 1991; Trivers 1971, 1985).

Despite the inappropriateness and erroneousness of an alleged incompatibility between the two viewpoints, however, a conflict between them persists, at least in the general intellectual and literary arena. The conflict or bias in effect pits nurture—the cultural, humanitarian, clinical approach—against nature—the evolutionary, impersonal, and disinterested approach. Its effect is to imply that varieties of psychopathology are concerns that belong in the cultural, clinical arena and not in the allegedly uncaring and disinterested terrain of biological evolution. The first part of this chapter outlines some of the properties of the two viewpoints that appear to sustain this opposition and frank bias. This is accomplished by presenting an exaggerated (and also idealized) formulation of the clinical, cultural approach to behavior and psychopathology and an equally hyperbolic formulation of the evolutionary approach. I then compare the viewpoints and indicate why and how an alleged opposition and inconsistency between them is untenable. This paves the way to the second part of the chapter, wherein I illustrate by means of hypothetical cases how both the cultural, clinical and the evolutionary, disinterested approaches to psychopathology actually complement one another.

The Clinical Perspective

For the culturally situated clinician, the individual and specific-named psychiatric disorders are the important units of analysis. Individuals are viewed as potentially vulnerable at any stage of life to developing a particular variety of disorder, each of which has distinct properties, causes, manifestations, temporal profiles or natural histories, and treatment schemes. This focus on individuals and named conditions of disablement and suffering is paralleled by the presence of healing practices and medical institutions in all societies and in the very rationale of the clinical perspective. The many varieties of psychopathology that are to be found are explained in terms of an individual's age and the phase of life at which they are likely to be manifested, as well as in their social circumstances. Subspecialties of psychiatry, for example, are defined in terms of life phases, and different specialists may address different types of psychopathologies.

In all human societies the presence of suffering is a hallmark of sickness and an occasion for healing (Fábrega 1974, 1975, 1997). Similarly, to a culturally immersed clinician, a psychiatric disorder causes distress and impairment, and the purpose of treatment is to alleviate and to mend. How the individual developed the disorder is the basic etiological focus. The psychiatrist explores circumstances in the patient's recent past or events during his or her lifetime to discover what can be done to alleviate suffering and restore a premorbid level of activity and function.

The cultural and clinical perspective endorses an explanatory model heavily influenced by an interactionist and, frequently, a mind-body dualism. Personal experiences with caretakers, peers, work mates, and family members throughout the life span can contribute to varieties of psychopathology, as can the material properties and functions of the individual's physical constitution and, especially, that of the central nervous system. Second, to explain how psychopathologies are produced over the life span, the cultural and clinical perspective endorses a developmental, linear concept of causality. Individuals may inherit a proclivity to develop psychopathology, a genetic vulnerability. During any phase of life, even during the embryonic stage, individuals are exposed to potential hardships and perturbations that can contribute to the onset of psychopathology by affecting brain organization and function, or adversely affecting behavior. According to the stress/diathesis model, genetic vulnerabilities and environmental factors, including physical, social, and psychological experiences, can affect the individual during any developmental phase and cause psychopathology. The sciences of developmental and clinical psychology explain how the stress/diathesis model operates with respect to the individual, and the science of epidemiology explains how individuals are affected by environmental and ecological factors.

The cultural perspective, under the banner of scientific medicine, focuses on individuals and their proneness or vulnerability to specific disorders, which are themselves conceptualized in individualistic terms. It applies a developmental, linear, causal, stress/diathesis model of psychiatric disorders. The cultural perspective is quintessentially value laden and also altruistic, since disorders are viewed as deviant, abnormal states that bring suffering, pain, disability, and impairment. The healer, whether shaman, folk practitioner, or psychiatrist, seeks to alleviate or correct these disorders. For this reason, the cultural perspective is usually thought of as humanitarian, altruistic, and compassionate.

The Evolutionary Perspective

The evolutionary approach to psychopathology is depicted in the general intellectual arena and especially in cultural, literary studies as altogether different from the cultural, clinical perspective. It is said to be predominantly concerned with how organisms manage to survive and reproduce, with a focus on genetic factors. The behavior of an organism is examined as to how it contributes to the organism's fitness (that is, its survival, longevity, and reproductive success) and its relationship to other organisms that share its genes in proportion to a measure of relatedness, its own inclusive fitness.

Despite its overriding concern with survival and reproductive success, and its explicit relevance to clinical psychiatry (McGuire and Troisi 1998; Stevens and Price 1996), the evolutionary perspective is criticized for allegedly not addressing the plight of individuals per se. Psychiatric disorders that need to be prevented, corrected, or eradicated are said to be ignored in favor of a network of factors. These may include genes, organisms, groups, and even species. While the researcher who undertakes the evolutionary approach may address the plight of individuals, he or she is said to examine these abstractly; for example, in terms of genotypes, phenotypes, social behavior, reproductive patterns, fitness, natural selection, and genetic drift. Vital concerns to the culturally oriented clinician—questions involving morbidity and suffering, for example—are allegedly underemphasized or neglected.

The evolutionary approach is said to focus, on the one hand, on populations structured in terms of ecology, geography, and evolutionary time; on the other hand, it focuses on forms of behavior and specifically behavioral variations that individuals display in relation to distinctive and changing ecological conditions affecting social organization and behavior. The content of behavior and especially its variation is again said to be viewed in terms of evolutionary time, so that one may distinguish between concurrent or proximal considerations and the long-term antecedents of behavior in ancestral environments, or ultimate considerations.

BEHAVIOR AS PHENOTYPE

Individuals of a group or population are described in terms of genotypes and phenotypes, as can be the groups or populations themselves. For our purposes, the important property of the phenotype is behavior that is conceptualized as a product of genetically inherited, innate factors, termed evolutionary residuals, and environmental influences coming to bear on organisms during various phases of the life arc—in particular, its period of development. In highly social organisms like monkeys and apes—and the earliest hominids, one can safely assume—and members of Homo sapiens, the influence of the genotype on the phenotype and behavior varies. Infants, especially those of highly social animals, are small, immature, dependent, and require close parental supervision, protection, and opportunities for guided trial-and-error learning.

THE RATIONALE OF BEHAVIOR: SOCIAL AND BEHAVIORAL ECOLOGY

The evolutionary perspective appears to eschew emotional and mental problems. Instead, it gives special emphasis to behavior that expresses the psychopathology. The rationale, function, and seeming strategy of behavior, viewed in relation to evolutionary time and changing social ecologies, are the important concerns. Like any other feature of the phenotype, behavior is conceptualized as the product of inherited programs and routines primed by the environment, in some instances during what appear to constitute critical or sensitive periods of development. Routines and strategies of behavior are seen as the product of natural—and in the hominids and varieties of Homo, cultural—selection, designed by past, selective environments of evolutionary adaptedness as a consequence of the survival and reproduction of genes and, by extension, of organisms and populations. The four dispositions of behavior invoked by evolutionary psychologists are pertinent here with respect to individuals.

The environment of principal concern to the evolutionist is said to be composed of social-ecological and, less often, cultural-ecological characteristics. In other words, the following factors are all judged as part of the current, proximal environment and were part of the EEA as well: (1) physical features of the habitat and its resources with respect to food and water; (2) size and organization of groups or populations; (3) family constellations and more extended networks of social relations with non-kin; and (4) general social practices and traditions that can include systems of symbols and their meanings. There is much debate about whether, and if so, when and how, cultural meaning systems and associated ritual and spiritual concerns may have entered the social and psychological behavior space of hominids and varieties of species Homo. Because they are viewed in relation to evolutionary time, these environments have changed and will continue to change in both predictable and not so predictable ways. Natural selection is viewed as having designed organisms and, by extension, shaped populations that can effectively track environments and consequently react to them in a satisfying way. The environment is thus a cardinal focus of the sciences of social and behavioral ecology, both of which provide a basis for the evolutionary approach.

Whereas the cultural, moral, and clinical approach sees the environment proximally—as engaging a phase or at most the lifetime of an individual in a caring humanitarian way—in the evolutionary approach, environment stretches back in time hundreds of thousands, and in some instances, depending on the behavior and the species, millions of years. Populations and species, then, are regularly and repeatedly undergoing challenges to their efforts to survive and reproduce; and to accomplish these basic goals, the behavior of individuals comprising them comes to be governed by distinctive mechanisms. More specifically, behavior involves strategies and routines geared to enhancing individual inclusive fitness (Hamilton 1964). This means that the fitness of the organism itself, and that of its relatives with which it shares genetic material, are important causes and functions of behavior.

THE RATIONALE OF BEHAVIOR: KIN SELECTION AND RECIPROCAL ALTRUISM
Traditionally, evolutionists have assumed extreme self-interest or selfishness as the basis of an organism's behavior. Individuals were viewed—and still are by hard-core evolutionists—as dominated largely, if not exclusively, by their quest for reproductive success and survival. Behavior seen as altruistic, namely, that which favors others at the expense of the individual, is on strict logical grounds something like an anomaly. An important feature of the modern synthesis of evolution is the concept of kin selection, which argues that individuals will sometimes favor, that is, be altruistic toward, genetic relatives, depending on the degree of relatedness (Hamilton 1964). This concept led to the adoption of the concept of inclusive fitness, mentioned earlier. This refers to the replication of the individual's own genes and those of its relatives that came about as a result of altruistic acts. Another important concept is reciprocal altruism (Trivers 1971, 1985), which stipulates that organisms can behave altruistically toward nonrelatives under conditions of high sociality when group mates know each other well and can expect to be paid back at some later time.

Reciprocal altruism, according to strict traditionalists and critics of evolution, simply gives a different slant to pure selfishness. However, in contrast to the unflattering picture painted of the evolutionary viewpoint, to other evolutionists reciprocal altruism implies an expectation of reciprocation and, in conjunction with succoring behavior, empathy, and sympathy, provides the biological, that is, naturally selected, raw material of mutual aid, care giving, and rule-following behavior. For many persons including myself, reciprocal altruism and related varieties of mutualism and social behavior constitute a biological foundation and have paved the way for the cultural elaboration of a sense of justice, fairness, morality, and possibly also spirituality and religion (Alexander 1979, 1987a, 1987b; Boyer 1994; de Waal 1996; Guthrie 1993; Rappaport 1999).

In the concepts of kin selection, reciprocal altruism, and inclusive fitness, evolutionists find the ultimate rationale and function of much of social behavior. Whereas they may understand and invoke these concepts differently, evolutionists agree that they explain the function, value, and durability of sociality and that its roots are to be found if not in the common ancestor of Homo and the apes, at least later in the line of evolution involving hominids, early varieties of Homo, and modern humans; in other words, in the EEA.

THE CONCEPT OF DEVELOPMENTAL PLASTICITY

The evolutionary perspective sees behavior as protean and mutable. The concept of developmental plasticity describes a phase of life, most especially infancy and juvenile periods, during which environmental influences significantly affect the shape, content, and function of behavior. Developmental plasticity is a product of natural selection that characterizes organisms born with immature organ systems, especially the brain, and consequently depend on parents for their protection and learning. In early development, during critical, sensitive periods, environmental influences can direct and shape the function of various organ systems. This eventually affects the content and structure of adult behavior. For example, genotypes program the developing organism to lay down redundant and variable synaptic typologies. This is the raw material on which neural selection acts so as to favor those synapses that are functional in the environment the organism is forced to survive and reproduce. This process, termed *epigenesis through selective stabilization* (Changeux 1985), describes how developmental plasticity allows for the shaping of adaptive behavior as a product of trial-and-error learning and play.

While a product of natural selection itself, developmental plasticity can also be a target of natural selection whereby new genetic programs or advantageous elaborations of others can be incorporated in the genome. These genetic traits are passed on and come to affect the phenotype of subsequent generations. The Baldwin effect (Baldwin et al. 1902) and genetic assimilation (Waddington 1959, 1961) describe processes that can actually select for new patterns of development that can produce behaviors advantageous in the environment. These behaviors may at first appear as natural variations of organisms, but if they enhance survival and reproduction, natural selection will come to favor genetic variants that more easily produce the behavior in question. The Baldwin effect and genetic assimilation, made possible by developmental plasticity, are both processes by means of which ontogenetic experience may affect phylogeny.

Developmental plasticity refers to events and processes interposed between the genotype and phenotype that shape and elaborate the behavior of the developing organism, ultimately culminating in adult adaptive behavior. The Baldwin effect and genetic assimilation describe how natural selection can modify developmental programs genetically so as to enhance adaptive behavior. Epigenesis, by means of selective stabilization, describes a neural basis for adaptive learning. All of these processes point to the pivotal function of developmental plasticity in shaping adult behavior in evolutionary terms.

In line with the contrast being developed here, developmental plasticity is what enables the organism to make the best of a "solitary, poor, nasty, brutish, and short" life wherein a disinterested, uncaring nature presides. In actuality, developmental plasticity, in conjunction with socialization, caring influences, especially from, but not limited to, genetic relatives, provide for the needs of infants and juveniles and lay the groundwork for equally caring socialization and essentially healing routines of adults (Chisholm 1988, 1996, 1999).

LIFE HISTORY THEORY

Organisms and populations show general traits and characteristics that in their totality describe how they have been programmed and how they behave so as to promote survival and enhance reproduction. Life history theory, which presupposes developmental plasticity, seeks to explain the links between genotype and phenotype in the population-centered evolutionary approach (Chisholm 1988; Lessells 1991; Stearns 1992). Again, the focus is on characteristics of individuals, but examined in relation to distinct demographic populations and to evolutionary time. Individuals and populations are conceptualized as constantly making trade-offs between seeking and consuming resources and allocating energy to components of fitness: survival, growth, maturation, storage, and repair, as compared to reproduction. Life history traits consist of an individual's age at first reproduction, number of offspring, interval between births, optimal rates and periods of growth, sensitive periods or prepared learning phases, and the timing of senescence and death. Because evolution is ultimately driven by differential reproduction, characteristics of the life cycles of individuals, as explained by life history theory, are conceptualized as reproductive strategies.

While these traits apply to individuals, they are often studied as properties of a population that has been molded by natural selection and is reacting to prevailing social ecologies. Traditional parameters of life history theory—for example, length of the estrous cycle, age at sexual maturity, age at weaning, interbirth interval—are best studied as features of demography, since variances between individuals can be mathematically controlled. It is the profile of a population with respect to its life history traits that constitutes the ideal basis for comparison with another. There is little value in analyzing the life history variables of one individual, since they could be influenced by any number of idiosyncratic factors. On the other hand, a population's measures afford a compelling statement of how it has adjusted to environmental factors.

By comparing life history variables across populations located in different habitats and environments, evolutionists are best able to appreciate the role of the environment in shaping fitness-related behavior. The theory predicts that in habitats with different physical conditions, food resources, disease agents, predators, and competitors, populations will display different life history traits. Theorists who study populations where social organization and behavioral ecologies resemble those of prehistory, and possibly that of the environment of evolutionary adaptedness, may make assumptions as to how and why evolution has inscribed in the genome of human populations distinctive dispositions of behavior with respect to survival and reproduction.

Developmental plasticity is a basic buffering mechanism evolved by organisms and populations to optimally adapt to the environment. As described earlier, by means of several processes, social organisms who are born immature and who depend on parental supervision and support during critical learning phases are able to learn routines and combinations of behavior that favor survival and reproduction. Among the critical behaviors that organisms come to acquire are those that pertain to various life history traits—for example, behavior surrounding age at first reproduction, affecting number of offspring. Through developmental plasticity, organisms and populations arrive at optimal values of life history traits. In their totality, the pattern inherent in these measures is the trade-off arrived at by organisms or a population, given their distinctive genomes, social and behavioral ecologies, and learning opportunities. Life history theory, then, provides one frame of reference from which the behavior of organisms can be analyzed and interpreted from the standpoint of the basic exigencies of the evolutionary perspective: survival and reproduction.

A strict interpretation of life history theory tends to suggest an impersonal, disinterested viewpoint about the individual. For example, mating and parenting, hallmarks of the cultural, clinical viewpoint, are reduced to questions of optimality about how best to track the environment and mates—for example, by sensing, categorizing, and responding—so as to enhance survival and especially reproduction, including appropriately investing in offspring. In actuality, it is the full picture of implications of life history theory with respect to individual behavior that is most important to one interested in drawing correspondences between the evolutionary and clinical approaches to behavior, particularly psychopathology. Remember that the tenets of life history theory apply to all phases of the individual life span, not just the period of early development when plasticity is most evident. It is true that, especially in highly social animals whose offspring are born immature and in which the distinction between genotype and phenotype is indeterminate, allowing for much learning, the effects of natural selection are most likely to be realized during early, critical periods. During this period of developmental plasticity, programs of behavior having a large innate, genetic component are most malleable and shaped by environmental factors. However, the effects of these programs on individual behavior are far-reaching indeed. Imperatives that promote survival and reproduction and maximize inclusive fitness clearly operate throughout the individual life span. These programs may be more vulnerable to being shaped and modified early in development, but they affect a range of behaviors that cover the whole life cycle—that is, behaviors that affect maturation, mate selection, fertility and fecundity, marital arrangements and commitments, parenting, health maintenance, and, finally, social activity and influence during senescence.

EVOLUTION OF THE LIFE CYCLE

A visible behavioral manifestation of the evolution of life history is that of the life cycle. It is divided into the stages following birth that describe major physiological and behavioral milestones of a species. Most mammals traverse from infancy to adulthood without intervening stages. Social mammals have three basic stages of postnatal development: infant, juvenile, and adult. It is customary to classify the human life cycle in terms of five stages: infancy (mothers provide all or some nourishment via lactation), childhood (after weaning, with feeding and protection still provided), juvenile period (prepubertal individuals no longer dependent on mothers), adolescence (marked by the onset of puberty or sexual maturity), and adulthood. This classification of the life cycle is applied to nonhuman primates and hominids.

Behavioral biologists point out that during hominid evolution complex changes have taken place in the timing and length of the various stages of the life cycle. These involve the duration and velocity of growth, general morphology including particular brain size, and changes in social and psychological behavior (Bogin and Smith 1996). The volume of the brain has increased. However, there are limits to fetal brain growth set by the size of the pelvis and this has necessitated greater postnatal brain growth. Behavioral immaturity and dependence of the young (altriciality) have resulted, placing special demands for learning and corresponding changes in the period of time involving infancy, childhood, the juvenile period, and adolescence.

During the course of human evolution, changes in morphology, reproduction, and behavior have accompanied changes in the life cycle. For example, because the period of infancy has shortened and that of childhood lengthened, the care of children being supplemented by older siblings, mothers have been freed to have more children. Also, adolescence has been associated with a significantly greater increase in the growth of virtually all skeletal tissue. Among apes, and presumably the earliest hominids, brain growth is rapid before birth and comparatively slow after birth, whereas in humans brain growth is rapid both before and after birth. "[R]elative to body size human neonatal brain size is 1.33 times larger than that of great apes, but by adulthood the difference is 3.5 times" (Bogin and Smith 1996, 706).

Brain size increased during the time of Homo erectus, and this is believed to be associated with an expansion of the childhood period so as to provide high-quality foods required by the more humanlike pattern of postnatal brain growth. Late Homo erectus showed an even larger brain size, and with this came a further expansion of childhood along with insertion of an adolescent phase, with its corresponding increments in body growth and social learning. Another significant evolutionary change in the life cycle involved the postreproductive phase and menopause. Apes continue to reproduce throughout the life cycle, but human females cease around fifty yet can live several decades more. What exactly selected for this menopausal phase is controversial, but it is generally assumed that one factor has been that grandmothers (and possibly grandfathers), by providing care for grandchildren and serving as repositories of group cultural knowledge, promote enculturation and learning in the offspring of their children and thereby increase their own inclusive fitness.

The evolution of the human life history program and life cycle means that the bio-behavioral structure that conditions the patterns of psychopathology has changed. The social ecology of hominid groups and the cognition and culture of their members have constituted factors that have affected how psychopathology was configured and enacted. Presently, it is the emergence of distinct stages and corresponding changes in their duration and function with respect to social and psychological behavior that need emphasis.

In general, the relatively automatic, uncomplicated, and seamless progression from infancy to adulthood found at the ape end of the evolutionary terrain has given way to developmental periods that were far more delicate and sensitively tuned to the behaviors and mental health of caretakers and group mates during the latter phases of human evolution. As the physical and temporal structuring of stages of the life cycle have changed, they have required different scheduling and programming of such things as maternal solicitation and nurturance, exchanges involving parental attachment and bonding, caretaking from kin, varieties of play with age mates, and supportive but also potentially stressful inputs from unrelated group mates. The greater dependence and immaturity of infants, children, and juveniles has increased their vulnerability to ecological pressures and challenges. The importance and vulnerability of adolescence as a phase of preparation for adult roles has come into prominence. And, finally, the postreproductive phase following adulthood has brought to the fore a whole set of potential problems involving emotional disposition, morale, and sense of integration with and worthiness to the group and the self.

To the social ecology, cognition, and culture that affect the configuration and enactment of psychopathology on the stage of human evolution one has to add the very concrete and material set of factors tied together and manifest in the life cycle as explained by life history theory. Changes in the causes, manifestations, and outcome of psychopathology linked to changes in cognition, culture, and associated social relations within the group, and as impacted by ecology, must be seen as conditioned by such specific, tangible, and mechanical factors as how well and how long infants are succored, how they are weaned, and who takes over their care, how many age mates are found in the group to provide the necessary stimulation and learning associated with play, whether male adults and especially fathers are present in the group, and whether postreproductive grandparents are available to promote enculturation and socialization. Varieties of psychopathology signaled by anxiety, depression, hostility, irritability, bodily ailments and pain, and social divisiveness change during human evolution not only as a function of such mental and soft items as language, cognition, emotionality, and culture, but equally importantly in terms of the comparatively harder ones stemming from the scaffolds that contain, constrain, and direct basic biological functions.

THE EVOLUTIONARY IMAGE OF PSYCHOPATHOLOGY

A particular combination of behaviors or a behavior routine that might constitute psychopathology to the clinician could be viewed as a possible variant of a behavior routine once designed to optimize survival and reproduction. Such behavior, for example, could be a reactive strategy to the challenges posed by the environment, or several environments, that populations have encountered over evolutionary time. Life history traits—for example, age at first reproduction, interbirth interval—constitute reproductively significant aspects of behavior produced by existing social environments affecting inherited genetic characteristics and behavioral programs, also termed psychological adaptations.

Because life history traits cover an individual lifetime, virtually everything that organisms do can be seen as a life history trait having a bearing on survival and reproductive success. For this reason, both culturally situated clinicians and pancultural evolutionists are concerned with much the same material. Both aim to explain and make sense of behavior of individuals. However, those emphasizing the incompatibility between viewpoints claim that what for the clinician is pathology or deviance is to the evolutionist perhaps merely a form of behavioral variation, or one of a range of possible strategies produced or set in motion as a consequence of the impact of the prevailing environment on an organism's or a population's attempts to survive and reproduce. In other words, to a strict evolutionist, psychopathology is not necessarily pathological, but must be examined in context. The behavior in question may instead constitute or reflect a strategy or routine that is tracking the prevailing environment, comparing it to ancestral ones, and selecting optimal strategies that work best given the exigencies of survival and reproduction.

AN INCOMPATIBILITY OF VIEWPOINTS?

According to intellectual critics, then, the cultural/clinical and evolutionary perspectives describe two different approaches to routines or syndromes of behavior. The first is culturally situated, personalized, concentrates on individuals and proximal circumstances, is value-laden because it is directed at abnormalities and deviance (undesirable and unwanted behaviors) and seeks to alleviate distressed and compromised functioning. The second is impersonal and objective, seeks to make sense of behavior of organisms and especially populations that have an evolutionary past and seek an evolutionary future, and examines behavior in relation to processes of natural selection and characteristics of life history theory, all of which provide the backdrop against which behavior makes sense. The routine or syndrome of behavior that interests the evolutionist is not likely to be accorded special clinical or diagnostic significance. To the evolutionist, psychopathology is not a relevant concept. The evolutionary approach can explain alleged disorders by means of alternative concepts such as disease, group conflict and social instability, negative selection, behavior strategy, or environmental variation. Nor is the behavior constituting a psychopathology of special interest. What is important is how the behavior fits into a sequence or pattern that in its totality explains how genotype, phenotype, environment, developmental plasticity, and the imperatives of survival and enhanced reproduction have united to design the behavior of organisms and populations constantly changed and challenged by the process of natural selection. Personal distress, suffering, and healing are irrelevant to the behavior routine in question because individual behavior has no essentialist meaning, as it does to the clinician for whom it can possibly represent a psychiatric disorder. I hasten to add here that this alleged polarity and incompatibility between viewpoints is one projected onto the respective sciences and disciplines and is belied by the efforts and accomplishments of evolutionary psychiatrists in particular.

James S. Chisholm (1988, 1996) has argued about the importance of life history theory for understanding human adaptive and maladaptive behavior in an evolutionary context. His starting point (1988) was the limitations of abstract, impersonal concepts such as inclusive fitness and many of those of life history theory for developing a science of human behavior. Chisholm outlined important principles of a "developmental evolutionary ecology" in his argument about the importance of the phenotype as against the pure genotype and of the pivotal influence of factors coming to bear on individuals during infancy and the juvenile period in relation to changing social ecologies. He saw this as important for the study of human behavior because "the relationship of the genotype to the phenotype is a developmental one of tradeoffs among the components of fitness, of phenotype-environmental dialectical contingencies and constraints and adaptive indeterminacy" (99).

More recently, Chisholm (1996) has argued that the ecology and organization of attachment behaviors provide the mechanisms in terms of which basic parameters of life history theory are shaped at the individual level. Bowlby (1969, 1973) had argued eloquently and forcefully that the behavior that infants and toddlers show in relation to caretakers constituted a primary, autonomous drive they are born with to be close to—to be attached to—their mothers. The attachment needs of offspring can elicit in mothers different varieties of maternal behavior—often equated with or said to stem from psychopathology—that in the infants and toddlers create behavior patterns—also equated with varieties of psychopathology—that are best seen as strategies about how to optimally take advantage of the prevailing amounts of resources and uncertainties in the social environment. Infants are born with a potential to adopt different strategies of behavior pertaining to reproduction as a consequence of environmental contingencies mirrored in their mother's behaviors toward them. This potential is said to consist of mechanisms of behavior designed in EEA, and it has far-reaching implications. In Chisholm's words: "The life history model of attachment . . . suggests . . . that the attachment process functions as an evolved mechanism for 'switching' developmental trajectories closer to a child's local optimal path [and that] the developmental basis for the difference between secure and insecure attachment might be part of a psychobiological mechanism for allocating resources contingently to optimize the trade-off between current and future reproduction" (Chisholm 1996, 22).

It is Chisholm's contention that the attachment strategy an infant, toddler, and juvenile adopt is a response to quality of mothering, which itself provides ecological clues about what can be expected from the environment. Not only does the strategy adopted early in life influence how social relations are conducted during infancy and the juvenile period; it helps predict future reproductive and parenting efforts when the individual grows to maturity.

For the argument being developed here, Chisholm's formulation of attachment organization as a mechanism in terms of which life history reproductive strategies are shaped early in life has two interesting implications. First, it blurs the difference between apparent maladaptive behavior or psychopathology and adaptive strategy or mechanism—that is, in both parents and offspring—in the process seeming to underscore the indifference and impersonality of how an evolutionary scenario depicts something that is important and vital—for example, distress, suffering—in a cultural, clinical viewpoint. This gets back to the postulated criticism of the evolutionary approach described earlier. Such an interpretation, however, is obviously illusory and also incorrect. Bowlby's general orientation, and that of Chisholm, is to promote a better understanding of factors affecting human development, personality, and ultimately adult behavior for the purpose of alleviating suffering associated with psychopathology. This is an explicit concern of evolutionary psychiatrists (McGuire and Troisi 1998; Stevens and Price 1996). In fact, and this is the second implication, Chisholm's formulation actually provides a good illustration of the complementary way in which the cultural, clinical as compared to the evolutionary viewpoints relate to one another. It shows how abstract, impersonal aspects of evolutionary theory, formulated as life history traits and demographic characteristics of populations, when examined at the level of individuals, bring into center stage important questions and problems pertaining to human behavior and psychopathology that beg clinical, humane answers and solutions best arrived at through an understanding of evolutionary theory.

Evolutionary Functionalism and Psychopathology

In evolutionary functional analyses, the physical and social environment is conceptualized as posing problems that must be solved if organisms are to reproduce and survive. The four general dispositions of behavior discussed in chapter 2, namely, attachment behaviors, temperament, psychological adaptations or algorithms, and personality traits, are the engineering solutions to distinct types of problems. They control and regulate social and psychological behavior so as to solve those problems, enhancing fitness. Individuals that live in close-knit, highly socialized groups, like nonhuman primates, the hominids, varieties of Homo, have evolved specific behavior programs geared to cope with specific challenges and problems. The following are some examples: (1) programs for recognizing the emotional states of group mates and emotions aroused within the organism itself; (2) sexual attraction or programs for choosing a suitable mate with whom to produce offspring; (3) programs for the care of offspring by one or both parents, depending on the species and phase of evolution, designed to protect, feed, and teach infants; (4) programs for assessing harm, to estimate the risks posed by predators, physical hardships, competition within the group or species, environmental hazards—each of which could entail a separate mechanism or adaptation; (5) programs for social reciprocation, to promote altruistic, mutualistic, or trust-building behavior in the organism and to enable it to determine which group mates return favors, which ones cheat, which ones pose no threat; (6) programs for detecting exploiters, to spot group mates who might take unfair advantage of the individual and/or the group, along with ways of dealing with them; (7) programs for arousing sympathy, which empathetically draw the organisms to the suffering of group mates; (8) sickness and healing programs designed to communicate and interpret states of disease, injury, or pathology in group mates and the self and to elicit healing actions; (9) consciousness-altering programs, to enable organisms to endure hardships by entering dissociated, heightened states of arousal and concentration that enable them to confront problems and resolve conflicts; (10) hope-building programs for confronting loss, separation, and harmful events or circumstances; (11) programs for arousing fear, anxiety, and hostility designed to elicit emotions that protect the organism from threatening situations or enemies; and, more generally, (12) mood-regulating programs that keep emotional arousal within boundaries so that it does not interfere with necessary activities, and regulate circumstances in the social and physical environment that generated the emotions. This is not a complete list—obviously, many other programs could be cited. The programs noted here can be viewed as evolutionary residuals, subsidiary to or manifesting the four dispositions discussed in chapter 2.

With this as background, one can construct a functional concept of psychopathology. That is, psychopathology consists of manifestations—signs and symptoms—of failures, breakdowns, or malfunctions of these programs. Each program mentioned above involves a scenario of social and psychological behavior and associated physiological functions elicited in response to specific social and environmental events and circumstances and directed toward survival and reproduction. When these programs fail, the individual is potentially compromised from a fitness standpoint and exhibits social and psychological behavior and related bodily symptoms that register its deficiencies, along with personal distress. Failures include persistent cognitive deficits such as inability to pay attention, recall events, solve problems; negative emotional states, such as fear, anxiety, anger, depression, irritability, impulsiveness; associated asocial or antisocial tendencies, such as extreme sexual jealousy, distrust, antagonism, social isolation or avoidance; neuromuscular, neurological, and general physiological reactions that register reactive or secondary bodily pain and dysfunction; and failure to regulate the operations of autonomic nervous system mechanisms registered as visceral and neurovegetative breakdown or impairment, such as the inability to sleep, loss of motivation, appetite, energy, or sexual desire.

Such failures are manifestations of psychopathology conceptualized in functional, evolutionary terms. While they are here framed in neutral and abstract terms, they would be given content and meaning depending on how the specific functional areas are realized in the culture of the group. In other words, in addition to impairments of general system functions, the thwarting of goals of a particular behavior program would have specific social and psychological manifestations, given a group's type of cognition and culture. Content-specific signs and symptoms can include sexual jealousy, parental neglect, fear of accompanying group mates during foraging forays, failure to share with siblings, failure to reciprocate favors, inappropriate emotional displays, avoiding open spaces because of exaggerated fear of predators, irritable or hostile behavior toward individuals whose support is needed for maintaining social rank and status, misuse of dissociative experience and trancing behavior, harassment of juveniles and individuals of lower rank, and, finally, prolonged mourning and inability to let go of the corpse of a group mate, sibling, or offspring.

Programs of behavior can be thwarted by genetic vulnerabilities, adverse developmental experiences, disease, injury, pathology, stresses in the social environment as reflected by food or water scarcity, competition between species, predators, stress caused by conflicts of rank, and competition from predators. The situations or problems that trigger the various content-specific programs may be impossible to resolve. Individuals may experience manifestations of impairment that include bodily pain, psychological distress, and visceral somatic, neurovegetative malfunctions, as well as more specific forms of distress and disability, depending on phase of evolution, linked to the specific content of the programs whose goals or functions were thwarted. In short, psychopathology as seen from the standpoint of evolutionary functionalism is a perturbation in the operation of behavior programs that brings about a breakdown in individual adaptation.

An evolutionary functionalist account of behavior programs also has an environmental focus. The behavior programs owe their rationale to and operate in the context of a distinctive habitat, its geography, and its inhabitants. Ecological parameters act as constraints on the behavior of individuals and groups and determine how effectively the programs function. In conjunction with a population's life history, demographic factors (a group's size, age structure, sex ratio), cognitive ability, including protolanguage or language proper, and strategies for learning and transmitting culture, ecological factors condition behavior and a group's level of cultural adaptation. The model proposed by Parker and Russon (1996) for distinguishing between the cultural potential of various species, from monkeys and apes to Homo sapiens—that is, preculture, protoculture, Ur-culture, or Eu-culture—emphasizes the functional interrelation of all of the parameters discussed here. The breakdowns of behavior that occur as a consequence of genetic vulnerabilities and environmental perturbations need to be interpreted in social and ecological terms. An evolutionary functionalist account of psychopathology, in short, is based on an appreciation of not only the function and malfunction of behavior programs but also the meaning, interpretation, and functions of psychopathology in an ecological context.

Points of Correspondence between the Perspectives

Part of the reason critics emphasize a polarity and incompatibility between cultural situated clinicians and impersonal evolutionists is that the latter can sometimes appear to focus less on the individual than on the group, less on behavior looked at from within and in the short run, that is, proximal factors, and more from without and over the long run, that is, ultimate factors. Not only does the strict evolutionary theory seem to place less emphasis on difficult-to-explain phenomena natural to the clinician—for example, individual suffering, deviance, eccentricity—but in emphasizing how groups smooth out conflicts and develop a modus vivendi of organization and integration, it seems to redefine and thereby exclude behavioral phenomena that the clinician holds paramount. The studies of Chisholm illustrate in a compelling way that group phenomena have profound relevance to an individual's circumstances as these pertain to the possibility of psychopathology.

There is a general reluctance on the part of evolutionary theorists to project actual varieties of psychopathology onto animals generally, especially hominid and earlier varieties of species Homo. This could bear a relationship to the alleged difference between a cultural, clinical compared to an evolutionary, detached viewpoint. Such a difference is belied by a cursory but careful reading of evolutionary theory. Nevertheless, the perception of polarity and incompatibility between viewpoints persists, especially in the general intellectual arena and literary cultural studies, including even cultural anthropology (Brown 1991). For example, the book by Power (1991), which is taken up later in more detail, reflects in part a critical view of ethology and evolutionary theory. The book by Wrangham and Peterson (1996) describes natural behaviors of chimpanzee males that raise pointed questions about morality and humaneness, not to say possible psychopathology. So also do the evolutionary writings of Smuts (1992, 1995) on behavior involving patriarchy and sexual coercion by males (but see her 1985). Finally, part of the rhetorical emphasis that M. S. Dawkins (1980, 1990, 1993) and others (Broom and Johnson, 1993) give to their account of the scientific relevance of the concept of animal suffering is based at least in part on the seeming indifference of evolutionary theory to this topic, which itself bears a relationship to the topic of psychopathology as pursued here.

Ironically, the forms of behavior that evolutionists might call pathological are precisely the ones that have been manufactured in the laboratory by scientists, steeped in evolutionary biology, working within the biomedical establishment to develop animal models of disease. Few evolutionists would quarrel with the notion that responses to extreme forms of maternal deprivation, the social avoidance and ineptitude shown by monkeys and apes with lesions in the cerebral cortex or amygdala region of the brain, and the abnormalities of vigilance and social communication produced by intoxication with various psychoactive agents are all abnormal or psychopathological behavior. Apparently, only by an intense manipulation of what is known to produce abnormality in humans and literally manufacturing what looks and behaves like human abnormalities are some evolutionists willing to embrace the viability of a concept of psychopathology as applied to animals. However, such abnormal behavior is likely to be viewed as unnatural or outside the expected behavior of members of feral communities. In short, there are grounds for claiming that strict evolutionists, as well as scientists steeped in its theories who work within the biomedical establishment, would support the idea that psychopathology entails qualitatively different behavioral phenomena compared to what is natural to animal communities. Yet, a psychiatrist's reductionist view that mental illness is based on significant and acquired neurobiological lesions, a bona fide medical disease, as it were, reflects a view that the environment is all-important in causing behavioral variation. Why should this not apply to hominids and earlier members of special Homo?

A potential compromise between extreme exponents of the two viewpoints lies in the concept of maladaptation. If one defines maladaptation as consisting of behavior of an organism that undermines its fitness, *broadly conceived,* one might accommodate the two perspectives and harness the power of each for understanding behavior in general and psychopathology in particular. An evolutionist would have to agree that an organism such as a monkey or an ape and, by extension, a representative of the earliest hominid group, that shows persistent behaviors which cause it to be marginalized in its group, lead to its ostracism, and undermine its efforts to survive and reproduce is showing behaviors that impair fitness (pathology). Examples of such behavior are excessive aggression or fearfulness, social avoidance, impulsiveness, selfishness, failure to reconcile conflicts, avoidance, neglect, and disregard of kin. Similarly, a clinician would have little difficulty in claiming that an individual who consistently, regardless of the social and cultural environment, behaves in ways that seriously hinder its efforts to work, look after itself, connect with the group, or form attachments and relationships with group mates of both sexes is showing evidence of psychopathology. To both the evolutionist and clinician, the bases of such behaviors, their causes, and their content would most probably be regarded as secondary. In other words, a genetic defect, a bad developmental environment, a brain lesion, or intercurrent stress might explain the behaviors in question. Similarly, how an organism or person shows its maladaptation might very well differ as a function of group, environment, species, and, in the case of human beings, society or culture. However, regardless of cause or content of behavior, it is easy to identify a maladaptation and/or psychopathology.

This provides a possible reconciliation between the two perspectives. Here, an individual's behavior routine or syndrome is judged pathological on clinical grounds—that is, it meets criteria for diagnosis of a disorder—and could also be judged as pathological or clearly maladaptive from an evolutionary standpoint. A picture of psychopathology or actual disorder that works against the fitness prospects of an individual would constitute the paradigmatic feature. Adolescents or adults who commit suicide after the breakup of a love relationship would meet the criteria for various diagnoses—major depression, adjustment disorder with depressed mood—and also be acting maladaptively from an evolutionary or life history standpoint. Other examples might be found in individuals who develop a change in personality and outlook as a consequence of situational contingencies or brain lesions and drop out of society or join a cult. The latter behavior could be construed as social suicide. From an evolutionary standpoint, they have forfeited their chance of passing on or contributing to the welfare of their genes. In all of these instances, the evolutionist might agree with the psychiatrist that such behavior is a manifestation of psychopathology.

Integrating Evolutionary and Clinical Perspectives of Psychopathology

The clinical and evolutionary perspectives about the social and psychological behavior of organisms are analytic frameworks imposed by the observer. In each perspective, a routine or syndrome of behavior possesses different characteristics, that is, ontological difference, and it is also interpreted differently, that is, epistemological difference. Yet the seeming incompatibility between the two frames of reference breaks down when one applies them to real or hypothetical groups of organisms located on the line of human evolution.

Consider four hypothetical groups of hominids, H1, H2, H3, and H4, located approximately 5, 2, 1, and 0.3 million years ago. In each group and in each setting one observes a routine or syndrome of behavior in a particular adult organism. We assume that the organism has shown the routines of behavior on previous occasions. The cause of the syndrome can be assumed to be a combination of genetic or temperamental vulnerabilities, factors that may have had a negative impact during key phases of development, or environmental or social stressors. In each scenario, we assume that the essential context and content of the syndrome is relatively similar and shared across time periods. However, at each subsequent phase of evolution, several forms of social and psychological behavior are added to the syndrome. These additions are consistent with presumptions as to the evolution of cognition and culture. In its final contextualization or construction, then, the syndrome would reflect phylogenetic continuity and yet it would have differed in significant ways over time, as can be expected because of evolutionary change.

SCENARIO H1

During this phase 5 million years ago, consisting of the very earliest hominids— for example, varieties of the Australopithecines—imagine a syndrome or routine of social and psychological behavior in an adult male who manifests behavioral changes over a period of several weeks, perhaps months. He is prone to fatigue, emotional timidity, and irritability, and displays overt anger and hostility toward lower-ranking males, females, and juveniles. The angry outbursts are assumed to be caused by frustration experienced during routine, everyday social activities linked to status and ranking, foraging, food sharing, scanning for and warding off predators, and mating. For example, during these syndromes, he engages in dramatic protest displays, during which he flings himself on and pounds the ground, and displays angry, hostile, harassing behavior when he cannot fulfill his needs, whether these are food and water, gifts from others that he begs for, or a secure, comfortable resting place. He may be responding to aggressive, imperious behavior on the part of those higher in status, inadequate displays of deference by males lower in status, or the reluctance of an ovulating female to mate. In this scenario, the male's frustration and anger are prolonged and go beyond the ordinary levels of everyday exchange. He detaches himself socially and emotionally, seemingly oblivious to group mates. Except for a few group mates who approach him and maintain social and physical contact via tentative attempts at grooming, he is essentially isolated. Attempts to comfort him and draw him into group activities are accompanied by vocal communications aimed at encouraging social participation. Such behavior is not unlike the angry, frustrated protests (temper tantrums) described by de Waal (1989, 1996) when a chimpanzee fails to achieve its goals, only in this example the organism's displays are repeated over a longer span of time and are accompanied by behavior that suggests weariness, distress, fatigue, and an emotional state one could describe as defeatist, discouraged, or alienated. A capacity for such behavior can be assumed to have been manifest among groups of the last common ancestor (LCA).

An evolutionist would explain this syndrome of behavior as a manifestation of frustration and anger at not being able to implement successfully a particular adaptive design—that is, one or several psychological adaptations designed by natural selection—which the organism has inherited. He would be described as illustrating individual variation in behavior and temperament, an example of the mix of genotypes and phenotypes found in natural communities. Organisms of this general disposition and temperament, it may be claimed, show a lowered fitness and are more likely to be selected against, that is, to fare poorly in the process of natural selection.

To the evolutionary psychologist, the behavior syndrome illustrates several types of psychological adaptations that could have been thwarted. Examples would include adaptations for searching for and processing food, cleverness and cunning in relations of social exchange or reciprocation, mating, and social cognition. The form and content of behavior would be explained as the consequences of a thwarted neuropsychological and behavioral adaptive mechanism. As an example, the orderly processes of cortically and subcortically activated mechanisms that connect to lower brain centers, mechanisms that mediate emotional and motoric displays and behaviors, might have been impaired and easily perturbed from a neurological standpoint as a consequence of acquired lesions or vulnerabilities caused by genetic or developmental factors. Alternatively, activation of cortically mediated social and emotional motor programs located in limbic/midbrain circuits ordinarily coordinated and kept in balance might have spilled over into one another, resulting in loss of control. These two explanatory schemes are obviously not mutually exclusive.

The evolutionist would maintain that how an organism adjusts its behavior in the above scenario is critical to its survival and reproduction. Whether the organism can elicit support and compliance for its needs and wants, whether it persists in tantrum behavior or escalates the level of hostility, and how long it remains detached and self-isolated all determine whether the organism will be able to accommodate to the exigencies of sociality and, by reaping the advantages of group living, continue to function adaptively in the group. Continued displays of protest, frustration, and self-isolation may result in its departure or banishment from the group, possibly along with kin. Alternatively, if an organism's needs are not met for a protracted period and it has tantrums that leave it physically exhausted, the organism's chances of survival and reproduction could be impaired, and it could become a weakened prey and vulnerable to attack.

The clinician's interpretation of H1 would differ sharply from that of the evolutionist. The psychiatrist or psychologist might see a distressed and behaviorally compromised organism (as would the evolutionist, but the latter would regard the clinician's theoretical concerns as falling outside the strict logic of evolutionary biology). However, the clinician would claim that the organism's behavior, if not neurotic or deviant, because it is repeatedly enacted, is seemingly maladaptive. The clinician might regard the routine or syndrome as an example of an emergent variety of psychopathology. The clinician would note the absence of consolation and accommodation offered by most of its group mates, offset by the perhaps tentative responses of a few group mates, perhaps siblings or a mother, offering support and comfort. Such gestures would reflect an awareness that the organism is in distress, suffering, and thus compromised if not impaired and disabled. Indeed, given their general (perhaps rudimentary) capacity for self-awareness and mental activity, members of the group might perceive the organism's "personality," behavioral dispositions, and repeated aberrant responses as deviant. However, emotional reactions to the organism would differ among subgroups as described.

The earliest hominids, the clinician would claim, must have manifested at least as much self-awareness, social cognition, and capacity for expression (although far from constituting full linguistic representation, it can be termed symbolic and communicative) as one observes among contemporary common chimpanzees and so-called pigmy chimpanzees (Savage-Rumbaugh and Lewin 1994). Because chimpanzees are commonly viewed as the best model of the social and cognitive abilities of LCAs, the clinician would assume that the very earliest hominids could have advanced beyond this level. Members of H1, then, probably manifested an embryonic capacity for comprehension and reference. They might have been capable of having invented or evolved a small set of symbols that referred to objects, events, and intentional states of group mates. Stored knowledge of this very rudimentary symbol system would constitute part of the group's culture and would probably be drawn on, expressed (behaviorally, emotionally, through displays and vocalized calls), and decoded. Thus, in the succor, consolation, and sympathy shown to the organism, however elementary, the clinician might infer the beginnings of an abstract conception that suggests a group awareness about the plight and behavior of the organism.

SCENARIO H2

In the hypothetical scenario of 2 million years ago, imagine essentially the same routine of behavior, only this time manifested in a member of a group of more evolved organisms. These could be either a late, robust variety of Australopithecine or an early variety of Homo, Homo habilis. However, in the H2 scenario, a more evolved level of cognition, culture, communication, and social awareness suggests that the organisms in question have evolved a more elaborated capacity for off-line thinking, symbolization, vocalization (possibly, although it is doubtful, a protolanguage). They appear to show the rudiments of a sense of group identity, even possibly awareness of differences in behavior between themselves and competing groups of hominids. For these reasons, the group could be said to manifest distinctive traditions of behavior or culture (Bickerton 1995; Deacon 1997; Donald 1991, 1993b; Parker and Russon 1996; Richards 1987, 1989a, 1989b).

Scenario H2 would include some features of social psychological behavior not present in H1. The organism, as an example, might be able to self-consciously identify its distress and frustration and be able to communicate about it as well. This would be accomplished by means of a representation of its mental, bodily, behavioral state, an associated mimetic routine, including facial displays of emotion, and vocalizations and gestures. These could be regarded as either relatively simple categories and forms of representation or as forms of communication that it acts out toward group mates—that is, the facial displays and vocalizations of emotion and meaning. If something like a protolanguage is present, albeit less tangibly and discernibly than in scenario H3, communication would consist of expressions that refer to an object, situation, goal, action, or thwarted action. The organism might emit sounds in association with gestures. Such behavior reflects basic cognitive capacities that involve symbolic comprehension and reference, an ability to produce and interpret vocalized utterances, and a capacity to carry out such functions as naming, requesting, and comprehending states of being and the import of events or circumstances. In other words, in H2 and in situations that involve routines and syndromes of behavior like the one depicted here, one would observe behavior reflecting capacities and abilities more advanced than those of H1—capacities and abilities, in other words, that will eventually culminate in language and speech.

A higher, more evolved sense of self-conceptualization would be associated during this phase with a more evolved capacity for social cognition. In other words, the organism's better developed mechanism for theory of mind thinking would allow it to better define its present state and better monitor and categorize the reactions, motives, and intentions of its group mates. In effect, with its enhanced capacity for cognition and communication, the organism should be able to express its emotional, self-conscious plight to the group at large or to selected members beyond its kin and allies. While in H1 the organism emotes and largely throws a tantrum of protest and frustration, in H2, given its more evolved level of cognition and shared culture, it may be said to be directing different messages toward different group mates or sets thereof. In short, a more directed and differentiated routine of expressive behavior is likely in scenario H2. Since a protosymbolic, protolinguistic representation of the organism's experience, behavior, and general distress entails interpretive and communicative actions, responses that reflect comprehension will likely be elicited from group mates.

These responses of group mates, as in scenario H1, might well include angry displays and denunciations, shows of indifference, vocalized exhortations geared to bringing the temper tantrum of the organism in line with group goals and demands, or accusatory gestures, protolinguistic routines condemning the organism's arrogation of power and resources, or its breaking the rules, whatever might have constituted the cause as they see it. For example, the organism's failure to comply with the group's routines and conventions concerning the procurement, processing, and consumption of food, as after a hunt, would be implicitly, perhaps even explicitly, communicated as the organism is socially chastised. Depending on ecological conditions and the group's comparative success in the habitat—which might foster a larger, more socially organized group, one with emergent shared conventions about behavior—the organism's attempt to mate with a particular female could be deemed a quasi moral violation subject to social chastisement. In other words, it is possible that participants in H2 would have realized a more evolved level of culture, cognition, and self-awareness.

The group might have more articulate categories of kin and social relations, in which case the organism's behavior would be implicitly mapped in relation to emergent social conventions. At any rate, any chastisements and condemnations would most probably be emotively but protolinguistically barked at the organism in question—embedded, of course, in expressive communications that involve emotive facial displays and gestures.

Allies, friends, siblings, and maternal figures and other relatives, depending on their number, size, strength, and relative rank in the group, might display against and/or threaten vocally members of the cohort that was critical of the organism or had thwarted its motives. The organism's visible frustrations or its retreat to a detached, silent, reverie-like mode, as evident, although less sharply etched in H1, is likely to elicit affiliative responses from allies, friends, and relatives. They may enact better articulated, focused responses in gestures and protolinguistic vocalizations that, in addition to reflecting understanding and sympathy, also offer consolation and support. If the organism could not be distracted and encouraged to return to group activities, allied group mates might provision it with food. Alternatively, if they decide that the precipitating incidents do not warrant such a response, or that the organism has committed behavioral violations, or at least that its behavior is faulty in conception, and/or they do not find social conditions in the group to be propitious for a full-scale confrontation with the dominant subgroup, they may join in chastising and denouncing it, albeit in a less strident way, in an effort to persuade, mollify, and regulate its behavior and experience.

If this behavioral scenario were not resolved, the organism could possibly be banished, or the group might break up if group mates of the thwarted and aggrieved organism maintain their alliance. As in H1, if the organism becomes physically exhausted or vulnerable to disease, it could become the victim of predators or attack from competitive neighboring groups.

Our hypothetical evolutionist would attribute such a routine of behavior to much the same causes as for H1: ultimate conditions involving innate, naturally designed psychological adaptations or mechanisms, the perturbation or thwarting of these as a result of proximal conditions, and the proximally evoked, natural responses of distress and frustration. Similarly, the enactment, course, and resolution of the H2 scenario present no essential conflict to the evolutionist, who can easily account for them in terms of traditional concepts of ethology: psychological adaptation, innate response patterns, competition for scarce resources, dominance status, social ecological contingencies, and natural selection.

The psychiatrist or psychologist would concur with the evolutionist's causal interpretation of H2 and would also largely concur with the latter's interpretation regarding its outcome. However, the clinician might argue that the organism's psychological and social identity has changed somewhat as a consequence of its difficulties and circumstances and is manifested in changes in personal experience and behavior. For one, the organism's distress—indeed, suffering—and obviously maladaptive behavior is more sharply profiled than in H1. The retreat into a self-absorptive state of detachment and social isolation is more marked in H2, visibly defining the behavioral angst of the organism. Indeed, given the advances in cognition and culture reflected in H2, the clinician might assume that selected group mates see these behavioral changes as expressive of limitations and constraints now linked to an awareness of the organism's mental and bodily condition and state of the self.

In short, in light of the more evolved level of cognition, culture, and social awareness of H2, group mates might come to see the organism as disabled. More specifically, kin, offspring, assuming a modicum of paternal connectedness, and allies would perceive that the organism is reacting in a compromised manner to intercurrent events and circumstances. The clinician would point to the subtle efforts at consolation, empathy if not sympathy, and support offered by relatives and friends of the organism, now better self-identified and identified within the group as a distinct entity with a network of affiliations. The clinician could then, with caution, maintain that the group's responses consist of condemnation and rejection on the part of the majority and an awareness of suffering and distress on the part of the minority. To the extent that actions of this minority involve attempts at grooming, sharing of food, and some accommodation to the threats to the distressed organism's ability to thrive, the clinician might even judge these social responses as constituting emergent forms of behavioral or mental health healing or even psychotherapy. In other words, it is possible to see in H2 responses of some group mates that reflect emergent conceptions that the individual whose plight is behaviorally expressed is actually compromised, impaired, disabled, and suffering; and about which special dispensations are appropriate (and might be required), given a shared body of past experiences and patterns of interaction in exchange networks that go back in time and entail future involvements. All of this, according to the clinician, would qualify the behavior described in H2 as an emergent form of socially, culturally recognized psychopathology.

SCENARIO H3

In this phase of evolution, approximately 1 million years ago, one observes a similar ensemble of behavior with a similar history of development. However, because the group in question is more evolved, consisting of organisms making up Homo erectus, one can assume advances in symbolization and protolanguage, perhaps, of what one can more confidently view as cognition and culture more generally. Consequently, we may expect that more situationally relevant content will be added to the scenario. In other words, the behavioral routine will reveal conceptual and pragmatic resources for describing, explaining, and acting upon the prevailing situation and circumstances. In short, the group's resources for social communication can now with greater confidence be thought of as symbolic.

In purely concrete terms, something like a protolanguage will be added to the vocalized strings of symbols and largely mimetic and gestural ensemble of H2. More expressive vocal communications would consist of more intricately, that is, more conceptually and linguistically, linked strings of utterances. Elementary phonemes, morphemes, and simple sentences could be uttered, now relating the plight of an individual, no longer an organism. As an example, the scenario of behavior might reflect frustrated attempts to obtain a resource. Facts about the situation—whether or not the individual might be entitled to it, given his status—would be explicit in the individual's and group mates' communications. Alternatively, the individual's communications may reflect a misunderstanding about the status or intentions of a group mate he infringed upon, about a rule or standard of group behavior, or about messages—vocal, facial, gestural, or emotional—from a female he accosted in an inappropriate mating attempt. In short, the group's resources will be better categorized and group needs will be more elaborately conceptualized. This would be explicitly manifest in how the event and circumstances were communicated.

If the individual were thwarted in his addresses to a potential mate and if he violated no moral rule, then his protest temper tantrum may be accompanied by allusions to unmet expectations, if not obligations. Social exchange, in other words, would be governed by psychological mechanisms to which are attached conceptualizations and representations that extend well beyond those described in H1 and H2. Behavioral responses to frustrations of social exchange would no longer consist of purely emotional, sensory motor representations or behavioral programs but should involve emergent conceptual, symbolic features conveying notions of relationships, obligations, and entitlements, however crudely they are conceived by the individual—that is, as thoughts, expectations based on exchange rules—or are publicly expressed mimetically, gesturally, and protolinguistically. In short, utterances that signal a temper outburst may in the H3 scenario be vocally or verbally communicated in a protolanguage with an elementary syntax, along with condemnations based on emergent notions of exchange contracts or obligations traceable to algorithms that are more or less consciously taken for granted (Barkow, Cosmides and Tooby 1992; Cosmides 1989). Or, by contrast, information explaining why or how resources may be properly earned, assertions that they will be shared under certain conditions of exchange, may all form the context of the behavior. The greater capacity for symbolization means that the temperamental and repeatedly frustrated individual has a better defined sense of personal identity and a more differentiated social identity in the group. Thus, the individual is capable of some degree of reflection and interpretation as to his behavior and experience, and group mates are better able to categorize not only the individual in question but also his present behavior and experiences.

Since during the H3 phase of evolution groups and individuals show a better developed sense of group identity and history, as well as a more evolved capacity for representing personal experience and behavior, we can begin to see the afflicted or deviant individual and his plight in relation to the dynamics operating in the group. Others might perceive the individual as a group problem, liability, or simply a nuisance. He might be regarded as a socially inept member of the group who must be ignored or passively tolerated. Still another possibility is that the individual is prone to suffering and distress and, as such, is disabled or compromised. These conceptualizations would affect the group's response to the individual and his behavior.

The resources of the environment, pressure and competition from other groups, the threat of predators, and the overall constraints on the activities of the group—its social and behavioral ecology, in other words—would all play a role in how group mates conceptualized and dealt with the problematic individual. He could be tolerated, commanded to leave or actively shunned, or allowed a somewhat marginal position in the group. The kinds of attachments that connected the individual to kin, friends, and allies would all play an influential role. Group mates might offer comfort and support, attempt to mediate the conflict, and alleviate his distress and heal him through social incorporation. The individual's ability to function and contribute to the group's adaptive needs and the willingness of kin and friends to assist and accommodate to his needs would also be crucial in how the incident of psychopathology played out in group.

The more evolved capacities for cognition, culture, and protolanguage in the H3 scenario, in conjunction with an emerging awareness of suffering and its bases, raises the question of whether at this phase of evolution a sense of cosmology, spirituality, or what has been termed the numinous might have emerged (Boyer 1994; Goodman 1972, 1988, 1990; Oubré 1997). Indeed, if group mates are able to identify certain behavior as revealing a form of affliction, malady, or sickness—a not unreasonable possibility, given the capacity for abstraction made possible by symbolization—they might ascribe the malady to forces outside the realm of ordinary, mundane events. In other words, some form of religious or cosmological attribution is possible (Boyer 1994). Similarly, to the extent that selected group mates categorize and interpret the behavior or the individual as afflicted, they might resort to some sort of ceremony, a playlike ritual response of gestures and mime in efforts to lift the spirits of the individual, engage his attention, remove him from his reverie by seeking supernatural intervention, and otherwise persuade him to reincorporate in the group's subsistence activities and social and family relations (Donald 1991). Chanting, singing, and dancing, hand clapping or other systematically repeated forms of aural stimulation and arousal may deepen the individual's detachment, helping him to enter an altered state of consciousness or trance in which he may be joined by selected group mates (Goodman 1988, 1990; Oubré 1997).

This phase of evolution, then, reveals a behavioral routine or syndrome that expresses personal distress and incapacity of an organism who can now be seen as an individual member of a group with a protoculture. His distress can now be made sense of in terms of a groupwide understanding of his responses and his frustrations. Consequently, one may discern the beginnings of social cultural identification and what a clinician could construe as enactments of routines of affliction and efforts to resolve them that involve the individual in some type of healing such as entering an altered state of consciousness or trance.

One sees in the H3 scenario a mixture of evolutionary-derived behavioral reactions and imperatives similar to those of scenarios H1 and H2 and group responses that in expressing concern, sympathy, support, and efforts at healing begin to suggest what, for lack of a better term, one may label as a clinical reaction and set of imperatives. The syndrome of behavior, its etiology, course, and natural history, are all rooted in inherited psychological mechanisms or adaptations that one can connect with patterns of behavior among members of earlier hominids; however, only in the present instance, an emergent clinical point of view is manifested in how the individual and group interpret and handle the situation.

At this point, then, the seeming incompatibility between the evolutionary and the clinical frames of reference appears to dissolve. Let me stress that in H3, the incident of emergent psychopathology still exists, ontologically, and can be explained epistemologically as a fact within an evolutionary narrative. In other words, its causes, manifestation, and outcomes can be viewed through the lens of fitness, natural selection, inclusive fitness, reciprocal altruism, and ideas about psychological adaptation. This is also true of an incident of psychopathology framed in a still more evolved context, H4. The point is that a clinical frame of reference is now depicted within the evolutionary frame of reference, as it were. In a rough way, the former explains how the group relates to maladies and sufferings of its members, whereas the latter frames the overall pattern of activities in a narrative that stretches back in time, has a basis in social behavioral ecology, and reaches into the future by a path marked now not only by biologically, genically stipulated selective forces but also by culturally stipulated ones.

In scenario H3, because a more or less discernible protoculture is made possible by a more advanced form of cognition and capacity for protolanguage, individuals are able to symbolize and abstractly configure, in group-relevant terms that acknowledge the individual's social identity and well-being, social happenings within the group. The behavioral features incorporated in H3 can be qualified in several ways; for example, as manifestations of a better-developed sense of morality and ethics (de Waal 1996), as showing a new sense of community and group integration and organization, as reflecting an appreciation of the agonies of personal suffering regarding which displays of a sense of compassion are natural, as showing an emergent form of spirituality and religion, as reflecting the emergence of a sense of medical consciousness in the group, and/or as showing the emergence of a culturally organized understanding and response to what we may describe as psychopathology.

SCENARIO H4

The behavioral syndrome described in H1, H2, and H3, when projected to approximately 300,000 years before the present era, would now be realized in a group of Archaic humans. This term is used to denote organisms who are placed phylogenetically somewhere between Homo erectus and Homo sapiens. The Neanderthals were very clearly in evidence during that period, although they are sometimes placed later. At this juncture, there were groups of individuals showing a more evolved cognition and culture. If a group of Archaic humans were severely stressed by the climatological and ecological characteristics of the habitat, its protoculture would likely be simple and unelaborated, and hence examples of psychopathology in H4 would have changed little since the time of H3. In physically stressed, ecologically impoverished circumstances, Archaic humans would show a more evolved form of language and cognition, but the inventory of symbols and their overall meanings would nonetheless be relatively impoverished in contrast to a comparable species situated in a more favorable social and behavioral environment. What follows is an interpretation of psychopathology in a comparatively more evolved hypothetical group of Archaic humans.

Assume that virtually all of the characteristics of behavior described in the earlier scenarios are also present in H4 and the earlier interpretations also apply to the hypothetical scenario of H4. However, in addition to, and as a consequence of, an evolved capacity for linguistic representation and more elaborate and abstract conceptions of personhood, group identity and history, and abstract metaphysical questions involving religion, cosmology, and death, one would likely find more detailed, elaborated, and organized pools of information and beliefs. Such bodies of knowledge, while small and relatively uncomplicated compared to those of today, would pertain not only to practical considerations—such as how to get and prepare food, related subsistence practices, natural history of the habitat, avoidance of predators, and sickness and healing—but also to social governance. One would find incipient myths and rituals that provided continuity and meaning, that sanctioned and explained emerging ideas about responsibilities, obligations, and contracts between individuals, and that validated standards of behavior toward others.

The evolving social expertise embodied in the system of symbols and their meanings that constituted the group's protoculture would more than likely address threats, possible calamities, and forms of sickness and healing. With respect to psychopathology, group members would likely share views about personality variations, behavioral styles, temperamental differences, relationship difficulties, and states of mental and behavioral competence. One assumes the existence of a primitive vocabulary that would allow group mates to categorize and make sense of the frustrated remonstrances of the individual displaying evidence of psychopathology. Moreover, there could be habitual ways of dealing with social problems, patterns of behavior, or emergent institutions for arbitration and resolving conflicts that could be brought to bear on the events surrounding the psychopathology. Among these response patterns would be scripts and scenarios of healing.

A likely feature of H4 would be a tendency to classify the psychopathology as a form of sickness for which healing is prescribed. The individual's distress, social isolation, alienation, and/or depersonalization might engender a more systematic, perhaps even standardized, syndrome of behavior in which an altered state of consciousness figured more prominently than in H3. The more evolved capacity for symbolization and for linguistic representation and the availability of symbols of affliction drawn from the group's social knowledge, stored in the adults' memories and transmitted to the young, would help the individual rationalize his experiences while under the sway of the group's appreciation of his plight. Indeed, if the group's system of symbols includes scripts and models of disability and despair, the conjunction of psychopathology and altered state of consciousness might conceivably lead to a culturally significant syndrome of affliction understandable to the group. If distress and suffering were visible or the syndrome of behavior compelling and culturally representative, it could silence those who had opposed or thwarted the individual. In short, in H4 categories of disability, sickness and healing could well have been components of the culture and cognition of the group. These categories could be used by group mates to explain the psychopathology, a capacity for altered states of consciousness could be expanded, models of syndromes of affliction could become resources of the culture, and as a result of these factors an incident of psychopathology could constitute a standardized, scripted form that is engaging and compelling to participants and the afflicted individual. Such syndromes may or may not have been qualified morally as pathological or simply as sickness, but if they aroused attention and concern among group mates and in some way validated the plight of the individual and exonerated him, such outcomes would follow from the logic of this and preceding scenarios.

To reconcile the evolutionary and the clinical frames of reference in interpreting the H4 scenario, one is further plunged into forms of analysis that complement those of H3. The group as a structure with a force of its own now plays an important role; not only in how psychopathology is configured and played out, but also in terms of its effect on the individual and the group.

To be sure, principles of evolutionary theory are still crucial in H4, as in all of the previous scenarios. First, explaining the ultimate reasons for the incident of psychopathology necessarily requires an appreciation of existing cognitive and cultural capacities conditioned by natural selection; the manifestations of the incident and its role in the group's social economy cannot be made sense of without consideration of social and behavioral ecological contingencies; the cost of the psychopathological condition to the individual still has to be calibrated in fitness measures; the motivations and behavior of the focal individual and of group mates in relation to incidents of psychopathology still depend on notions of inclusive fitness and reciprocal altruism; and, finally, an interpretation of the incident as a factor of the population in which it is found still depends ultimately on considerations of life history as well as social and behavioral ecology. The H4 phase of evolution (like the H3 scenario) may be interpreted in light of both the clinical perspective and the evolutionary one. The cognition and culture of the group reflect directives of both perspectives. A clinical perspective becomes a developed characteristic of the group as an institution, with a structure and momentum whose impact is calibrated in terms of the group's history and social identity. Possibly the emergence of abstract notions about experience would affect how the incident of psychopathology is configured and played out and what significance it has within the group. Stated more directly, with the advent of beliefs, narratives, myths, rituals, models about behavior and deviations thereof, and practices for dealing with medical or quasi-medical rents in the social fabric, incidents of psychopathology would now constitute cultural traits or facts. A clinical perspective now operates as more than a tangible, real, on-the-ground fact within biological evolutionary space, that is, in the group of Archaic humans constituting H4. In addition, it represents a new order to which features of this clinical perspective can be ascribed, namely symbolic and cultural ones, with a power to select, influence, govern, and determine how incidents of psychopathology are exhibited and understood. Whereas in evolutionary terms incidents of behavior that a clinician would label as psychopathology are played in an objective and impersonal economy governed by impersonal processes like natural selection, in the evolved situation of H4 psychopathologies and pathological behavior are group phenomena governed by culture and influenced by tenets that reflect a clinical perspective. In short, a greater degree of assimilation between evolutionary and clinical points of view has taken place.

Comment

In studying psychopathology on the basis of evolutionary criteria, we need to ask three separate questions: How would a clinician define and interpret a syndrome of psychopathology in a group of early hominids or an individual member of the group? Criteria here are complex but would include signs of personal distress, manifestations of psychological and social behavior, as well as bodily functions, that deviate from the usual or typical, and some impairment in social functioning. Second, how might an evolutionist define and interpret a syndrome of psychopathology? Here, these criteria are no less complex but, as indicated earlier, would involve aspects of animal suffering and welfare and decreased fitness. Third, how would members of a hominid group themselves define and respond to a syndrome of psychopathology? The scenarios described in this chapter suggest that early in the long process of evolution, exploitation, and pursuit of selfish, individualistic needs and opportunities would predominate, whereas later in evolution group mates might desist from responding in this way and instead choose to respond altruistically by offering support, comfort, and perhaps attempts at healing. In other words, the latter group of hominids could in effect manifest their own clinical or moral criteria about psychopathology that may or may not coincide with those of a modern clinician, or with those of the evolutionist, for that matter.

The scenarios presented earlier illustrate the interplay among these considerations. Somewhere along the line leading from early hominids to modern human beings, a point was reached at which evolutionary residuals and imperatives to seek personal advantage in competitive situations came into juxtaposition with an emerging clinical, moral disposition to support, comfort, and heal. A need arose to balance or resolve differences between alternative orientations, dispositions, and outright strategies. Social and environmental circumstances and the group's organization and resourcefulness would influence which response would be adopted—that of the strict evolutionary as opposed to that of the clinical observer.

The fact that groups of later hominids might have evolved strategies and directives designed to balance strict evolutionary residuals against evolving clinical and moral imperatives toward afflicted members does not mean that criteria and definitions of psychopathology in prehistoric species were unambiguous and easily applied. In other words, because individuals might have forgone opportunities for exploitation in favor of providing support and comfort to those behaviorally compromised does not mean that there was a clear-cut difference between an evolutionary and clinical idea of how psychopathology was to be defined. Nor did one perspective supplant the other. From an evolutionary point of view, psychopathology should be defined as entailing suffering and impaired social functioning or fitness, that is, a potential diminution of the capacity for reproduction and survival. However, this does not mean that when more advanced hominids chose to desist from exploitation and instead provide comfort and healing to a behaviorally compromised group mate that psychopathology was then defined in a clinical framework, no longer an evolutionary one. In many instances, both perspectives would apply: an episode of psychopathology would entail suffering and impaired fitness, as well as an appeal for support, healing, and comfort. In certain instances, the evolutionary criteria may not have been met, so that the hominid group might have taken a clinical approach—or vice versa. Prevailing circumstances in the group and the social ecology would influence whether an evolutionary or a clinical response predominated, as well as how conflicts in strategies were worked out.

We have to allow for another set of possibilities: the external observer's clinical criteria for defining psychopathology need not agree with that of the evolutionist, or with those we can impute to the hominids themselves. Differences between modern clinicians and native, hominid subjects are easier to deal with. Simply put, a clinician's definition of psychopathology or psychiatric disorder may not conform to how even fully evolved hominids would respond to individuals whose behavior met the clinical definition of psychopathology. This is hardly surprising. It is well known that although some varieties of psychopathology may be interpreted and handled similarly across cultures, societies differ markedly in their criteria for psychopathology. Differences between normal and abnormal behavior remains an elusive and ambiguous cross-cultural boundary, for example, and scientifically trained clinicians see problems differently, even if they acknowledge the fact of cultural relativism.

Similarities and differences between the modern clinician's and the modern evolutionist's points of view are harder to deal with. Ideally, an evolutionist's definition of psychopathology should be consistent with that of the clinician acquainted with evolutionary concepts. The clinician's criterion of personal distress would seem to overlap with the evolutionist's concern for animal welfare and suffering. Similarly, what for the clinician is social impairment or disability is not so different from the evolutionist's criterion of potentially decreased fitness. Of course, both sets of traits may be difficult to measure, but here we are primarily addressing analytic questions. In earlier chapters I have formulated psychopathology broadly so as to accommodate both points of view. As long as we hold to these descriptive criteria, we are on comparatively safe ground regarding diagnosis, allowing for the influence of prevailing circumstances in the group and social ecology in how behavior syndromes were defined and, more to the point, acted upon by a group of hominids.

Because an evolutionary perspective brings into play natural selection and the possible design of behavior mechanisms, this creates additional problems about differences between the clinical and the evolutionary definitions of psychopathology. Since psychopathology, like psychiatric disorder, implies something negative and undesirable—a behavioral breakdown, anomaly, or dysfunction—there is a tendency to explain it as resulting from a breakdown in the mechanism that produces normal, adaptive behavior. This leads to the view that psychopathology might be defined as a breakdown or malfunction of the mechanism itself and, by implication, of what the mechanism is supposed to produce, namely, adaptive behavior. There is in fact considerable controversy as to whether a clinician's idea of psychopathology conforms to evolutionary criteria about adaptive and maladaptive behavior, and if so how (Lilienfeld and Marino 1995; Wakefield 1992a, 1992b). It is not at all clear that clinical definitions can be made to conform to evolutionary ones, since the circumstances to which criteria are to be applied, as well as the criteria themselves—that of the evolutionist and especially those prevailing in a group of hominids—change across time, especially evolutionary time. To anticipate later discussions, I simply add that criteria for what is adaptive, maladaptive, functional, and dysfunctional differ in relation to prevailing circumstances in the life of a group and across phases of human evolution. As groups evolved in cognition, culture, and language, they must have developed different sets of values according to which they themselves applied emerging conceptions of psychopathology. In short, given such fluid and changing circumstances regarding what constitutes psychopathology and how best to define it, the clinician's and the evolutionist's perspectives may not be made to correspond.

- A clinical and humanistic perspective about psychopathology emphasizes cultural conceptions about medicine, addressing questions of meaning, suffering, social deviance, moral values, and help giving or treatment.
- An evolutionary perspective is alleged to be comparatively objective and impersonal since it addresses questions about survival and reproduction seen in relation to requirements of ecology and constraints of natural selection.
- Characteristics and implications of the clinical/cultural perspective compared to the evolutionary/objective perspective on psychopathology are discussed and analyzed.
- The evolutionary compared to the clinical perspectives are generally viewed as highly contrastive, addressing very different sorts of concerns of organisms, from strict matters of survival and reproduction to aspects of suffering, meaning, and value, respectively.
- When the evolutionary perspective is examined carefully in terms of phenotype and genotype, kin selection, reciprocal altruism, develop-

mental plasticity, and life history theory, it is shown to incorporate concerns of the clinical, cultural perspective.
- From a purely analytical point of view, the two perspectives are not incompatible at all. Indeed they are complementary since they both examine behavior from the standpoint of biological functions, although the contexts where these play out obviously vary.
- Hypothetical scenarios of behavior and psychopathology during phases of human biological evolution are formulated that illustrate the continuity and overlap between a clinical, cultural focus and an evolutionary, allegedly objective one.

Chapter 4

Accounting for the Universality of Psychopathology

IN THE BROADEST SENSE, psychopathology refers to specific types of behavior problems found in human societies. Such problems may be defined in different ways. Psychiatry and psychology formulate psychopathology in terms of specific disorders, each with its own causes, manifestations, course, prognosis, and distribution or prevalence. Psychopathology, viewed in a social context, that is, in social groups, has a genealogy that stretches back into earlier periods of human biological evolution.

By means of scientifically informed hypotheses, I shall attempt to trace the characteristics and implications of antecedent varieties of psychopathology, viewed descriptively in terms of prevailing social ecologies and phase of evolution. One could do this by concentrating on each of the psychiatric disturbances described in the contemporary scientific diagnostic system. However, a basic disadvantage here is the culture-bound and tentative, changing status of what are called psychiatric disorders. How disorders are defined, and even their actual number, changes. Moreover, such an attempt, requiring specific evolutionary narratives for each disorder, would be cumbersome and redundant in many ways. It would be inefficient to select abstract features of each disorder and project them backward through every phase of human evolution. Moreover, because the orienting theme would be a complex of signs and symptoms, such a method would be less likely to provide a descriptive picture within which to discuss general theoretical themes. The advantage of adopting a general definition of psychopathology is that it allows one to construct a single evolutionary narrative. Because of the generality of the definition, the narrative can be long and detailed.

Contributions from Studies on the Genetics of Psychopathology

Certain types of psychiatric disorders, most notably Huntington's disease, show a simple Mendelian pattern of inheritance stemming from defects in a single gene (Plomin and McClearn 1993). Even in this disorder, the responsible gene is held to cause different types of malfunction in the central nervous system, involving involuntary motor movements, incoordination, intellectual deterioration, and eventually dementia. Several other varieties of psychiatric disorders, including phenylketonuria, fragile X syndrome, and a familial version of Alzheimer's disease, also have a one-gene disease pattern of causation. However, the most challenging and more common disorders that are explained in terms of genetic factors involve the interplay among environmental factors with multiple genes, each of which has varying effects and none of which may be a necessary or sufficient cause of the disorder. Genes that contribute to genetic variance of traits quantitative in nature are called quantitative trait loci. Although the traits in question are distributed dimensionally or quantitatively in a population, they are additively implicated, together with environmental factors, in causing a disorder qualitative in nature—in other words, a psychopathology that to a clinician depends on a yes or no decision involving diagnosis. Such genes are assumed to be implicated in a number of psychiatric disorders and a variety of complex behaviors, such as paranoid schizophrenia and violence or hyperactivity.

Because genetic factors are implicated in personality traits, more than likely they are responsible for related personality disorders. An example is provided by traits listed in the California Psychological Inventory (Carey and DiLalla 1994). Most of these traits are positive attributes that refer to complex behaviors—for example, self-control, well-being, and responsibility. For personality traits more directly implicated in psychopathology, studies reported in Eaves, Eysenck, and Martin (1989) demonstrate genetic variance attributable to genetic factors on the traits labeled impulsivity and neuroticism. A number of theories relate personality factors to psychiatric disorders and postulate complex relationships of cause and effect. Personality traits reflecting psychopathology have been explained as resulting from psychiatric disorders, suggesting a reverse causality. The relationship between genetic, environmental, personality, and psychiatric disorders is therefore very complex. (See Bailey 1997, 1998; Carey and DiLalla 1994.)

Genetic factors are thus implicated in a variety of contemporary, medically understood forms of psychopathology. Since genetic factors are inherited attributes of human populations that have a basis in the history of these populations, and possibly their evolution, we may ask what role evolution may have played in causing psychopathology. As noted earlier, Stevens and Price (1996) and McGuire and Troisi (1998) discuss the appropriateness of evolutionary theory as a basic science for psychiatry. These textbooks summarize the main subtheories within evolutionary biology that can usefully illuminate aspects of selected psychiatric disorders. They point out that many psychiatric symptoms and even some disorders may represent exaggerated responses to innate, prepared fears of biologically significant stimuli from the environment. Various types of phobias, for example, can be explained as stemming from inherited tendencies to react with fear to stimuli that might have been threatening in the environment of evolutionary adaptedness, such as phobias regarding snakes, strangers, thunder, or heights. Similarly, many emotions and forms of behavior labeled as psychopathology, and which in fact are listed as symptoms of psychiatric disorders, such as separation anxiety, sibling rivalry, sexual jealousy, sadness, strong angry responses toward social cheaters, and certain forms of antisocial behavior may, when viewed in the light of evolutionary psychology, reflect the operation of behaviors that are adaptive and natural because they were once designed to enhance fitness (Stevens and Price 1996).

Many aspects of psychological and interpersonal behavior and conflicts that are part of psychopathology may in fact constitute normal variation. Traits for such characteristics as selfishness, self-interest, a propensity to fearfulness, wariness of strangers, and suspicion of being cheated can all be viewed as having an innate basis but obviously not as pathological. In terms of optimality theory of evolution, these traits could have been naturally selected, providing advantages to those who inherited them. In this view, the psychopathology becomes a relevant clinical diagnosis as a result of an unusual aggregation of genetic material in a particular individual that causes an aberration in behavior.

The discussions by McGuire and Troisi, by Wenegrat (1984, 1990), and by Stevens and Price reinforce the view that psychopathology has a phylogenetic pedigree. As a concept, psychopathology is applicable to human populations considered as pools of genetic information as well as conglomerates of phenotypes living in cultural and historical communities. The formulations of researchers such as these make it apparent that there is an interplay between genetic and environmental factors in the production of complex behaviors of all types, including seemingly pathological and maladaptive ones. These researchers explain how genetic causes of psychopathology can be maintained in human populations, despite the seeming winnowing effects of natural selection.

Most forms of psychopathology, including some bona fide psychiatric disorders, do not negate an individual's capacity to reproduce, either because they are not totally disabling psychologically and socially, or because they have their onset well after the reproductive phase of life. Since forms of psychopathology most likely reflect the additive effect of quantitative trait loci, a full spectrum of such genes may be needed to produce pathology per se; furthermore, it is not unlikely that some of the responsible genes, when not enhanced or facilitated by others, may contribute to adaptive behaviors because of heterosis. In instances when homozygosity is required to cause or predispose toward a rare disorder, individual genes may be moderately common and distributed in the population, with the heterozygote condition bestowing some advantage. More general expositions of the influence of evolutionary factors in psychiatric disorders and behavioral disturbances are found in D. R. Wilson (1992, 1993), Zorumski (1988), Saku (1992), Smith (1993), Cates et al. (1993), Epstein (1987), and Gottesman (1991).

K. G. Bailey's (1987) work on human paleopsychology gives an equally general and comprehensive formulation of psychopathology in terms of evolution. Bailey's concerns go well beyond the strict accounting of human psychopathologies as conceptualized in clinical theory and practice today. Bailey deals with broad classes of human psychopathologies, seeing them in motivational, emotional, and behavioral terms; he roots psychopathologies in hierarchically placed and organized systems of brain structures and processes that are a product of the evolutionary process. But whereas other researchers are primarily concerned with explaining psychopathologies (largely conceptualized as psychiatric disturbances) per se, Bailey's frontier is far broader. He is concerned with formulating and thus understanding the structure and organization of normal and psychopathological behavior not only in evolutionary terms but also in terms of the organization of an apparatus (the central nervous system) that was designed, layered together, and integrated by natural selection.

The concept of phylogenetic regression is centrally important in Bailey's formulation. He uses this concept to explain how psychopathologies are produced; namely, how it is possible that evolved, fully enculturated, neocortically governed members of Homo sapiens can show a developmental reversal, hierarchic disintegration, and a behavioral dedifferentiation that leads to psychopathology. According to Bailey, below the surface of human behavior lies an animal, beastly character of man, potentially capable of spilling over into psychopathological behavior, under circumstances that can loosen higher nervous systems of control. Bailey not only presents a theory about evolutionary aspects of psychopathology, like other researchers, but also gives explicit form to a research tradition (see Laudan 1977) about the character of human behavior, including its psychopathological varieties.

Evolutionary Aspects of Selected Human Psychopathologies

Price (1988) presents a detailed formulation of depression as an example of a psychiatric disorder that was selected for during human evolution in order to handle conflicts and stresses associated with establishing and maintaining asymmetric social relations. In his formulation, he draws on knowledge from communication theory as well as ethology to explain how organisms or individuals may have coped with low social rank and the hostility of dominants in the social hierarchy. He explains forms of depression as subroutines of behavior that evolved as yielding components of ritual agonistic behavior. In his words, "Two related but distinct yielding subroutines are postulated. One subserves social *homeostasis* in the form of maintenance of low rank in spite of motivation to rise in rank, and is a chronic and relatively mild condition that may provide a phylogenetic template for *neurotic depression*.... The other yielding subroutine subserves social *change* in the form of fall in rank, and is a severe but self-limiting condition that may provide a phylogenetic template for *psychotic depression*" (1988, 190; emphasis in original). Price offers a very tightly reasoned argument that explains how and why depression is found in the human species, using evolutionary and ethological concepts involving ritual communication. Stevens and Price (1996) also seek to explain the origins and reasons for the prevalence of a number of specific psychiatric disturbances from an evolutionary standpoint (see also Wenegrat 1984, 1990).

Gilbert (1992a, 1992b) also offers a thorough exposition of psychiatric disorders from a general behavioral sciences standpoint, including evolutionary biology. Gilbert focuses especially on psychiatric disorders also classified as forms of depression. His formulation bears a resemblance to that of K. G. Bailey (1987) with respect to its evolutionary basis, comprehensiveness, and integration with contemporary theories, but Gilbert does not anchor his view of depression in ideas about phylogenesis as Bailey does. His main concern is to use evolutionary theory in order to better understand distinctive forms of psychopathology; and how to make sense of their manifestations, content, and subjective/objective parameters to better understand human nature and suffering (1992a). Besides relying on ideas of key evolutionary biologists who articulate the modern Darwinian synthesis, Gilbert uses ideas of Carl Jung, the psychoanalyst; Paul MacLean, the neurobiologist; Michael Chance, the ethologist; Russell Gardner, the evolutionary oriented psychiatric theoretician; Aaron Beck, the cognitive psychotherapist/psychiatrist; and Timothy Leary, the clinical psychologist and interpersonal personality theorist. Gilbert develops a highly detailed reversed pyramid formulation of behavior, relying on general biological and evolutionary principles pertaining to innate behavioral tendencies; varieties of biosocial goals, needs, and imperatives; evolved animal and then human social strategies viewed in terms of group relations, interpersonal competition, and striving for status and prestige; and finally personality, interpersonal, and social systems factors—all of which bear on the topic of human social adaptation, human nature, and suffering. Gilbert (1992b) applies his evolutionary, social-psychological synthesis to varieties of depression.

Gardner (1982) also concentrates on specific psychiatric disorders as distinct entities in developing his theory about the role of evolutionary factors in the etiology of bipolar affective disorders. He holds that animals have "distinctive neuropsychological 'organismic' states that underlie behavior patterns correlated with social position" (1436). He emphasizes innate, genetically canalized programs of behavior associated with social positions of very high and low rank. He argues that animals are innately disposed to associate distinctive behavioral routines, in themselves and conspecifics, with positions of dominance and leadership, as compared to submissiveness, passivity, and banishment. Given social realities in the group involving the mix between social organization and behavioral ecology, on the one hand, and individual differences in behavior and temperament—such as those based on strength, size, agility, cleverness, intelligence—on the other, animals display the appropriate manic (dominant, confident) and depressive (avoiding, withdrawing) routines or responses. Animals, in short, are programmed to notice, respond to, and display communicative behavior that mark their positions in a social hierarchy. In humans, the respective bipolar psychiatric disorders represent poorly regulated, exaggerated responses of this wired-in routine independent of the social reality of the individual. "Because inheritance factors cause a regulatory abnormality in vulnerable persons, these organismic states are too easily triggered and rigidly maintained" (1436). "This instability-rigidity complex most likely reflects an altered regulatory system, probably neuronal, for example, one exhibiting some kind of biased set point" (1439). Like other researchers who study anxiety disorders, Gardner gives an important role to the well-described, seemingly invariant behavioral profiles of bipolar disorders. Gardner, in other words, relies on the alleged ontological specificity of behavior, viewed in behavioral, phenomenological terms, displayed by psychiatric patients diagnosed with the respective disorders. He uses the well-delineated profile of behavior of bipolar patients, together with their subjectivity and the resemblance of all of this to the syndromes associated with extremes of social rank, to make sense of the bipolar disorder from an evolutionary standpoint.

Anxiety disorders have been studied by Öhman (1986, 1987), Marks (1987), Hofer (1995), Marks and Nesse (1994), and Blanchard and Blanchard (1990), who use the general fear response of animals to danger stimuli to interpret human emotions and disorders involving anxiety and phobia. Fear is understood as a protective or defensive response of an organism that enables it to avoid dangerous predators, threatening conspecifics and anomalous, poorly understood environmental circumstances that pose danger. Marks (1987), in particular, illustrates how fear responses function in relation to stress and danger, arguing that fear and animal analogues of anxiety have served adaptive functions. He interprets contemporary anxiety disorders in human beings as consisting of many of the reflexive physiologic and behavioral responses of animals exposed to threats and dangers, with cognitive and subjective issues as the more evolved expressions of the fear response gone awry. Marks sees disorders characterized by anxiety, obsessions, compulsions, and phobias of many varieties as rooted in the evolutionary past. Hofer (1995) even argues that unicellular and multicellular animals provide prototypes or analogues of fear responses and anxietylike behavior to aversive stimulation, illustrating how sensory stimulation is linked to motor neurons that set in motion neural propagation resulting in elementary, primitive fear/anxiety reactions (see also Kandel 1983). Hofer (1984, 1987) also develops the concept of relationships as regulators, not only of social behavior but also of internal states. Hofer contends that separation anxiety and bereavement, for example, can be explained in terms of altered relationships of attachment that have wide-ranging physiological and emotional correlates. (See also Bowlby 1969.)

Crow (1995) relies also on evolutionary theory to explain the etiology and persistence of a well-described psychiatric disorder or syndrome of behavior: schizophrenia. He builds his theory on three facts pertaining to schizophrenia: its genetic basis, its rather high prevalence, and its apparent universal distribution in all known human populations. Crow asks: "The problem is an evolutionary one—why have the manifest disadvantages associated with the psychosis genes not ensured their elimination from the human population?" He concludes: "The genetic variation is likely to be species specific, an interesting special case being that in which the variation is present in relation to the particular characteristic by which speciation has occurred" (1995, 12–13). Here Crow seems to raise the possibility that psychosis/schizophrenia is purely a disorder of the human species, has no prevalence or analogue among our evolutionary ancestors, and is tied to presumably key evolutionary transformations such as the emergence of language and the rich emotionality that has made possible forms of modern human cognition and culture. After reviewing the possible evolutionary bases for psychosis/schizophrenia, Crow develops his view that the prevalence, age of onset, and sex distribution of psychosis are linked to foundational aspects of human biology involving the plasticity of the brain at birth, hemispheric specialization, the centrality of social cognition/intelligence, and the empirical observation of sex differences in cognitive style. The processes of natural and sexual selection are both held to play a role in producing and distributing psychosis in the human species: "Some extreme variants are associated with the deviations of psychological function that we describe as psychosis. These states are seen as boundaries of the distribution of personality variation, including the capacity for language and emotional expression. In particular, those with the earliest manifestations (i.e. schizophrenia, Asperger's syndrome and autism) have the greatest impairments of communication and social ability, and also demonstrate a failure to develop anatomical asymmetry" (1995, 24).

In their approach to the sociobiology of disorders said to involve psychopathy or chronic forms of antisocial behavior, Draper and Harpending (1988) draw on generalizations about the biological and social sciences, encompassing evolution, game theory, developmental psychology, and learning theory. In essence, they hold that a genetic predisposition toward antisocial behavior is found normally distributed in all human populations. Cheating and deception, in short, are by-products of a cunning, Machiavellian intelligence that has evolved in a context of intense social competition. Presumably, kin selection and reciprocal altruism have constrained the full expression of psychopathy, thereby selecting for mutualism, cooperation, and playing by the rules. Nevertheless, anyone can be predisposed to antisocial behavior, given appropriate environmental conditions. Males are held to show a higher genetic loading for psychopathic behavior, a factor explained by their generally greater variance in reproduction (compared to females), which leads them to adopt more risky life strategies. As a consequence of selection to fill a small frequency-dependent, evolutionary niche, a percentage of individuals at the extreme of the distribution pertaining to psychopathy will be deemed morally insane in any culture, while others, located at less extreme positions of the continuum, and in response to specific environmental conditions, will pursue a life strategy of antisocial behavior.

COMMENT

In the neo-Darwinian synthesis, social behavior is considered an attribute of the phenotype associated with genetic variation and an object of natural selection (Hull 1988; Mayr 1982; Plotkin 1988, 1994). Since it can influence fitness and adaptedness, this phenotype constitutes an integral part of the natural selection syllogism. According to Endler (1986), if (1) a population shows variation among individuals in some trait, (2) fitness differences are consistently related to different measures of the trait, and (3) the trait is inherited in offspring from their parents, then trait frequencies will differ among age classes of the population, and trait differences between parents and offspring will be predictably different. Endler defines evolution as "any net directional change or any cumulative change in the characteristics of organisms or populations over many generations.... It explicitly includes the *origins* as well as the *spread* of alleles, variants, trait valuers, or character states" (1986, 5; emphasis in original).

In analyzing how psychopathology is affected by the process of evolution, one should begin with a description of the content of psychopathology, the phenotype, that includes either specific items of behavior or a whole ensemble. As an example, psychopathology might equal social impulsiveness, behavioral timidity, social avoidance, and/or stereotypies of motor behavior. Additionally, one needs a description of the genotype or genetic features contributing to psychopathology—for example, the location, number, and penetration of the genes contributing to the heritability of the behaviors, as well as the frequency of psychopathology in the population. The degree of density of the behaviors in question in the population would provide possible clues about frequency-dependent selection. Also needed is a description of how psychopathology affects the fitness and adaptedness of organisms, animals, or individuals under prevailing living conditions, such as whether or to what extent the behavior impedes mating ability, fertility, and parenting, for example. One would want to know why or how psychopathological behaviors might have been naturally selected for or against. For example, psychopathology manifested as timidity or impulsiveness might reduce mating ability and parenting in the case of negative selection, whereas psychopathology as heightened aggressiveness, cunning or deviousness, and bullying of females could actually enhance mating ability and be positively selected for. Knowledge about the absolute and relative frequency of the psychopathology phenotypes or component behaviors, or the tendency to display them, in the succeeding generation or in offspring would also be needed. Finally, one should know whether there is any directional, cumulative change in the prevalence, incidence, or salience of psychopathology in the population over many generations.

Researchers who study the genetics of officially recognized disorders are primarily interested in explaining their variation, distribution, etiology, manifestations, and pathogenesis. Evolutionary factors, when they are invoked at all—for example, bipolar disorders, psychoses, antisocial personality disorder—play a largely secondary role, since the emphasis is placed on proximate factors that may aid in understanding and treatment. Although they may draw on important aspects of mammalian and especially primate behavior and evolution, their primary concern is with psychiatric theory and practice. As an example, McGuire and Troisi, Gilbert, Stevens and Price, Marks, Hofer, Crow, Gardner, and Draper and Harpending all draw on important generalizations about human evolution. They all project what is known today about the descriptive content and etiology of a contemporary, official psychiatric condition back onto an evolutionary scenario so as to connect realistically with it but not necessarily to illuminate it. They do not address the possible role of psychopathologies in the evolutionary process, what circumstances might have precipitated psychopathology in a social group, how it might have been dealt with, and how it might have become part of an emerging social economy involving sickness and healing. In the case of anxiety disorders, antisocial personality disorder, and bipolar disorders, the roots of the putative psychopathology are said to go far into the phylogenetic past, whereas comparatively more recent factors are considered to be critical for psychoses, such as those involving the speciation of Homo sapiens. In most instances, the researchers project whole syndromes onto the evolutionary past and are not interested in examining component psychopathological behaviors in terms of specific biosocial functions, behavioral ecology, or social organization; nor do they examine how the syndromes might have been affected by changes in cognition or how they might have affected interactions in specific groups.

An exception here is the compendium of McGuire and Troisi (1998), who conceptualize psychopathology as involving a composite of impairments of major infrastructural systems of behavior keyed to the pursuit of biological motivations and goals. However, this exception does not contradict the general point: while their analysis implicates all facets of psychopathology and is, thus, comprehensive, it does not go into detail about specific evolutionary scenarios, nor does it systematically consider how these ultimately caused behavior systems played out in contexts of evolving cognition and meaning.

Psychopathology as a Waste Product of Coping and Adaptation

A prominent feature of the environment of mammals and especially nonhuman primates (and, one assumes, human ancestors) is competition over basic subsistence and reproductive needs, as well as cooperation in various altruistic or politically expedient ways to meet selfish needs. Accommodating to challenges and threats posed by group members is held to have played a central role in the evolution of human cognition (Humphrey 1976). In general, striving for and maintaining status in the group is viewed as having posed intellectual challenges and obstacles and promoted social skills involving alliance building, deception, and mind reading (Byrne and Whiten 1988; Whiten and Byrne 1997). While these conditions are analyzed primarily in terms of cognitive demands, scholars have not given enough attention to the fact that they also entail stress and wide-ranging psychological changes that can thwart the realization of biological functions and social goals.

Abundant evidence from general biology, social epidemiology, and clinical medicine shows that disease and injury/pathology are precipitated, if not produced, by emotional, cognitive, and physiological strains resulting from an individual's attempt to compete effectively. Psychopathologies, then, can be viewed as potential if not inevitable consequences of the stress-inducing selective conditions that have led to the evolution of human cognition and culture. Many other factors only partially related to stress per se can also produce psychopathologies. Behavioral anomalies produced by embryological imperfections, later developmental failings that are a product of unwholesome and noxious interactions between genes and the environment, or physical lesions interfering with or damaging the functioning of central nervous system—all can also give rise to striking forms of behavioral breakdown. To this we must add more subtle forms of interference in social behavior and psychological distress resulting from developmental deprivations or conflicts and stresses arising from local contexts and situations that favor the evolution of cognition (Chisholm 1999). All of the resulting psychopathologies can be seen as constituting dysfunctional or negative outcomes of attempts to cope with environmental challenges manifested in measurable changes in social and psychological behavior. In evolutionary theory such behavioral disturbances can readily be explained as resulting from proximate stressors and strains.

Viewing psychopathologies from this perspective is to equate them with other negative consequences of external environmental hazards and challenges, such as food scarcity, infections, injuries, organ malfunctions because of nutritional deficiencies, or attacks by predators. Indeed, many instances of sickness and healing can themselves be viewed as the natural consequence of exogenous stressors or hardships and not genetically entrained or based. It seems incontestable that some forms of psychopathology should be viewed in the same way. Just as a tree is injured by lightning or a virus, organisms and individuals can be similarly passive victims of pathology, injury, and disease (Mehmet and Edwards 1996).

Viewing psychopathology not as genetically determined but as a negative consequence of coping with exogenous factors in no way denies that many of the behaviors that make it up—their content, their pathogenesis—are a consequence of genetically canalized programs regulating normal behavior that have been disrupted or simply injured as a result of anatomical-physiological changes set in motion by environmental factors, including stress and coping. While responding to stress in the form of psychopathology may indicate the perturbation of innately determined general programs of behavior, psychopathology itself or its components do not figure in the natural selection syllogism per se. In other words, the components or ensemble of behavior qualified as psychopathology are not actual traits with systematic frequencies, fitness differences, or inheritance patterns with cumulative change.

Psychopathology as a Negatively Selected Trait

Vulnerability to psychopathology may result from genetic traits that have a negative impact on fitness and that are being eliminated through natural selection. If so, this raises a number of questions: First, what sorts of psychopathologies might have been removed during the evolution of the human line? And when, why, and how did this occur? Second, what are the forms of psychopathology that remain in the genome? And how does one explain their persistence?

It seems incontestable that during human evolution some, to us unknown, varieties of psychopathology were naturally selected against. One can visualize this by constructing a three-step process: As a result of such factors as genetic copying errors, such as mutations causing sexually deviant actions, lack of emotional control or social insensitivity; or because of founder effects, such as the emigration or isolation of behavioral deviants who could not cope with predators in the new environment or could not integrate and mate; and/or owing to striking changes in paleoecological conditions that made a certain individual's genetically based behavioral differences less socially functional, populations of organisms with genetically based behavioral vulnerabilities came to manifest forms of psychopathology that diminished fitness, and therefore the traits were eliminated from the population.

Continuing with this argument, what components or forms of psychopathology might possibly have been eliminated, and under what conditions? Many possibilities can be entertained. Some behaviors may have helped the individual to avoid predators or to resist hostile incursions of competing hominids of the same or different species; or some behaviors may have enhanced reproduction and parenting under certain conditions, such as those favoring low investment in mating and parenting. But, then, during a subsequent phase of evolution, when different social ecologies prevailed, the behaviors or traits may have worked against mate selection or parenting and were thus eliminated by natural selection—high investment strategies were preferred. Selection against psychopathology may have benefited conspecifics: psychopathologies that were eliminated by natural selection, or even genetic drift, relieved stressful, challenging conditions of group mates, thereby giving them selective advantages that eventually contributed positively to the evolutionary process. Any number of behavior handicaps (psychopathology?) may have occasioned responses that selected against the victim. Special ecological conditions, such as population bottlenecks, may have selected against behavior ensembles or routines, thus constituting them as psychopathologies, thereby enhancing the social and psychological functioning of group mates.

Conversely, many phylogenetically older forms of psychopathology may have ceased to be selected against as social ecologies subsequently changed and became more favorable. The traits for these may lie buried within or layered under more evolved programs of behavior. For example, following the argument of Donald (1991, 1993b), one could say that modes of relatively evolved cognition and culture (termed by him "mythic") may conceal, have modified, or have replaced older forms of cognition and culture termed mimetic, and that the latter exist as genetically determined behavior routines or mimetic eccentricities manifested in members of contemporary populations whose coping abilities are inadequate. Mimetic varieties of psychopathology that were never fully eliminated may surface when special conditions are created in contemporary populations that resemble conditions prevalent during the evolutionary past.

In this context, it is appropriate to recall the formulation of K. G. Bailey (1987) involving the paleopsychology of man and the concept of phylogenetic regression. Bailey developed this from the writings of MacLean (1970, 1973, 1982a, 1982b) and Valzelli (1981) regarding the hierarchical organization of the nervous system. Implicit in Bailey's thinking is that hierarchically lower systems of neural mechanisms and routines, most likely paleomammalian, though possibly even reptilian, incorporate forms of genetically canalized behavioral routines that can, in the context of stress, disease, pathology, or injury, produce what could be called released, disinhibited forms of psychopathology in vulnerable members of contemporary populations. Bailey explains such properties of psychopathology as involving developmental reversals, hierarchic disintegration, and behavioral dedifferentiation. Viewed in light of prevailing conditions in the environment of evolutionary adaptedness, the primitive, instinctual, explosive, and highly emotionalized behavioral routines Bailey has in mind might have been adaptive. However, even if they have always been maladaptive, given changed social ecologies, such behaviors have been not only restrained but also overridden, appropriated, and reintegrated into routines that themselves could be genetically determined or culturally elaborated and neocorticalized, as both Bailey and Plotkin and Odling-Smee (1981) suggest. When phylogenetically older behaviors are released or inhibitions removed, they may prove highly maladaptive in a different social environment with different forms of social organization and culture, that is, environmental mismatch.

During the phases of human evolution that came before language and culture, hominids and early varieties of Homo undoubtedly relied on some as yet unclear social, emotional, gestural, and viscerosomatic system of cognition, culture, and communication. Such a system might have resembled the one Donald (1991) calls mimesis and is explicitly invoked by Nelson (1996), who analyzes how infants parse phenomena of the world into meaningful units, event construction, that provide a basis for language learning. Mimetic cognition and communication in early hominids obviously required the integration of motor programs for making gestures, facial expressions, and vocalized signaling. Moreover, it may have revealed aspects of human paleopsychology that K. G. Bailey (1987) contends are still regressive in contemporary peoples. Given such an evolutionary context, certain forms of psychopathology would have been produced by neurodevelopmental delays, neuromuscular rigidities, or failures of coordination among motor routines. If they proved selectively negative, they would have been eliminated from the human behavioral expressive capacity altogether. However, if the underlying neurologic traits were merely reassembled, then they might provide the regressive pull that Bailey contends still prevails in the human genome.

Individuals in the process of developing a mimetic form of communication and culture, and, what is more important, consolidating a sense of group identity, would have needed social skills for self-conceptualization, social attribution, social politicking, emotional regulation, emotional restraint, subtle forms of social expressiveness, tolerance of behavioral differences, and close bonding with group mates. However, some neuropsychological and neuromuscular programs that favored effective responses to ecological contingencies involving not only purely physical but also behavioral aspects of such skills as foraging, hunting, avoiding predators, or tolerating climatic extremes may have actually clashed with and been rendered less effective by routines favoring responses that contributed to social and psychological integration. For example, routines favoring emotional intensity, behavioral persistence or rigidity, capacity for sustained attention and vigilance, comfort with social isolation, ability to postpone or ignore adverse stimulation or hunger, extreme emotional guardedness, auditory or olfactory sensitivity, and inability to control desires and appetites may all have favored ecological survival but militated against survival in a dense social group where mimetic abilities were especially adaptive. Along the behavioral line that selected for social intelligence, one might have found suites of behavior favoring ecological intelligence that under current conditions were maladaptive or pathological and, because they impaired fitness, were eliminated by natural selection in later social ecologies. Alternatively, the suites of behavior that were pathological and maladaptive and gradually being selected against, given new environments, might have proven adaptive.

Fodor (1983), Nelson (1996), and Mithen (1996a) offer a different interpretation of social intelligence and ecological intelligence in their theory of the modular mind. They propose that the human brain is composed of modules that operate quickly, automatically, and separately, and that the mental component that promotes general intelligence, general problem solving, self-conception, and social awareness has comparatively little access to how sensory data are processed and to the mechanism of cognition of these separate modules. Presumably, only the products of cognition carried out in these modules are available to the individual, late in the chain that produces behavior. Indeed, a feature of more evolved human mentation is said to be that an adult has greater access to the modes of information processing or reasoning in the various specialized modules and can better integrate their modes of operation and their products with general intelligence and self-awareness through creative reasoning and behaving (Karmiloff-Smith 1992). According to this scenario, perturbations in the functioning of any of the modules would have produced selected deviations of personal experience and emotional, motivational expression in relation to natural history undertakings, technical or manipulative pursuits, and interpersonal relations. Such deviations of modular thinking could result in behavioral ineffectiveness or social behavioral anomalies classifiable as psychopathologies. For example, using Mithen's (1996a) modular interpretation of prehistoric mentalities, one could say that deviations in the integration of cognitions of the various mental modules, or how they combined with general intelligence, would have produced ancestral, prehistoric forms of psychopathology. Any of these could have impaired fitness and been selectively discarded. Such a traditional, mechanical formulation of prehistoric psychopathologies, when "warmed over" by formulations provided by K. G. Bailey (1987), can suggest a better picture of psychopathology among earlier hominids and varieties of Homo.

The idea that psychopathology had genetically deleterious effects in the evolutionary past can be interpreted differently. Keep in mind that abnormal homeostatic response patterns denoted as hyposensitivity, hypersensitivity, failure to habituate, to accommodate, to assimilate, and to integrate are universal in animal forms, certainly in mammals, and especially primates. This must have been the case with hominids. Such response patterns mark boundaries and thresholds that enable adaptive behavior, but all organisms sometimes "normally" cross or fail to cross them. The psychophysiological systems that mediate these responses vary from one individual to the next. Given these considerations, how would extreme variations in the distribution of such psychophysiological response patterns be manifested? In short, one can attempt to conceptualize the character of psychopathology in prehistoric populations by visualizing the effects of genetically based deviations in how psychophysiologic routines promoting basic homeostatic responses were realized and fared in the prevailing social ecology.

The potential for losing the ability to control and coordinate motor, sensory, vegetative, and social responses is universal in higher animals and undoubtedly produced maladaptations and psychopathologies during every phase of human evolution. Each type of cognition and culture that described prehominids, hominids, and Archaic humans favored some forms of personal experience and some routines of social behavior; and each type of cognition and culture also disfavored other forms. The former were selected for and now constitute the neuropsychological and psychophysiological machinery that shapes human behavior and adaptation today; the latter were selected against and may be regarded as lost or silent forms of maladaptation or psychopathology that may still constitute unwanted potentials or possibilities in the human enterprise.

Formulations of the interplay between psychopathology and evolution make apparent the necessary link between psychopathology and prevailing modes of social organization, cognition, and culture. An evolutionary bias, in other words, determines what constitutes psychopathology: how maladaption is calibrated. This evolutionary bias must be distinguished from our contemporary cultural bias that leads us to interpret certain forms of behavior as deviant or undesirable purely on the basis of prevailing symbolic conventions. In the contemporary situation, many forms of behavior that are handled as psychopathology may not in fact impair fitness (that is, they are not maladaptive). Varieties of true psychopathologies that may have existed in ancestral environments and become manifest under special circumstances today, such as phylogenetic regressions, may or may not constitute psychopathology, depending on whether one uses a social, cultural, or evolutionary mode of reckoning. This formulation is predicted by the theory of environmental mismatches. In short, one must keep in mind different criteria about psychopathology when evaluating its character in the phylogenetic past, during prehistory, and today. Our biases about pathologies—what we are familiar with and have come to discredit—shape what we can conjure up as possible forms of psychopathology during the phylogenetic and prehistoric past, like our biases about forms of psychopathology in nonindustrialized societies today and how we interpret them. To paint a more realistic canvas of what psychopathology might have looked like during evolution, we must separate out notions of behavior, social action, and forms of psychological experience that are shaped by our contemporary pictures of the world, behavior, morality, and normality.

Psychopathology as a Positively Selected, Adaptive Trait

If psychopathological behavior diminishes fitness, by definition it is maladaptive. Yet its genetic basis and comparatively high frequency in human populations implies positive selection, which would make it an adaptation—an obvious contradiction. However, there are several ways in which a set of genes predisposing to or causing a maladaptation could actually be positively selected. First, consider a quantitative trait. A large number of genes that are additive may predispose an individual to psychopathology, but fewer may actually confer advantages. Second, when expressed or overt, psychopathology may impair fitness, but when latent and covert, the genes responsible for it may promote fitness. This could also be the case with qualitative traits, given that the appearance or emergence of a psychopathology may require a mix of environmental factors or precipitants. In these instances, aspects of the phenotype conferring advantages may or may not be part of the psychopathology itself: they could involve social psychological functioning—as, for example, cunning, ability to better conceal one's motives, ability to form coalitions; psychophysiological functioning—such as limbic brain activation patterns that produce dysphoric states that can act as spurs to problem solving, heightened sensitivity to enter states of dissociation, tolerance of ingested brain toxins; or something quite separate from behavior all together—such as a pattern of DNA structure that enhances digestion.

Such positive, fitness-enhancing correlates of genes otherwise predisposing to psychopathology could come to prevail for different reasons. For example, the phenotype may, as a consequence of exaptation (the process by which a biological trait that has evolved for a particular function, or that has none, is used by natural selection for a different or altogether new and special function [Gould and Vrba 1982; Gould 1991]), enable the incorporation of a novel, advantageous behavioral routine; or the genetic-developmental material that is expressed as a variety or component of psychopathology may itself be incorporated or elaborated into material that encourages behavioral strategies favoring fitness. Furthermore, it may enable those vulnerable to emotional perturbations to find better ways of foraging, hunting, migrating, or solving social behavioral problems. Much as Gardner (1982) theorizes, quantitative genes may under normal conditions confer clever patterns of status seeking, which certainly can benefit fitness, but given other conditions (or too heavy a load of the genes in question), it may lead to reckless behaviors costly to fitness. As Price (1988) formulates it, a trait for depression may be a small price to pay for the ability to adapt to situations in which one's status or rank is diminished.

Finally, what today we regard as psychopathology could result primarily from adverse social, behavioral, ecological, or cultural conditions that simply did not exist in the environment of evolutionary adaptedness—in other words, the environmental mismatch theory. For example, a propensity to abuse sugar and gain excessive weight (Bateson 1988), to abuse alcohol and other addictive drugs (Dobkin de Rios and Winkelman 1989), to display extreme narcissism or suspicion, or to become depressed at the loss of important prestige objects must be assumed to rest on (a) a genetic complex that makes sugar attractive, mind-altering drugs enjoyable or useful for rituals, celebrations, or problem solving; (b) personality traits that enhance and guard the self through social relationships, and (c) programs of social and psychological behavior that protect individuals when their status is diminished.

In the evolutionary environment, the traits just described may have been components of physiological and psychological adaptations. In other words, they may have been designed in a dialectic of social ecology, behavioral ecology, social organization, culture, and personal awareness and experience, helping to contain, restrain, or channel behavior and the psychophysiology of early hominids. However, given the very different conditions of modern life, the previously adaptive traits have led to a number of disorders. For example, overeating and taking drugs are encouraged in a technological setting that produces concentrated sugars and mind-altering agents; personality disorders arise in a setting of dense public networks in which social display, achievement, and high status are rewarded; and mood-affect disorders are inevitable in a setting where a plethora of social markers, objects, and emblems are used to calibrate personal esteem. According to the theory of environmental mismatch, all of these disorders could be called contemporary forms of psychopathology (Fábrega 1974, 1975, 1997). Also keep in mind the formulations of Plotkin (1994), Odling-Smee (1983, 1988), and Gould (1991) regarding the complex interplay between genetic information, cultural information, genetic inheritance, ecological inheritance, activities of niche-constructing phenotypes, and the brain's flexible exaptive capacity to make use of genetic potentials while responding (positively or negatively) to challenges posed by new social conditions, some of which were the result of phenotypic activities in the first place.

Coping with a harsh, dangerous, and unpredictable environment and encountering setbacks and defeat can cause anyone (mammal, primate, early hominid, or archaic human) to become disheartened, avoidant, and meditative, as well as more responsive if not sensitive to the incongruities and misfortunes of life. Somewhere along the line of evolution stretching from early protoculture to culture, human ancestors began to reflect upon the self and its experiences and to evaluate social circumstances affecting kin, associates, and the group as a whole. In a context of adversity, loss, and deprivation, negative reactions could readily escalate to demoralization, which is associated with personal distress, social inefficiency, and neurovegetative-psychophysiological disturbance (Gilbert 1992a, 1992b). Such conditions create a psychosocial complex that could nurture the creation of myths and rituals about social origins, social identity and bondedness, and social renewal. Demoralized but creative members of the group would probably be led to develop new programs and routines for coping with demoralization and calamity. A feature or component of this crucible of existential inquiry and rumination might lie at the roots of creativity, hope, and spirituality, traits that undoubtedly proved advantageous. This is the positive side of the trait complex. Its negative side is demoralization, despair, defeatism, and a failure to find meaning and value in group activities and shared cultural programs that could bring subjective renewal and welfare.

Hyland (1990) offers another scenario for the emergence of psychopathology as a necessary concomitant of an otherwise positive or adaptive solution to paleoecological living conditions. He hypothesizes that contemporary psychosomatic ailments, forms of experience, and behavior that are undoubtedly universal (Fábrega 1974, 1975, 1997), are the burdensome residues carried by certain individuals as a result of their capacity to psychologically "point" and react with "psychosomatic" ailments—that is, through psychological alerting mechanisms that spread from internal systems of regulation to affect the body—to problems and conflicts in group settings in the evolutionary environment. In other words, nomadic groups of hominids fared best when they were highly bonded and interdependent, since this allowed more effective planning, coordination, and sharing of activities. However, coping with environmental challenges and hazards could lead to interpersonal conflicts. Such conflicts could interfere with foraging and other livelihood activities by undermining the social and political economy and the group solidarity necessary to survive. A predisposition to sensitively monitor and react to conflicts in the group's economies caused by a breakdown in the selected internal economy of an individual because of psychosomatic ailments could have provided help and healing for the individual in question, thus bringing relief and perhaps status to one seen as vulnerable but who manages to overcome sickness and contribute to the group's continuation. In the prevailing group's experience, such individuals and the healing responses they generated would have provided opportunities to renew and reestablish relationships of trust and cooperation in a group threatened by conflict. According to Hyland (1990), being vulnerable to psychosomatic—and, one must add, somatopsychic—ailments, thereby signaling imminent social disorganization as a result, had the effect of securing fitness benefits and also helping the group. In short, the ailing individual received attention and satisfaction because healing, besides mending torn social fabric and relieving group tensions, restored the status of the sick individual. Here again, then, certain psychological response patterns can actually be positively selected for, despite the burden they carry of proneness to psychopathology. On balance, a maladaptive trait, vulnerability to Hyland's psychosomatic ailments, if infrequent or brief, may have actually conferred fitness advantages to the individual while also militating against group disorganization. Hyland's interpretation, it should be noted, implicates frequency-dependent selection of individuals but flirts with group selection.

- A term like *psychopathology* facilitates the formulation of a unified and consolidated picture of behavior problems across the evolutionary terrain. To better understand how these fared, we must examine the genetics of specific psychiatric disorders.
- Single gene disorders provide compelling examples of genetic causes, but disorders resulting from additive effects of many genes in conjunction with and influenced by environmental factors were probably more influential during human evolution.
- Many contemporary psychiatric disorders are explained as exaggerated responses to environmental factors underlying behavior routines that are otherwise adaptive and were positively selected for in ancestral environments.
- To illustrate how evolutionary psychiatrists explain disorders, five were discussed: depression, anxiety, bipolar disorder, schizophrenia, and antisocial personality.
- Many psychiatric disorders are best explained as examples of wear and tear of basic neuropsychological mechanisms and social behavior routines.
- The evolution of hominid populations probably negatively selected out many psychiatric disorders because by definition they compro-

mised fitness. Hypotheses about why and how this might have occurred were discussed.
- That something detrimental like a psychiatric disorder might have actually been positively selected during evolution seems counterintuitive, yet some of the reasons why this may be plausible were discussed.

Chapter 5

An Active Role for Psychopathology in Evolution

THE PROPOSITION THAT psychopathology has a genetic basis in human populations raises the question of its possible adaptive value. The comparatively high frequency of psychopathological traits is viewed as evidence of positive selection. Because psychopathology constitutes an ensemble or routine of social and psychological behavior, we may seek to clarify the mechanism by which it may have operated as an active, positive influence in natural selection.

Niche Construction, Coevolution, and Psychopathology

Plotkin (1988, 1994) and Odling-Smee (1988, 1994) draw on ideas pertaining to evolutionary biology, community ecology, human social behavior, and cultural evolution to develop a multilevel model of the coevolution of genotypes, phenotypes, and environments. Basically, they propose that phenotypes play an active role in the evolutionary process. Plotkin's and Odling-Smee's concepts are consistent with the contemporary neo-Darwinist synthesis (Mayr 1982, 1988). They broaden the traditional interpretation, which holds that evolution operates in a relatively closed, deterministic way—purely in terms of the nonrandom process of natural selection and the random, chance-based processes of genetic drift and mutation pressure. Plotkin and Odling-Smee propose that aspects of the phenotype, not necessarily genetic in the traditional sense, can influence the evolutionary process. However, they adhere to the primacy of genetic inheritance as the main pathway that produces change in populations and organisms, insofar as it constitutes "the only route via which organisms can contribute anything to descent. So, if a phenotypic trait is to affect evolution at all, it can only do so via genetic inheritance, regardless of whether the trait itself is genetically determined or not" (Odling-Smee 1994, 163–164).

Odling-Smee and Plotkin contend that evolution entails the operation of several nested processes that culminate in the acquisition of information and knowledge by individuals, groups, populations, and species. These processes include, in addition to the nonrandom process of genetic inheritance by natural selection, several additional nongenetic, nonrandom processes that affect phenotypes and populations (or cultures). The latter include the processes of development, immunology, individual behavioral learning (including imprinting), protocultural processes such as social enhancement and imitation (arguably present in the higher apes), and all the ways in which cultural information is acquired and transmitted in culture-bearing organisms (Boyd and Richerson 1985). Rather than viewing these supplementary processes merely as statistical, passive influences and/or as merely assisting organisms, that is, phenotypes, to pass on genetic information, Plotkin and Odling-Smee hold that through phenotypic modifications of an individual's behavior, the processes actually modify the selective environment of themselves and other organisms and populations, including succeeding generations.

Plotkin and Odling-Smee use the terms *niche construction* and *ecological inheritance* to refer to the effects produced by the activities of phenotypes in this multilevel coevolutionary process. (Durham [1991] has provided a full discussion of coevolution in relation to human societies and cultures.) They refer on the one hand to the capacity of phenotypes to alter, thereby to reshape or construct, the autonomous environment via actions determined by genetic inheritance and previous learning experiences; on the other hand, they note the fact that the reshaped environment operates as a form of ecological inheritance insofar as it is a new, modified selective environment.

Organisms and populations, then, participate in a coevolutionary process linking the phenotype and the environment whereby they acquire genetic and cultural resources and information, as well as ecological resources shaped by niche-constructing activities of phenotypes. Organisms and populations also pass on information and resources to descendants. The Plotkin–Odling-Smee model describes two types of information flows, genetic and ecological, which would include cultural, on four levels. On level 1, the population is the information-gaining unit insofar as genetic information is held to tightly determine phenotypes in an obligate manner, yielding phenotypes that are viewed as nonlearning. On levels 2 and 3, organisms are the information-gaining and information-using units or entities; they receive genetic information from the lowest-level information-gaining entity, that is, the population and its genome, and modify their developmental and learning tasks and behaviors. In this instance, the phenotype undergoes facultative, that is, environmentally open as compared to genetically closed, development. On level 4, culture-sharing individuals and populations—note that allowance is made for group selection—constitute the information-using entity, and here a modification of cultural tasks takes place. Level 4, then, involves organisms pursuing facultative developmental processes and culminating in learning phenotypes that also use cultural knowledge, a repository of information that grows and evolves and that organisms acquire and make use of. Furthermore, at each level of the coevolutionary, phenotype/environment process, information-gaining and information-using units—populations at level 1, organisms at levels 2 and 3, and organisms and cultures at level 4—are actively engaged in changing the environment—that is, niche construction. In altering the environment, they create new selective environments for themselves as well as other organisms, and, by extension, other populations, thereby contributing to what Plotkin and Odling-Smee term ecological inheritance.

One can apply this perspective to hominid populations and their environments. In particular, one may use it boldly to conceptualize the potential role played by psychopathology during different phases of human evolution. The genetic material for the production of psychopathology resides at level 1, in the genomes of populations of early varieties of hominids, Homo, Archaic humans, and anatomically modern humans during the transition from the Mid- to Upper Paleolithic Era. As an example, given what is known of the behavior differences within groups of chimpanzees and bonobos, one can anticipate similar variation in the behavior and personalities of members of groups of the last common ancestor of the pongid/hominid split. Genetic factors would have made some individuals vulnerable to certain varieties of psychopathology.

Social relations among the earliest hominids were undoubtedly complex and selected for higher levels of social intelligence and cognitive flexibility. Cooperation among different temperaments or personalities required toleration of individual differences in behavior and personality, or social attribution. While genetic factors were influential in much of this behavior, on level 1 of the Plotkin–Odling-Smee model, higher levels came into play as learning phenotypes gained information through social experiences. In short, in communities of evolved social hominids, perhaps operating only at levels 2 and 3, although aspects of level 4, culture, were no doubt emerging, individual experience and learning must have modified and fine-tuned psychological mechanisms underlying behavior. These modifications of social behavior, given the challenges and stressors emanating from the physical and social environment, might have produced cognitive and emotional conflicts. These would be manifested, for example, in selfish or mutualistic pursuits, suspicion of cheaters, mate selection, sexual jealousy, social participation or withdrawal, loyalty versus deception, avoidance of irritable, belligerent, or impulsive group mates, and the like. Behavior designed to deal with complex social interactions and attempts to cope with myriad other group-centered conflicts could have given rise to psychopathology in populations of early hominids hindering conflict resolution.

Viewed in terms of the Plotkin–Odling-Smee model, genetic information contained in the population's genome and the niche-constructing activities of hominids, or varieties of Homo, Archaic humans and others, led to alterations of the selective environments of existing and succeeding populations. Psychopathological incidents led to conditions that might have altered the environment in ways that created entirely new forms of ecological and cultural inheritance, eventually shifting the mix of selective influences and selecting for or against certain varieties of phenotypes. As an example, a few aberrant, eccentric, withdrawn, moody, but creative introverts may have engaged in activities that reshaped (examples of niche-constructing phenotypes) the behavioral ecology of the group, thereby creating an environment that selected for or against a particular type of behavior or a personality style that came to be over- or underrepresented in succeeding generations.

The Plotkin–Odling-Smee model, then, offers a way of conceptualizing how niche-constructing phenotypes, or behaving individuals, who may have displayed signs of psychopathology can be fitted into the evolutionary transformations that culminated in Homo sapiens. During human evolution, genetic information, in conjunction with information acquired by social learning—that is, niche-constructing phenotypes capable of contributing to ecological inheritance— led to programs of flexible behavior calculated to minimize social conflicts. However, under certain circumstances, these programs may have been subverted, thereby contributing to the emergence of varieties of psychopathology. An important implication is that niche-constructing phenotypes, that is, real organisms, through individual or social learning, social imitation, and cultural learning, affected the selective environment of other individuals in diverse ways. This ecological inheritance constituted an altered environment that, in turn, selected for or against phenotypes that were themselves produced by a combination of canalized genetic influences and those affecting developmental, learning, and cultural tasks. Varieties of psychopathology, as products of genetic information and individual and social learning, can be understood as making up the pool of niche-constructing phenotypes that contributed to the ecological inheritance of succeeding groups. This interactive process could have had diverse influences.

To say that varieties of psychopathology during human evolution may have played a positive role in the evolutionary process appears to contradict the generally accepted idea that it is a maladaptive form of behavior, that it contributes to a lowering of fitness. Recall that so-called positive functions attributed to psychopathology may have taken place during selected, limited phases of evolution, bottleneck periods, when unusual actions or strategies were effective and positively selected for. Furthermore, psychopathologies are hardly the product of obligate development. However, as the Plotkin–Odling-Smee model indicates, they can incorporate exaptive learning influences, that is, niche-constructing phenotypes, that can alter the effects of psychopathology on behavior, and consequently its impact on the group. Moreover, the concept of psychopathology does not refer to complete behavioral breakdown and ineptitude but admits of many varieties of behavior. While it is maladaptive, strictly speaking, and causes personal distress and suffering as well as impaired fitness for the individual, over a shorter period it can have beneficial effects, if not on the individual at least on kin and the group. An important implication of Plotkin and Odling-Smee's model is that the niche-construction activities of phenotypes that display a variety of psychopathology do not exclude the possibility that while psychopathology may ultimately be negative for the individual, it can have salutary consequences on group mates. Psychopathological behavior can often contravene or push beyond the perimeter of normal, standard group conventions. By selectively adapting, modifying, or negatively reacting to psychopathology, group mates create new ecological inheritances that can lead to the selection of adaptive traits in succeeding populations.

Development as the Setting Where Psychopathology Affects Evolution

The behavior of organisms can create changes in the environment that come to play an active, selective role in the course of evolution. In effect, individual behavior can indirectly influence the process of natural selection and evolution. This raises the logical possibility that psychopathology, considered as a form of behavior or a trait of the phenotype, could have played an active, positive role in the evolutionary process. In other words, instead of simply representing a secondary consequence of environmental perturbation or a primary but negative change in the genotype, psychopathology could have sometimes played a primary and positive role in evolution.

But how to explain the emergence of psychopathology in the first place? When and how might have its characteristic features arisen and developed in the organism? Specifically, one may consider the following questions: What sorts of conditions, mechanisms, or processes could have led to the emergence of psychopathology and cause it to play a more active role in human evolution? This section pursues the analysis of the origins of psychopathology at a different level, specifically, a microdevelopmental level. How does the organism fit into its environment? What is the role of environmental changes in differentiation of the embryo and development of the organism? Psychopathology, as a negative change in the behavior of the phenotype, can arise as a result of environmental change, of migration into a new ecological zone, or, indirectly, simply of change in niche preference as a consequence of microbehavioral changes without necessarily involving geographic migration. Any of these factors could have changed the relation between the organism and the environment. The long-term effects could consist of changes in individual development by creating alternate life strategies and life history parameters, such as those involving mating, parenting, competition, survival, and, eventually, the adaptability and adaptation of an entire population. This can culminate in heritable modifications of social and psychological behavior, no less than morphology and physiology, that are hallmarks of the evolutionary history of a species. Psychopathology may have figured importantly in this whole process, but in order to explain this possibility, we need to understand how psychopathology becomes an integral and especially an innate part of the organism.

DYNAMICS OF AN ORGANISM'S DEVELOPMENTAL MANIFOLD

In studying the interplay between genetic factors, behavioral embryology, development, the origins of novel behavior, and evolution, Gottlieb (1992, 1995, 1997) argues that populations are capable of enormous behavioral modification and plasticity, or adaptability. This adaptability is a feature of the population's genotype, a reservoir for the production of phenotypic behavioral variability that is tapped as the relationship between the organism and the environment changes. Significant environmental change initially affects the embryology and development of the organism, thereby changing behavior patterns, or morphological and physiological systems, that are generally thought of as innate and genetically determined. In subsequent generations, the changes in behavior introduced during development can become heritable components and part of the population's genome.

Gottlieb explains that the initial change in behavior as a result of environmental perturbation is not genetic in the strict sense in which it is generally construed in the modern synthesis, heavily influenced by population genetics. This synthesis stipulates that mutation and recombination of genes, selection, drift, and changes in gene frequencies are the essence of evolution, with these parameters often examined in abstract, formal, and mathematical terms. The population genetics model leaves out of focus how actual variation arises in concrete circumstances of embryology and development—although, of course, innumerable experimental studies have documented factors that cause mutation and influence genetic variability. Gottlieb argues that the initial impetus for evolutionary change lies in what he calls the organism's developmental manifold. This consists of a complex, nonlinear, interactional complex of influences that characterizes the normal behavioral embryology of organisms, as well as their development and maturation more generally, that is, their developmental experiences. Genes, of course, are an integral component of all of this: "The *gene* [DNA] is the ultimately reduced unit in the ever-expanding developmental pathway" (Gottlieb 1992, 142). However, Gottlieb emphasizes a range of influences that are supra- and extragenetic and that involve interactions affecting the products of genes and the environments in which genes and their products exist during phases of embryonic and subsequent development.

The development of an organism, considered broadly—that is, during its embryonic and early postnatal phases—is a complicated process that consists of influences that take place at the same and different levels of the multiple systems that come into play following fertilization and differentiation of cells during the formation of the embryo. The systems incorporate activities within cells (nuclear DNA-RNA-cytoplasm), between cells (involving tissues), between organs, and between organism and environment, including social interactions between organisms that affect circulating hormones that affect cells and their DNA. The development of an organism is a complex amalgam of influences that move forward, backward, and across levels and systems; in other words, displaying directional, coactional, hierarchical integration. By studying the modification of the late embryonic experiences of mallard ducklings, which affect their behavioral relationship with their environment, Gottlieb was able to modify the seemingly instinctive, innate preference for their species' maternal calls.

Gottlieb (1992, 1995, 1997) interprets the implications of behavioral changes caused by changes in the early developmental experiences of vertebrates. He points out that much of the potential for behavioral plasticity brought about as a result of new developmental experiences and influences on the behavioral embryology of individuals cannot be strictly accounted for by the traditional, populational model that explains the roles of genes and undergirds the modern evolutionary synthesis. Some of the potential resides in changes in the cytoplasm of cells that affect behavioral development, and some of it resides in genes that are activated or turned on during development but are not in a strict sense structurally in control of the processes and behaviors that result, as postulated by the gene-trait model of population genetics.

The behavioral potential released by environmental change, a part of the behavioral adaptability of members of the population in question, can persist over generations, provided the same environmental conditions affecting development that initially caused the changes are maintained. Subsequently, as a result of processes such as genetic assimilation (akin to the Baldwin effect) and stabilizing selection, an actual change in gene frequency can occur, resulting in a more direct, structural control of the behavior in question. At this point, stage 3 of his evolutionary pathway, termed changes in genes, the process of natural selection as ordinarily conceptualized in the modern synthesis, has come into play. Gottlieb emphasizes that in reality evolution has already occurred by stage

2, when nongenetic influences on development that bring about the new behaviors are operative. His formulation of the evolutionary pathway also includes stage 1, changes in behavior—for example, new behavioral shifts encouraging new environmental relationships—and stage 2, changes in morphology—for example, new environmental relations that bring out latent possibilities for physiological change.

The basic feature of Gottlieb's formulation involves changes in the developmental conditions of organisms, the activation of quiescent genes, and induction of behavioral plasticity realized as exploratory behavior and problem-solving activities. The release of this behavioral potential as a consequence of perturbations of late embryonic, perinatal, and postnatal experiences brings the organism into a new relationship with its environment: The behavioral change can prompt the organism to change its existing environment, propel it toward a new one, set in motion modifications of inherited programs that in time can lead to alternative life strategies, and can in general lead to a shift in the organism-environment equation altogether. Gottlieb emphasizes that what he terms "probabilistic epigenesis" and the creation of "behavioral neophenotypes" are most likely to occur in vertebrates (mammals and birds), who, compared to invertebrates, are capable of greater neuropsychological and behavioral change. In other words, changed behavioral patterns are more likely to take place in organisms with larger brains and that depend on variation of early experiences for behavioral development and maturation.

Gottlieb emphasizes that a species whose young show long periods of dependence and social learning, along with eventual behavioral variation and plasticity, is especially likely to undergo the changes he describes. Consequently, one can assume that hominid newborns and infants (possibly juveniles) were potential candidates for probabilistic epigenesis and behavioral neophenogenesis. As a result of general phylogenetic changes, hominids acquired large brains and their young depended on adults for their development. The character of the environment, the stimulation provided by its physical and social characteristics, and prevailing developmental experiences could increase dendritic and axonal differentiation and connectivity as the brain matured. All of this modified behavior and made it more complex, thus making hominids good candidates for behavioral neophenogenesis as formulated by Gottlieb. In short, hominids were capable of enormous behavioral variability, being able to show adaptability and plasticity in the event of environmental change both before and after birth.

Gottlieb describes a variety of scenarios that can bring into play behavioral neophenogenesis. His examples are lower mammals, but his formulation is equally relevant to nonhuman primates and hominids. Moreover, his research focuses narrowly on late embryonic development and the period that immediately follows, thus emphasizing certain environmental factors that can affect the developmental manifold. However, in species that produce altricial young and whose period of development stretches well beyond infancy and even into puberty or adolescence, influences that can induce behavioral neophenogenesis are equally relevant. Examples of changes that could affect experiences during development and contribute to behavioral neophenotypes would be severe changes in climate and the amount of daylight time, changes in nutrition and in the availability of key mineral ingredients (factors necessitating radical changes in living space and density of population), and changes creating radically new living conditions that required greater vigilance, living conditions that demanded more activity, emotionality, and risk taking.

In summary, Gottlieb outlines a line of causation starting with the inherited genetic potential for adaptability and moving through (1) environmental change, (2) change in conditions of ontogenesis, (3) changes in experiences during development, (4) realization or release of potentials for plastic changes in behavior, (5) the creation of new behavioral responses, that is, behavioral neophenogenesis, (6) changes in behavioral coping or adaptiveness associated with the neophenotype, (7) transgenerational maintenance of changed behavior, and (8) natural selection for genes actually controlling the behavior, which as a result can be conceptualized as an adaptation, at least in the short run. A key functional relationship exists between step 1, environmental change, and step 5, the creation of new behavior phenotypes.

IMPLICATIONS FOR PSYCHOPATHOLOGY

The behavior pathologies of adult chimpanzees resulting from altered environmental conditions, such as those reviewed by Power (1991) in her critique of the work of Goodall (for example, Goodall 1986) and the laboratory manufacture of stress on maternal behavior of rhesus monkeys and its effects on infants (see Harlow and Novak 1973; Schneider 1992b), reflect the kinds of changes that adversely affect relationships of the fetus and infants with the environment. Such changes could bring about behavioral deviations or psychopathology. Some of these, of course, are catastrophic stresses that may overwhelm the resources of the developing organism and preclude its survival. However, exposure to milder degrees of stress and deprivation could cause changes in the organism-environment relationship that could lead to the creation of behavioral neophenotypes. Such experiences could have been followed by periods of stability during which infants could benefit from positive and even therapeutic relations with maternal figures and siblings, if not others. In other words, psychological stressors and associated changes in the social environment during development could have led to the creation of behavioral neophenotypes that in selected instances, given conditions similar to those outlined by Gottlieb, could have culminated in heritable forms of psychopathology in populations of early hominids. This supports the niche construction idea.

Gottlieb's ideas about behavioral neophenogenesis are consistent with and complement the formulations of behavioral and biological anthropologists who are interested in evolutionary theory and who have speculated about the causes of contemporary human psychopathologies. For example, Draper and Harpending (1982, 1987, 1988) postulate that later human reproductive strategies are influenced by sensitive periods of development. They argue that children raised in households without a father present have a very different development experience from children raised in households where fathers are present. They form very different perceptions about the availability of resources, how they should be spent, and how they should be obtained. Boys and girls appear to have different reproductive strategies that are assumed to be inherited and rooted in the evolutionary past, depending on whether a father was present or absent in the home. These strategies have long-term consequences for psychopathologies that affect both men and women. This is also consistent with the niche idea.

Draper and Harpending, then, argue for something that is rooted in the open developmental manifold of Gottlieb. They imply that conditions during evolution may have selected for various developmental programs that became part of the inheritance of hominids. They then apply this formulation to contemporary populations, focusing on infancy and childhood.

Formalizing the Role of Psychopathology during Human Evolution

The idea that psychopathology played an active, positive role during human evolution can be put in more theoretical and formal terms. According to Goodwin (1985), in an article aptly entitled "Constructional Biology," psychopathology is an example of a paradigm of constructed and constructive behavior change. This change is aimed at restabilizing the individual in its setting, correcting changes introduced by environmental perturbations, or simply solving (within a set of boundary constraints entailed by the behavioral manifold) a particular adaptive problem signaled by a perturbation.

Goodwin presents a general perspective described as a structuralist biology, or a structuralist view of organisms:

> Structuralist biology attempts to render the biological realm intelligible by giving an account of how organisms are possible in a manner similar to the way in which Newtonian mechanics gives an account of how celestial motion is possible. Each actual organism is seen as an instance of the possible, a specific manifestation from a potential set defined by the invariants which characterize the entity "organism."

A field description of organisms is an example of a structuralist analysis which seeks to describe organismic morphology and its transformations in terms of the solutions of particular field equations which define the potential set of developing forms in a particular developmental process. . . . As far as morphology is concerned, organisms are then seen as transformations of one another within the group defined by changes in initial and boundary conditions of the field solutions describing their form, which delimits the set of possible generative processes available to organisms as structures of a particular type. From this perspective, the major problem of biology is to articulate and develop a theory of biological organisation in these terms, attempting to define the transformation rules which underlie embryonic development and the relationship between different species as generative transformations of one another (that is, transformations via developmental trajectories). *Behavior also needs to be understood in these terms, extending to the rules of transformation which underlie developmental cognitive processes.* (1985, 56; emphasis added)

Goodwin's explanation is consistent with those of Plotkin and Odling-Smee and Gottlieb. He relies on ideas about self-organizing systems, feedback and regulator theory, and organism-environment interactions as these relate to evolution. While Goodwin emphasized changes in morphology, he makes absolutely clear that the same principles apply to changes in behavior. He shows that general, nonspecific physical environmental perturbations of low information content can mimic or copy changes in morphology, and, by extension, behavior produced by genetic mutations, that is, phenocopy phenomena: "Thus, environmental perturbations whose 'information content' is low, and hence which have low specificity, have the same causal status as genetic perturbations. . . . The two categories of perturbation are radically different in this case. . . . We are thus led to the conclusion that genes are not specific causes of morphology . . . since their role can be duplicated by nonspecific perturbations. Rather, both genes and environmental stimuli are acting upon a self-organizing morphogenetic process, classically referred to as a morphogenetic field, and it is the (limited) set of possible responses of this field which is responsible for the phenocopy phenomena" (1985, 49). One might adapt Goodwin's formulation as follows to underscore its relevance for behavior and, by extension, the origins of psychopathology: "The developing organism is to be understood as a field (or set of primary, secondary, and tertiary fields with defined spatial boundaries within the embryo [the developing brain of an early hominid]) whose solutions define the potential set of morphologies [behavior programs of strategies] which it can realize" (1985, 50).

Goodwin first reviews some of the ideas of Piaget (1980) about relations between cognitive development and evolution—for example, how organisms assimilate aspects of the environment and equilibrate with it by developing an internal model of it. Goodwin cites Piaget's claim that "not only are children stimulated to produce appropriate operative schemes by specific environments, but . . . they also spontaneously create new schemes of behavior for which appropriate environments are then realized if possible. In the first instance there is a 'copying of an exogenous formation by an endogenous one,' which Piaget calls phenocopying in the broad sense, in the second, there is the discovery or creation of an exogenous formation which 'copies' an endogenous one, a process which Piaget does not consider" (1985, 53). Goodwin continues: "If we now translate these into biological terms, then we see that appropriate (genetically adapted) organisms can be generated by either a response to a new environmental challenge, a hereditary state arising by some means and resulting in an appropriate organismic form for that environment; or spontaneous reorganisations within the hereditary constraints can occur, procusing [sic] organisms with new morphologies and behavior patterns which must then either discover or create appropriate environments" (1985, 53–54).

Using ideas about information processing and feedback systems drawn from abstract automaton and regulator theory, Goodwin keys these to the insight that organisms operate in terms of generative principles of biological form and transformation as they internalize exogenous formations and develop new structures of knowledge, function, or morphology. In this way, Goodwin is led to develop ideas about an "internal model principle." He uses this principle to help explain how organisms are able to internalize and transform exogenous information to solve problems posed by the environment: "Thus, although some degree of adaptation to the external environment is clearly a requirement for all species of organism which persist, the internal model principle shifts the emphasis from mere survival to a notion of harmonisation or dialectical stability between organisms and their environments, and between different levels of order within organisms" (1985, 60–61). It is this scheme, then, that accounts for progressive or directional evolution: "There is a general direction of organismic change, a transformation asymmetry, which makes it legitimate to talk about evolution as 'progressive,' or at least directional" (58).

How can Goodwin's ideas be used to explain how psychopathology could have played a positive, active role in human evolution? First, consider psychopathology as a form of behavior that is keyed to, is in harmony with, or exists in a dialectical relationship with a social environment. The organism involves an overarching system consisting of connected and hierarchical systems that takes in, processes, and acts upon information about the social environment—actually, a set of relationships directed to solving adaptive problems for the organism and group mates. Both the organism or system and the social environment have exhibited change during evolution. (Goodwin defines *system* and *environment* in relation to one another and defines *organisms* as hierarchies of such relationships [1985, 60].) Hominids have evolved language, cognition, and culture, while social environments could be said to have evolved different demographic patterns, population structures, patterns of social exchange, and transactions with the ecology. Moreover, during the process of evolution, the two sets of changes, internal and external, were keyed to one another in a temporal, transformational asymmetry. In the process, aspects of the environment are incorporated in the hominids' internal model principle, entailing something like a cognitive and genetic assimilation of the environment. Finally, it should be clear that perturbations affecting either organisms, such as genetic mutations, developmental traumas, physical harassment by group mates, or the social environment, such as predator pressures, interspecific competition, group fissioning, migration, and emigration, bring about changes in the organization and functioning of their respective domains that also affect the relationship between the two, leading to a new cycle of change and assimilation.

Such an approach to behavior and evolution avoids many of the weaknesses of either genetic or environmental determinism and is not inconsistent with ideas about behavioral programs, behavioral strategies, and psychological mechanisms or adaptations promulgated by evolutionary psychologists, sociobiologists, and behavioral ecologists. Psychopathology fits easily into this formulation: it is a form of potentially maladaptive behavior marking a new state or condition of the hominid organism that is prompted by perturbations arising from either within itself or the environment, most likely both. The internal aspect of psychopathology, pertaining to the organism, reflects the behavioral responses of a system made up of organizers and regulators, structures that internally represent the outer world so as to maintain order and stability—or postpone disorder, disintegration, chaos. These goals, if successful, bring satisfaction, comfort, and safety; moreover, they promise a new, possibly even better state of organization for the organism. The outer aspect of psychopathology, pertaining to the social environment, consists of a set of altered social relations and behavior routines that could theoretically promote or allow the inner aspect a measure of stability, order, comfort, and purpose. These routines of the social environment include routines or management systems involving repair and growth of the apparatus, physical safety, rank and status, reproduction and parenting, interspecific competition, group identity and history, and eventually spirituality, a sense of one's place in the cosmos. Thus, one can ascribe to psychopathology an active, positive, creative role in human evolution by seeing it as part of a process whereby organisms replace external patterns of influence by patterns of internal contingencies.

Comment

The model proposed by Plotkin and Odling-Smee of niche-constructing phenotypes, the mechanism of behavior neophenogenesis proposed by Gottlieb, and Goodwin's formulation about a constructionist biology provide three related and complementary ways of visualizing the role that psychopathology could have played during human evolution. All three theories propose the development of a new form of behavior that has a positive impact on the organism-environment relationship and, given environmental contingencies, becomes incorporated in the genome and thus becomes heritable. In subsequent generations, under changed conditions, the same behavioral pattern may impair fitness and be seen as psychopathology. Alternatively, a negative, maladaptive behavior routine resulting from an environmental perturbation may, despite its short-term fitness costs, persist in the population transgenerationally and later, under changed conditions, prove adaptive and be selected for. The advantage of Gottlieb's perspective is that it provides a window through which one can visualize the developmental dynamic that induces behavioral change, even psychopathology, that proves important in the evolutionary process.

From the standpoint of evolutionary biology, psychopathology is a pattern or syndrome of social and psychological behavior that is brought about by and that depends on underlying neuropsychological and psychophysiological processes with a genetic basis. Overt changes in social and psychological behavior constitute the surface picture of psychopathology, whereas changes in how the brain handles and responds to elementary components of experience and social action form its inner, hidden core. The overt, manifest syndrome involves perceptions, attitudes, feelings, cognitions, beliefs, and altered social relations and social functioning, all in the context of the group's cognition and culture. This is the expressive character of psychopathology. On the other hand, the hidden, covert set of changes in neurophysiology and neuropsychology are caused by cerebral processes molded by genetic, embryological, and developmental influences in relation to the environment. In other words, as Gottlieb argues, by acting on the hidden, covert component of behavior the environment releases new overt behaviors by influencing the developmental manifold, and some of these may prompt microhabitat changes in the old environment, migration to new environments, general changes in the organism-behavior relationship, or, as Draper and Harpending suggest, differing rearing environments of infants and children that may prompt alternative reproductive strategies that have implications for the development of psychopathology.

The new behavioral phenotypes can be viewed as the niche-constructing phenotypes of Plotkin and Odling-Smee and conform to dynamics of the internal model principle of Goodwin, all of which incorporate changed environments into the evolutionary equation. These changed environments should be conceptualized broadly so as to include altered social experiences and relationships and ways of coping, as described by Draper and Harpending. Thus, insofar as psychopathology is a feature of the phenotype—more specifically, a selected subset of the social and psychological behavior of individuals—it can be said to have played an integral role in the process of evolution, like virtually all other components of the phenotype.

In this scheme, a syndrome of psychopathology is understood in terms of how it enables individuals to cope with the environment and manage their affairs. While distress, suffering, inability to cope, and impaired fitness are relevant criteria of psychopathology, these effects are not timeless and universal, but relative and conditional. Changes in environment and social and behavioral ecology more generally can change the nature of experience, function, and fitness associated with a particular pattern of behavior. The cerebral substrates that constitute the hidden component of a behavioral syndrome, and the syndrome itself, may have persisted and been selected for under favorable ecologies, only to contribute to the genetic load of the species at a later phase of its evolution when conditions of living had changed.

- Positive selection of traits causing psychopathology illustrates the way in which symptoms and signs might have affected behavior of adult victims residing in natural communities.
- An evolutionary explanation of psychopathology is more compelling if one explains why and how underlying behavior mechanisms were naturally selected in EEA. We need a broad evolutionary framework that integrates genotype, phenotype, and environment.
- Ideally, such a framework should encompass behavior, biological evolution, and social activities and traditions including culture, thus helping to integrate the human sciences.
- Niche-constructing phenotypes and ecological inheritance refer to the effects of organisms' behavior on their environment, which then alter the conditions of natural selection of group mates, essentially helping to codirect subsequent biological evolution.
- Characteristics of Plotkin's and Odling-Smee's broadly integrated evolutionary framework involving niche construction is as a mechanism that could in part explain positive selection of psychopathology.
- A concrete and graphic explanation of positive selection is obtained if the mechanisms and processes that mediate it can be linked to developmental periods when individuals are more plastic, such as during late phases of embryology, infancy, and adolescence.
- Gottlieb's empirical work on the evolution of behavior of newborn mallards led to the derivation of new concepts and a dynamic model

integrating effects of genes, behavior, and environment termed probabilistic epigenesis and behavioral neophenogenesis.
- Gottlieb's model, which describes how environmental change releases new, nongenetic behavior that subsequently changes gene frequencies, becoming part of the organism's genotype, helps explain the positive selection of psychopathology.
- Goodwin's structuralist biology about automata and regulators addresses how biological forms internalize environmental information to develop new systems of knowledge and function, providing a still more abstract model of positive selection of psychopathology.

Chapter 6

On the Limits of an Evolutionary Conception of Psychopathology

AN EVOLUTIONARY EMPHASIS has entered the study and treatment of human psychopathology. Psychiatrists and clinical psychologists interpret disorders that seem connected to contemporary society and explicable in terms of modern medicine and social science as having ancient origins. Differences are to be noted in their formulation of psychopathology. McGuire and Troisi (1998) seem primarily concerned with explaining the evolutionary basis of psychopathology in terms of breakdowns, deficiencies, or perturbations of fundamental organizational systems of behavior (termed by McGuire and Troisi as infrastructural) that were naturally designed ("ultimately caused") in relation to pursuit of fitness-enhancing biological motivational goals, including that of survival, reproduction, and associated social relationships among kin and nonkin. I have referred to these aspects of human behavior as "evolutionary residuals" to emphasize their roots in the process of biological evolution. Imperatives of adaptation to past environments—for example, EEA—and adaptiveness to current ones condition the character of psychopathology and help explain its manifestations in the McGuire and Troisi formulation. A specific difference between current environments and ancestral ones and, particularly, EEA is not a dominant feature. In essence, they appear to hold the view that many of the important systems of behavior, the breakdown of which produce varieties of psychopathology today, may predate EEA and, moreover, play a dominant role in the production of psychopathology in all environments, including the contemporary one.

On the other hand, while Stevens and Price (1996) explain contemporary psychopathology in terms of the importance of basic evolutionarily centered or rooted parameters of social, psychological, and behavioral functioning, they seem more wedded to the incongruity or discontinuity, that is, environmental mismatch, between EEA and contemporary modern environments (Crawford 1998a, 1998b). They attempt to explain psychiatric disorders by studying phylogeny, early ancestral environments, and evolved mechanisms underlying human behavior. Like many evolutionary psychologists, these investigators give central importance to the character and ancestral bases of human adaptations and behaviors. However, as indicated, they emphasize why and how these adaptations and behavior patterns malfunction in contemporary environments to explain psychiatric disorders.

A central assumption in evolutionary studies of psychopathology is that there are many enabling conditions that account for its prevalence. Some of these include genetic factors such as pleiotropy, heterosis, behavior polymorphisms, and deleterious genes not yet selected out of the genome. To this must be added straightforward variations of traits related to behavior and physiology, a point emphasized by McGuire and Troisi (1998). The list would include developmental abnormalities of behavior resulting from infection, stress, or psychological trauma, acquired defects affecting behavior, effects produced by defenses imposed by intercurrent stress, and the wear and tear of imperfect design features of the human brain (Nesse and Williams 1994). In emphasizing that the roots or ultimate causes of psychopathology reside in naturally designed structures of behavior that have gone awry, evolutionary psychiatrists provide support for the idea that it, that is, psychopathology, constitutes a harmful dysfunction.

Is Psychopathology a Passive Reaction or a Positive Response?

The topic to be pursued here involves the implications of handling a contemporary variety of psychopathology as though it constituted a remnant of mechanisms or adaptations that were selected for in EEA. This entails viewing a variety of psychopathology not simply as enabled by evolutionary factors—for example, pleiotropy, trait variation, behavior polymorphisms, deleterious genes—a result of a lesion or stress, or a passive, mechanical reaction of inherited routines in relation to environmental problems. Instead, this chapter examines the question that psychopathology should be construed in purely positive terms: its manifestations are the result of trade-offs, entailing costs of implementing behavior routines that were beneficial in EEA but that create special problems in a contemporary society because of a mismatch in environmental conditions or simply intercurrent stress. In other words: how might psychopathology make sense as a purely adaptive response? This question is worth asking and attempting to answer even though it evokes features of the dilemma arising from differences between the cultural, clinical compared to the strict evolutionary viewpoints and limitations of some of the conceptions of psychopathology implied by evolutionary theory. Its analysis will point to dilemmas and paradoxes inherent in the fields of evolutionary biology, psychology, and clinical psychopathology that cannot easily be resolved and which, as a consequence, pose intellectual challenges to and place some limitations on the enterprise of an evolutionary psychiatry.

The unorthodoxy of the exercise carried out in this chapter needs to be emphasized at the outset. As discussed earlier, a hallmark of what was described as the cultural and clinical approach is that phenomena thought of as psychopathology are by definition undesirable and unwanted. Psychopathology is thought to reveal a negative condition that should be corrected, to indicate an individual in distress who is socially and behaviorally impaired. All societies have institutions and healers for this. This cultural, clinical view of psychopathology is clearly not independent of an evolutionary one that sees fitness as at least potentially compromised and that seeks treatments in evolutionary principles. Indeed, as mentioned earlier, all textbooks of evolutionary psychiatry are based on a premise that the perspective in question offers solutions to the plight of the mentally ill. Moreover, a leading, although controversial, approach to psychopathology conceived as disorder is that it constitutes a harmful dysfunction viewed in terms of evolution (Wakefield 1992). In other words, psychopathology as disorder is thought to involve a breakdown in the functioning of a system or adaptation, a failure to perform the function it was naturally designed to perform. All of these claims imply that psychopathology is not only negative but contra adaptation, as it were.

In this chapter, this formulation of psychopathology is in some respects inverted. Psychopathology is examined as though it actually constitutes an adaptation; in other words, as embodying something positive. The manifestations of psychopathology are held to be part of an adaptive mechanism that natural selection has designed in order to effectively cope with distinctive strains and stressors associated with social living in an individual's behavioral environment. There may be associated costs and penalties inherent in psychopathology, to be sure; especially when environments deviate too far from that of EEA. But the penalties and costs of psychopathology might be construed as the necessary, perhaps more immediate, short-term, trade-offs exchanged for other, perhaps more long-term, benefits. At any rate, the basic claim to be examined is that not only does the syndrome of behavior so designated constitute the individual's adaptive response to an undesirable situation; the response pattern itself, that is, the psychopathology, was naturally designed to cope with recurring stresses and strains associated with the environment of evolutionary adaptedness.

Causal and Functional Questions Pertaining to Psychopathology

There are two basic ways of studying any trait of an organism from a general biological point of view (Curio 1994). Here I will focus on behavior generally and on social and psychological behavior specifically. To understand behavior from a biological point of view, one can concentrate on (1) causes and mechanisms, which depend on proximal questions, how things function or malfunction in the organism now; or (2) biological, adaptive functions and goals, which bring into play ultimate questions, such as the origins of behavior in the EEA in response to natural and sexual selection. Because psychopathology is manifested in social and psychological behavior, as well as psychophysiological symptoms and signs, we should examine both types of considerations in relation to one another.

Contemporary psychiatry is overwhelmingly preoccupied with causal, mechanistic, and proximal questions, because it is quintessentially a service profession pursuing practical ends; its goal is to relieve suffering. The psychiatrist approaches a psychiatric disorder by identifying the mechanisms that bring it about, such as defective or disturbed neurotransmitters or neurotransmitter systems. Alternatively, the psychiatrist seeks to clarify what factors in the physical or social environment could precipitate a clinically relevant instance of a disorder. If adverse developmental experiences have predisposed a patient to psychopathology, these factors would have to be identified, as well as later, precipitating ones. And, as a final example, a psychiatrist would seek to clarify the mechanisms and processes that intervene between the precipitating cause or vulnerability factor(s) and the outward picture of psychopathology. This would include its pathogenesis, its premorbid phase, its clinical emergence, the connections between perturbation and manifest signs and symptoms, their progression and course, and eventually the end points of the disorder with or without appropriate intervention.

By contrast, a functional, adaptive approach, which considers ultimate questions as to origins, would attempt to clarify why a particular routine or program of behavior, including that of a psychopathological syndrome (or vulnerability to it), is part of human nature in general; and why vulnerability to varieties of psychopathology is part of the genome of human populations in the first place. An evolutionary functionalist wants to know, for example, why a behavior strategy that could include a variety of psychopathology is present in human populations and why and how it might have even come to be naturally or sexually selected.

A functional, adaptive analysis does more than ask why, however. For example, such an analysis can involve determining how the syndrome in question, or at least certain features, might have proven advantageous in the EEA. If a functional analysis is viable, then a critical examination of psychopathology should indicate what goals or functions it could have possibly fulfilled in the evolutionary environment. By unpacking the features that make up the psychopathological syndrome, we may be able to understand its adaptive, functional aspects.

For example, Price et al. (1994) present a social competition hypothesis to explain depression. By using ideas about ethology and evolutionary biology—such as sexual selection, ritual agonistic behavior, social hierarchy, and resource-holding potential—they provide a compelling analysis of the apparent functions of many manifestations of depression. In other words, the descriptive attributes of depression, its symptoms and signs, fit in with the social logic and adaptive rationale whereby depression arose as a response to recurring circumstances and problems in the EEA.

Reemphasizing the Complementarity between the Two Approaches

Clearly, the causal, mechanistic approach and the functional, adaptive approach are not mutually exclusive (Curio 1994). Depression, for example, can be caused or triggered in vulnerable individuals by any number of events or social circumstances. We may assume that analogous circumstances in the EEA operated to select depression as a potential response pattern to recurring adaptive problems that this particular syndrome of behavior attempted to solve. The syndrome could in theory be the manifestation of only one general performance mechanism. In other words, one or more cues about some feature of the social environment could activate the same mechanism and associated responses, such as physiological processes or neurotransmitter release, that together make up the response or performance system that sets in motion the behavior syndrome called depression. Also, cues and stimuli from the environment that cause depression, that is, that set off the causal mechanisms in the individual, could have different effects, perhaps because pathways to the one response mechanism have differing thresholds or because several of these response systems are related in some hierarchical way.

While this type of analysis has a causal, mechanistic orientation, it could be broadened by the functional, adaptive approach, which seeks to know what the syndrome of depression is responding to and why. Thus, by clarifying the evocative potential of various cues, or the ecological correlates of differing thresholds of physiological response, all causal, mechanistic questions, one might learn the motivational, functional, and adaptive implications of these cues for individuals in the evolutionary environment. In other words, insights about what depression might have been designed or selected to solve could be clarified. Seeking causal mechanisms, in other words, leads naturally to functional considerations, and both types of analysis may be able to explain how depression is produced and why.

With respect to the mechanism of depression, alternative formulations are possible. A syndrome of depression may consist of breakdowns in one or more homeostatic mechanisms, such as those regulating neurotransmitters in the brain, hormone balances throughout the body, visceral physiological functions, and/or diurnal rhythms. These account for a cluster of descriptive, behavioral manifestations, each associated with its own faulty mechanism. Alternatively, the syndrome could consist of only one basic mechanism complemented by others that can be recruited by associated cues or motivational factors (temperament, appetitive predispositions, or state of physical health).

Ideally, understanding depression as having an evolutionary basis involves clarifying the processes and neurobiological changes associated with each mechanism, clarifying how the latter are linked neurobiologically and how all elements are activated in relation to ecological cues that in theory initially designed and later came to set off the mechanism, or several internally linked or physiologically independent mechanisms, in order to cope with a certain social or ecological problem. In short, a causal, mechanistic account is bolstered by asking functional questions. For example: What are the adaptive implications of, or what kinds of adaptive problems are signaled by, the various ecological clues that can evoke the mechanism(s) that produce a syndrome of depression? Did the functional, adaptive characteristics of an evoking ecological factor or mechanistically conceived cause (in the EEA) bring into play alternative internal mechanisms that, combined with the depression syndrome mechanism, explain its varying manifestations? Here again, causal, mechanistic concerns lead naturally to functional, adaptive ones, and vice versa. Both are needed to examine the evolutionary origins and character of depression. Moreover, these two approaches can be mutually beneficial in the study of virtually any form of psychopathology whose origins and character one would like to understand from a general biological point of view.

Nesse (2000) has recently offered a critical analysis of the idea that the depression spectrum constitutes an adaptation. Based on much earlier work involving the evolutionary function of emotions and the biological basis of responses linked to general medical disease, he offers a summary of the possible functions of low mood—states in the common range of normal experience—and depression—severe states of negative affect, usually pathological. He sees these as pleas for help, the elicitation of help from group mates, and also as a communication designed to manipulate others to provide resources and then conserve them. Depression is part of a motivational package to plan and reassess a course of action with a possible view to change or alter goals. Even some conditions of frank clinical depression, Nesse implies, can be explained as serving evolutionary functions. However, his analysis and experience lead him away from explaining depression in terms of one function and instead to view the spectrum as states shaped to cope with a number of unpropitious situations.

Psychopathology as a General Defense System

There is no realistic likelihood that depression or any other variety of psychopathology could be adequately and fully explained as a positive adaptive response pattern. The idea being proposed appears to commit the worst sins of the Panglossian paradigm as formulated by Gould and Lewontin (1979). It is also vitiated by the claim that varieties of psychopathology constitute complex ensembles of disturbances of several overlapping behavior systems that signal deficiencies and impairments in solving one or several of the many functional motivational goals of an organism (McGuire and Troisi 1998). Finally, many varieties of psychopathology, as well as disease, result from defects in the apparatus, lesions, or infection and in no easy way can be conceptualized as adaptive or an actual defense system of behavior (Nesse and Williams 1994). The general proposition being pursued here, then, is not only counterintuitive but also somewhat simplistic in a contingent, empirical sense, largely resulting from an analytic supposition that psychopathology constitutes a unitary object or thing. Yet, the adaptive formulation, as illustrated earlier, is premised on positions too often handled in a contrastive, independent way from the standard mechanistic clinical account that, logic compels, should be integrated. Furthermore, an adaptive formulation is not without some validity and has been proposed in a fruitful way to explain selected varieties of psychopathology.

If one accepts that the causal, mechanistic approach and the functional, adaptive approach to explaining behavior should ideally be interdependent, then one could choose to examine virtually any form of psychopathology as consisting of manifestations of a general defense or response system that was selected for in the evolutionary environment. Varieties of psychopathology would of course be based on different neurobiological routines and connected to or triggered by different ecological conditions. The overall, common evolutionary basis is what is emphasized. From the standpoint of strict evolutionary biology, this implies the existence of species-specific mechanisms for the control of behavior, such as psychopathology, that were selected for in the EEA because they enabled individuals to survive and reproduce. In other words, forms of psychopathology have evolutionary significance because conditions in the EEA made them significant, and therefore individuals evolved neuropsychological mechanisms to produce them. Because such evolved mechanisms contributed to the reproductive success of individuals and their genetic relatives, the mechanisms continue to operate in current environments. Evolved mechanisms for psychopathology can have very different results today because of environmental mismatch.

We can imagine that early hominids lived in a constant state of threat and potential deprivation because of the harshness of the environment. This necessitated the design of a complex, multichannel, polyvalent, and general defense or response system to confront hazards, solve problems, survive, and reproduce. This applies to morphological and physiological responses such as disease or injury, and behavioral responses, such as to stress caused by the social and psychological context. Because even early hominids and certainly later varieties of Homo were social creatures with an evolving symbolic culture, possessed a social intelligence consisting of a sophisticated capacity for mental attribution and theory of mind thinking, all threats and challenges posed by the social environment in particular were by definition framed in some type of social-psychological idiom (Bickerton 1990, 1995; Donald 1991; Foley 1987, 1995a, 1995b).

The evolutionary hypothesis is that hominids and varieties of Homo interpreted threats to security and responded to them adaptively, in light of the social and psychological implications they attributed to those threats. It could be that the behavioral responses designed by natural selection—avoiding predators, seeking food, detecting cheaters, resolving social conflict—and by sexual selection—competition over or selection of mates, reproductive success, decisions over parental investment—were embodied in largely unconscious behavioral routines and strategies inherited from earlier varieties of species Homo, that is, evolutionary residuals, that once had been responses to social threat and security, some of which led to psychopathology. However, the expression of some of these largely nonconscious routines must have been accompanied by some form of social and psychological awareness and goal directedness. This is based on the assumption that culture and cognition have an evolutionary history. Here we cannot pursue the question of how the evolution of cognition and culture, such as mental attribution, problem solving, social cunning, became integrated in the effects of behavioral routines designed by natural and sexual selection—what one could term *pure* evolutionary residuals.

The behavioral environment during protocultural phases of human evolution was self-conscious and socially focused (Hallowell 1960). Events were calibrated in terms of their implications for organisms as persons, kinfolk, and social beings with a rudimentary sense of morality, group awareness, the group's historical placement and even spirituality (Boyer 1994; Guthrie 1993; Rappaport 1996, 1999). This includes what Donald (1991) describes as the evolution of the modern mind and Deacon (1997) describes as a capacity for symbols. A protocultural phase overlaps with the evolutionary biologist's abstract EEA. Moreover, it is important to emphasize that it spans an indeterminate expanse of evolutionary time, for this provides the context for the linkage between ecology/environment and language/culture/cognition—that is, for the design of psychological adaptations or algorithms. Threats to the security of such organisms, whatever the stimuli that activated their response-defense systems, were necessarily registered in a social-psychological frame of reference. In line with the argument that psychopathology has evolutionary significance, one can claim that syndromes of psychopathology made up one type or set of social and psychological responses to perceived social threats. If this reasoning is scientifically valid, then the response-defense system triggered by environmental challenges and hazards that involves syndromes of psychopathology during evolution was a defense system designed to combat social frustration and social threat, or to maintain security.

Let us pursue the idea of psychopathology as a general response defense system. Assume that as a result of developments in the EEA, individuals inherited (1) a perceptual system for identifying ecological hazards or social conflicts that pose threats to security; (2) psychological mechanisms and behavioral strategies designed to cope with those contingencies, whose expressions are varieties of psychopathology; (3) a system for evaluating the danger posed by the contingencies; (4) a selection system for choosing the appropriate response to address the contingencies; and (5) a monitoring device or regulator, most likely part of (4), which assesses the efficacy of the chosen response mechanism to meet the threat or to return the individual to relative safety and security.

I will provide some examples of how psychopathology might have originated as a general response defense system. It is generally assumed that natural and sexual selection were the two main forces that shaped behavioral routines in evolutionary environments, intertwined with problems of survival (see Buss 1999; Crawford 1998a, 1998b; Geary 1998; Miller 1998). Threats to physical security such as male sexual coercion, predator pressure, aggression from a stronger competitor, or an environmental catastrophe would lead to anxiety, panic, and a phobia response. In addition to fear, other social and psychological responses could be aroused, such as those aimed at securing support, understanding, and a change in living circumstances. The regulator of the latter algorithm would presumably monitor the results of the psychopathological syndrome and the effectiveness of the response system in procuring physical and psychological safety.

In the same way, social threats to self-esteem such as the failure to protect a mate and/or offspring, failure to provide food, lack of resourcefulness and cunning in group decision making, and loss of face because of failure to meet a social obligation, would lead to behavior designed to restore reciprocal relations and repair self-esteem. In addition to physiological fear, the individual might display behavioral responses that communicated an explanation of failure and sought to ward off threats to social standing. (The following examples are framed anthropomorphically and do not imply conscious deliberation, but rather cognitive solutions in light of evolving culture and cognition.) These behavioral responses might include a sickness response, evidence of bodily malfunction; a response suggesting possession, a state of mental dissociation, with behavior suggesting control by a powerful spirit; a paranoid response, impugning the motives and trustworthiness of a rival; a quasi-delusional response, claiming revelations about the future or malevolent groups nearby. A perception of actual or impending loss of social rank could lead to a depression response, retreat, withdrawal, capitulation, self-derogation, and passive ineffectiveness; or, contrariwise, to a manic response, a burst of energetic self-assertiveness, a relentless quest for markers of accomplishment, and a frenzied attempt to persuade and manipulate others. Finally, a motive to rationalize a desire to freeload, in response to real or imagined loss of vigor and/or resourcefulness, could produce a somatoform response, a syndrome of sickness and illness with bodily manifestations. Such examples of psychopathology as a response defense system are attempts to return the individual to a position of safety, to enhance self-image, to improve efficacy, and/or to rehabilitate social status.

The efficacy of the chosen response syndrome must be carefully monitored. In other words, the individual manifesting psychopathology in response to a threat to social standing or rank must possess a regulator, presumably a part of the algorithm that elicits the psychopathology as response system in the first place, whose function is to carefully weigh the effects of the response on the behavior of group mates in relation to the self. Such a regulator need not operate with the individual's conscious awareness. Rather, the strategy merely reflects the implementation of designs and mechanisms that evolved in the EEA, which, so to speak, reflect the wisdom of natural and/or sexual selection as applied to behavior generally and psychopathology specifically. In effect, regulators of a strategy of psychopathology are required to examine the efficacy of manifestations of psychopathology as ways of avoiding danger, avoiding stressors and burdens, restoring or improving social standing and rank, securing support and protection, procuring resources, and rehabilitating social identity. The regulator must evaluate whether the behavior syndrome succeeded in persuading, manipulating, or recruiting significant group mates in a way that benefits the individual. Because many if not most stressors and crises that give rise to psychopathology are rooted in social and interpersonal relationships and because the syndrome response itself endures across time with varying effects on others, the regulator of the algorithm monitoring the response is an integral part of the whole psychopathology mechanism and is actively engaged throughout implementation of the response syndrome.

A regulator and monitor of the efficacy of a defensive psychopathology response would presumably also need to weigh the costs of engaging in the behavior in the first place and continued costs incurred throughout its implementation. In other words, in the above examples of possible syndromes of defense, the individual incurs social costs and risks further loss of social status because others may discover the aim of the stratagem, achieving comparative fitness advantage, and may recognize the psychopathological syndrome as an attempt to take advantage of nongenetically related group mates. In other words, a syndrome of psychopathology may undermine or negate ordinary, acceptable forms of social and psychological behavior. Consequently, it carries a risk of further loss of status for failing to comply with rules of reciprocal exchange, seeking advantage over others, avoiding group responsibilities, and ignoring appropriate behavioral conventions or shared values. Stated baldly, when measured against the values and standards of group members, an evolutionarily significant syndrome of psychopathology carries by definition a negative value. Because syndromes of psychopathology, or responses to social threats, are intended to restore or rehabilitate individuals with compromised, devalued social identities at the expense of others, their manifestations can alienate or exploit others and undermine the individual's credibility or bring shame and discredit. Consequently, the system that regulates and monitors the efficacy of psychopathology as a defense response system must consider not only efficacy of removing or diminishing danger and threat but also the social and psychological cost entailed in selecting the syndrome in the first place. That is to say, the regulator's assessment of the costs and liabilities of implementing a particular syndrome of psychopathology entails an assessment of the syndrome's social stigma; furthermore, since the whole threat defense system was designed in the EEA, this implies that psychiatric stigmas (Fábrega 1987, 1990a, 1990b, 1990c, 1991) must have evolutionary roots and must have been laboriously calibrated or at least taken into account during the evolution of the modern mind (Donald 1991).

Tensions between Psychopathology as Adaptive Response and as Failed Strategy

A functional, adaptive conception of behavior, and by implication, psychopathology, may be presented in two ways. One view, already described, is that psychopathology is a package of time-bound, discrete, carefully targeted, and circumscribed responses of a threat defense system. It is a directed syndrome of behavior evoked by challenges to the organism's security and safety. Thus psychopathology is a form of instrumental behavior designed to do something, to overcome a threat or problem. Since psychopathology is a phenomenon found in highly social species, with evolved social intelligence, mental attribution, language, and culture, it functions through communication and persuasion.

Another version of psychopathology as a defense response system holds that the principal functional, adaptive mechanism that can lead to psychopathology is developmental and occurs early in life. Hence, a syndrome of psychopathology is a consequence of a lifelong strategy adopted during sensitive periods of development as a result of exposure to specific ecological conditions (Belsky, Steinberg and Draper 1991; Draper and Harpending 1982). The evolutionary environment presented myriad adaptive problems and threats that forced individuals to design a range of reproductive strategies for coping with them. This continues to be true today. A specific constellation of ecological conditions or clues operates as stimuli that infants and children respond to by adopting more or less lifelong social strategies for coping in environments containing these clues. For example, it has been posited that infants and children raised in a home without a father present and without adequate resources grow up to adopt a lifelong reproductive strategy as a way of coping with instability (Draper and Harpending 1982). The strategy embraces not only mating behavior per se but also cognitive, interpersonal, and parenting behaviors, and it can have deleterious effects, especially in contemporary society. If the environment changes or the long-term goals of the strategy are not realized, psychopathological symptoms can result (Belsky, Steinberg and Draper 1991; Mealey 1995). Thus infants and children in both nonindustrial societies and some industrial societies raised in poverty without a father present, or without kin and extended group socialization, respond to these ecological clues and come to follow a course of behavior, specifically mating and parenting, that can be equated with psychopathology in the long run (Mealey 1995).

Compared to the earlier formulation, in this instance the response defense system is not in itself psychopathological, nor is it necessarily directed at changing specific or immediate environmental problems. It is a lifelong social and psychological behavior strategy geared to coping with what appears to predict future reproductive constraints and limited resources. By definition, the adopted strategy is designed to cope with the environment that selects it. If the goals and functions of the strategy fail when the environment changes or the individual moves to a different social and cultural environment, psychopathology may be the result.

Two conceptions of the hard evolutionary view of psychopathology have been presented: (1) it is an elicited syndrome of social and psychological behavior that is geared to solving an immediate problem; and (2) it is a long-term behavioral strategy geared to coping with circumstances that may lead to future adaptive problems and thus result in psychopathology. Although analytically these patterns of behavior appear to be reasonably independent, in practice they are confounded. Consider this question about the first definition: How long must an immediate response endure before it is said to fail and on what grounds? That is, when does a response become maladaptive or psychopathological? With respect to the second definition: If the adaptive strategy counteracts local deficiencies that may produce benefits in the current environment, but endures and is costly, when does it become a failed adaptive strategy and on what grounds? Dilemmas posed by these two questions underscore obvious tensions between the adaptive response approach and the failed adaptive strategy approach. Many syndromes of psychopathology are by definition chronic and some are permanent, particularly so-called personality disorders. Are these varieties of psychopathology examples of positive adaptive responses or of adaptive strategies that can fail and be costly? In a clinical setting, the degree of stress and social maladaptation, or impairment, would alert the clinician to the correct diagnosis and solution. However, over the span of evolutionary time, these parameters are blurred and difficult to establish.

These complexities may be merely hypothetical, a matter of semantics. However, given the difficulty of measuring fitness in highly social species, and especially in relation to behavioral patterns and strategies that endure, conceptual problems can naturally arise. Moreover, the complexities are inevitable, since the settings and circumstances that give rise to them involve groups of individuals in the process of evolving culture and cognition. In these circumstances, clashes between the clinical and evolutionary points of view are likely to be more evident. Complexities such as these, and others that could be introduced, come into view when one searches for the origins of psychopathology by examining hypothetical behavior that may have occurred in the evolutionary environment. As noted earlier, these messy details are often, perhaps wisely, ignored by evolutionary psychiatrists. What was termed their intellectual conservatism—in other words, their tendency to focus on modern humans as completed evolutionary products and to merely accept that psychiatric disorders have an evolutionary basis—allows them to develop theories about the evolutionary origins of different psychiatric disorders without providing realistic and detailed examples and mechanisms.

Seeing psychopathology as an immediate response to a social ecological problem implies that alleged signs and symptoms constitute adaptive, functional, that is, positive, behavioral characteristics that are part of the evolutionary logic of behavior, particularly of psychopathology. By contrast, psychopathology as a consequence of a failed adaptive strategy implies that signs and symptoms may become costly, negative consequences when an environmental mismatch supervenes. In essence, the first approach minimizes clinical considerations in favor of evolutionary ones, whereas the second allows the incorporation of clinical concerns and thus permits one to integrate the evolutionary and clinical points of view.

The Evolutionary and Cultural/Clinical Points of View Revisited

Defining a behavioral pattern as a functional, adaptive response system to a threat to security is to define the syndrome of behavior that is produced as positive. Its function is to mitigate problems in the individual's environment, to communicate, change, and cope with an adaptive problem. To the evolutionist, the social and psychological behavior response pattern, a clinician's psychopathology, is analogous to other response defense systems rooted in natural and sexual selection—for example, for dealing with predators, for dealing with competitors, or for finding a mate and producing offspring. These are positive endeavors designed to promote survival and reproduction, that is, to produce fitness benefits. Whatever personal distress and associated signs and symptoms the response system may include would be trivial costs entailed in the otherwise positive response pattern. To a clinician, on the other hand, psychopathology by definition involves personal distress, social inefficiency, and social impairment. It implies disorder and suffering. Evolutionary psychiatrists, of course, integrate the two viewpoints.

There is an obvious tension between the content and meanings of the evolutionist's explanation and that of the culturally based clinician. They cannot be reconciled without acknowledging the intellectual dilemma inherent in studying psychopathology from an evolutionary point of view: How can the concept of psychopathology, a quintessentially human, moral, and value-laden construct that has a cultural context, be applied to animals, including hominids and earlier varieties of Homo? Evolutionary psychologists and psychiatrists attest to the viability of both the evolutionary and the clinical perspectives on human behavior and psychopathology. However, evolutionary theory explicitly handles members of Homo sapiens as fully cultural beings and uses evolutionary theory as a basic science to explain how they function. Evolutionary psychology and psychiatry never confront the question of the logical and moral status of the concept of nonhuman, animal psychopathologies. Given the basic phylogenetic continuity of behavior and psychopathology, arguing that hominids were capable of psychopathology contravenes basic assumptions about the alleged uniqueness of human culture. Conversely, to deny that psychopathology could have occurred among hominids reveals a bias of speciesism. Finally, it should be apparent that since culture and human moral systems, where the concept of psychopathology originates, have an evolutionary foundation, at some juncture during human evolution both the evolutionary and clinical frameworks about psychopathology are describing the same process of development. In other words, it would have been possible for protolinguistically, protoculturally informed varieties of Homo to reflect an objective, impersonal, individualistic, and selfish, that is, strict evolutionary, conception about psychopathology as well as manifesting an emergent personal, culturally rooted, and moral/evaluational, that is, strict clinical, one.

Psychiatric Disorders as Harmful Conditions

Evolutionary theory has recently been used as a basis for a more active approach to the study of disease and the general clinical care of patients (Nesse and Williams 1994). As part of this application to medicine, a disorder has been defined as a harmful dysfunction (HD), with dysfunction viewed as a failure or breakdown of an internal mechanism to perform its natural function (Wakefield 1999a, 1999b). Harmfulness is described in general terms as a condition that is painful and/or detrimental to an individual's well-being and functioning. Harmful conditions have many causes and may be based on environmental happenings or social arrangements that conflict with biological imperatives; however, only those traceable to dysfunctions of natural mechanisms are applicable to the evolutionary argument.

The HD slant on a disorder is compelling. On the one hand, it has general resonance: a natural function and a failure of it are conceptualized intuitively, that is, in terms of common sense, as what persons ordinarily mean when they think of disease or disorder as involving something that has gone wrong or is not working properly. On the other hand, it also has a formal philosophy of science basis. It rests on the classical theory of categories and invokes a scientific epistemology: a failure of a naturally designed function. While the HD approach has general medical implications—for example, diabetes, hypertension, kidney failure—it has been systematically applied to psychiatric disorders.

The various psychological adaptations, mechanisms, or algorithms singled out by evolutionary psychologists to explain behavior biologically are examples of natural functions. Hence, the HD analysis holds that true or scientifically valid psychiatric disorders are based on dysfunctions of psychological mechanisms or algorithms that produce harm in an individual. Since these mechanisms were naturally designed, sculpted to solve recurring social problems during human evolution, the HD analysis of a psychiatric disorder reinforces the link between psychiatry and general medicine (D. F. Klein 1978, 1999).

An evolutionary conception of psychiatric disorder is an example of an essentialist or classical approach to the definition of a concept. This is so because the definition of HD stipulates two individually necessary and jointly sufficient defining features that provide psychiatric disorder with its meaning. The HD analysis has been the target of critical analysis by both psychiatrists and social scientists. Whether the concept of psychiatric disorder is Roschian and conforms to the prototype theory of concept formation—that is, any one condition qualifying as a disorder because it approximates or resembles an ideal or prototype—

has constituted one line of attack (Lilienfeld and Marino 1995). It is clear that exponents of either an evolutionary or Roschian conception can provide cogent reasons why their position is valid, sometimes basing their arguments on the same disorder.

Many harmful dysfunctions of psychological mechanisms, that is, disorders, qualify as conditions that should be treated, but so do other conditions that may not be disorders in the strict sense (Cosmides and Tooby 1999). A treatable condition is the product of a decision based on values and conventions, either that of the individual, significant other or the reference group in society. The environment—for example, ancestral (evolutionary) or contemporary—wherein a condition of behavior is situated is critically important in distinguishing between disorder and treatable condition. Some treatable conditions may arise because a function naturally designed in an ancestral environment and operating naturally in the current one nevertheless causes impairment or suffering (sexual jealousy). The converse is also true: natural functions may be dysfunctional—for example, repeated sensation seeking and dangerous risk taking—yet produce behavior in contemporary environments that is satisfying and not impairing—for example, rock climbing. Finally, many conditions of behavior that are treatable may have no relationship to a naturally designed mechanism but simply result as a by-product of one or are due to simple human variation.

In short, in the evolutionary conception, human conditions of potential psychiatric import constitute points of intersection for three logically separate considerations. The first is who does the evaluation and how—for example, a positive or negative condition, the person or a reference group. The second is whether the condition results from a naturally designed mechanism that is doing its job, that is, is evolutionarily functional or dysfunctional, or is simply a by-product of a mechanism. The third is the environment in which the condition is situated, that is, the ancestral or the present one. As suggested above, some conditions may be harmful and disvalued to the individual in that they pose hazards and may cause suffering and even misery, they may be disvalued by others and yet may be evolutionarily functional in that they promoted reproductive success in ancestral environments (e.g., sexual predation). "In general, then, individuals can be expected to seek to increase the valued effects of useful functions, dysfunctions, by-products, and idiotypic genetic or environmental variation and to decrease the[ir] disvalued effects. . . . To the suffering person, it does not matter whether the condition is an adaptively designed outcome, damage to an adaptation, and unwanted side-effect, or simply an entropic accident" (Cosmides and Tooby 1999).

Echoing a treatable condition perspective, Kirmayer and Young (1999) point out that the HD analysis is particularly vulnerable to the challenge that it is not fully impersonal and objective; in other words, that it rests on implicit positions of value and that it totally disregards the importance of cultural conventions. An emphasis on the importance of treatable conditions in a psychiatric nosology implies that a defect or breakdown of a natural function is insufficient: it involves an implicit criticism of the HD formulation of disorder.

Sadler (1999) has emphasized that the HD analysis started out as a descriptive and prescriptive formula for definition of disorder as compared to nondisorder, thereby functioning as a basis for the conceptual validity of disorder. Most recently, Sadler claims, its exponent is concerned mainly with descriptive questions (why and how generally held psychiatric disorders conform to an HD analysis) and less so with prescriptive ones that clarify what and why a condition qualifies as a disorder—for example, whether hyperactivity, premenstrual syndrome, or paraphilic rapism should or should not constitute a disorder. It is clear that in the debate about prescriptive questions regarding a particular condition, one can point to sticky and fuzzy concepts about what constitute natural functions and, thus, whether the condition constitutes a disorder. Sadler, like Kirmayer and Young as well as Lilienfeld and Marino, then, makes clear that despite the seeming rigor of the HD formulation, its application can entail messy questions of politics, values, and conventions and standards about normality, deviance, adaptation, and natural functions.

There are several additional reasons why an HD prescription for psychiatric disorder cannot be expected to neatly serve the needs of diagnostic systems, at least in the foreseeable future. Most diagnoses that have emerged from the traditional history of psychiatry do not conform in any regular, point-for-point way with failures or breakdowns of a natural function. Entities like depression, schizophrenia, anxiety, and somatization disorder embrace many levels and layers of social and psychological function, and there is little evidence that they can be reduced to or equated with failures of one or even a few adaptations or mechanisms (D. F. Klein 1972, 1999). Most embody complex behavior phenomena that are the outcome of failures of several natural functions, mechanisms, and different levels of analysis.

Furthermore, many of the functions or mechanisms governing pathological behavior in prototypical psychiatric conditions are conceptualized as involving the interplay of hierarchically arranged levels of functions. Perturbations and dysfunctions in one level can be propagated up and down the hierarchy and at different levels may be subject to positive or negative feedback. When a hierarchical view is used to conceptualize how individuals function and also to define what can or should constitute a disorder, that is, what are natural functions, the neat, elegant, and precise solutions that an HD analysis promises break down and are rendered opaque and fuzzy.

Many so-called psychological adaptations are really descriptions of domains of biologically significant but highly complex social behavior. They may have promoted the solution of biological problems, for example, mate selection, acquisition of rank, and social competition; however, they do not readily map on to specific psychological mechanisms, other than tautologically, nor can they be equated with conditions or disorders as classified in psychiatry. Other adaptations, while certainly fundamental in promoting fitness and adaptation, really refer to rather narrowly defined, that is, content-specific, cognitive/perceptual functions that serve or contribute to the solution of many biological functions. For example, mate selection, achievement of high social rank, solution of subsistence problems, resolution of social conflicts and/or ability to avoid predators in the hominid environment of evolution were dependent on or required adaptive or good functions in many areas of perception, cognition, recognition of emotion, linguistic and/or emotional communication.

Both the molecular, hardware-related functions or adaptations and the more molar, socially complex or integrated biological functions implicate behavior. Moreover, both could figure as a natural function—or, more realistically, a *cluster* of natural functions—that can fail and break down. Some of the evolutionary arguments that have been developed for psychiatric disorders—depression, schizophrenia—embody whole packages of maladaptive behavior that can be reduced or fitted in to an HD analysis only with great difficulty. There is also the associated problem of the functional specificity of psychological adaptations or mechanisms and of what should count as a biological function in the first place.

There are reasons to be cautious with respect to the proposed evolutionary conception of psychiatric disorder generally and on the harmful dysfunction formulation in particular. While the classic theory of categories that supports the HD formulation is theoretically compelling and aesthetically pleasing, its use for deciding whether any one condition of psychiatric import is, is not, or should be defined as a disorder raises numerous problems. Nevertheless, evolutionary biology and psychology generally, and the HD analysis of disorder specifically, embody insights that should be included in a science of psychiatric diagnosis and classification.

Advantages of the Hard Evolutionary Conception

It is appropriate to ask what the advantages and disadvantages are of understanding psychopathology as an evolutionarily designed general response or threat defense system. The main advantages of a hard evolutionary conception are that it provides an elegant, integrated formulation, with pragmatic, clinical concerns anchored in the theoretical tenets of evolutionary biology. Moreover, it creates research opportunities by broadening the scope of scientific study of psychopathology and potentially leading to therapeutic benefits. Following is a brief discussion of these issues.

THEORETICAL INTEGRATION AND ELEGANCE

A hard evolutionary conception of psychopathology integrates a diverse body of information in the social and biological sciences. Psychopharmacology, molecular biology, and psychiatric genetics, all germane to questions of proximal causes and mechanisms, require knowledge not only of evolutionary biology writ large but also of clinical psychology, epidemiology, sociology, and cultural psychiatry. These are the sciences that, with respect to human populations, elucidate aspects of the causal/mechanist view, for they shed light on the kinds of stressors and events that can set in motion complex, lower-level changes in intra-organism response systems.

Conceiving of psychopathology as having evolutionary significance requires giving attention, in the mechanist's model of psychopathology, to how the brain takes in characteristics of the social ecology. Specifically, the causal/mechanistic–oriented analyst would need to know how individuals perceive and respond, such as neurochemically, neurovegetatively, to environmental threats such as ecological crises, resource depletion, loss, and other stresses and strains. Because psychopathology is keyed to happenings in meaning-rich settings, upheavals in the environment affect personal experience by threatening the well-being and security of themselves and others.

The functional/adaptive view is also enhanced by adopting the strong conception that integrates proximal mechanisms and causes, for example, pathogenesis, with factors tied to ultimate, adaptive solutions to problems in the EEA. One would need to examine and integrate laboratory research about micro-level processes and field studies about social environmental causes and mechanisms. To understand the functional and adaptive implications of a syndrome of psychopathology, one should take account of (1) ecological clues in triggering the response defense systems, (2) the various neurobiological and psychophysiological responses that might signal how the defenses aim to cope with the threat, and (3) the signs and symptoms of psychopathology and how they are communicated to group mates. Understanding how syndromes of psychopathology provide security in the face of threats requires understanding the social and psychological implications of the mechanisms perturbed by ecological crises and the possible communicative and adaptive significance of the behaviors produced.

Taking into account the significance of external, releasing causes enhances the causal/mechanistic understanding of how the individual's response systems work, just as the functional/adaptive view is helped by considering how internal mechanisms are keyed to biological goals and functions posed by the environment that selected for psychopathology. Thus, a hard evolutionary conception is a unifying directive to make sense of psychopathology, both its causes and mechanisms and its functional/adaptive aspects, in light of how individuals respond to their social environment. While the hard conception stipulates that circumstances in the EEA selected the evolved mechanisms producing psychopathology, extrapolation to later phases of human evolution provides a window through which one may explain psychopathology in contemporary settings and develop useful therapeutic interventions that rely on evolutionary insights.

DEVELOPING AND TESTING EVOLUTIONARY HYPOTHESES

The study of psychopathology has heretofore been dominated by psychology, sociology, and anthropology, and by sciences associated with neurobiology such as chronobiology, psychopharmacology, and psychiatric genetics. The value of their contributions notwithstanding, all of these sciences provide hypotheses and theoretical frames of reference that are in competition with evolutionary ones.

It is difficult to test evolutionary hypotheses. The task involves adopting presuppositions from evolutionary theory, extending the theory through an appropriate research program so that it applies not only to human and nonhuman primate behavior but also to aspects of psychopathology, and then formulating a conceptual framework that incorporates a characteristic of psychopathology in a way that meaningfully engages theoretical premises (Holcomb 1998). In other words, a research program should subsume psychopathology under the umbrella of evolutionary theory so that findings pertinent to psychopathology are integrated into a conceptual framework that makes its evolutionary relevance explicit.

A two-pronged research program would adopt these guiding premises: Psychopathology results when individuals experience a threat to their security. Threats activate evolved behavior mechanisms that are keyed to survival and reproduction, depending on the character of the threat. It is assumed that the social implications of the current environment resemble those of the EEA. Consider four types of threats and response syndromes: (1) If an individual's evolved mechanisms pertaining to gaining social rank and status are threatened, a psychopathology syndrome characterized by behavior that reflects a loss of status, or striving for higher social status, will be the result. (2) If mechanisms evolved to protect physical safety are threatened, the individual will display a syndrome characterized by behavior that reflects personal vulnerability. (3) If evolved mechanisms pertaining to a sense of social and intellectual expertise are threatened, the individual will develop a syndrome that projects social and intellectual incompetence. (4) If evolved mechanisms pertaining to social reciprocity and trust are threatened, the individual will respond by ascribing malevolent motives to group mates.

These situations could lead to two different courses of action. From a therapeutic standpoint, a clinician could advise a patient about how the symptoms developed, using evolutionary explanations, to find a way to cope with the evoking circumstances. In therapeutic sessions with the patient and/or family members, the clinician would try to help a patient find solutions to problems. It would help the patient simply to learn that what appear to be unique, overwhelming circumstances can be fitted into a valid conceptual framework.

Alternatively, one could devise an epidemiological study that collects data on a sample of acute psychopathologies and their causes. One might test one or more hypotheses about the effects of various adaptive problems with respect to types of psychopathology. One might learn more about adaptive problems by concentrating on differences in sex and social economic status, assuming threats to social safety and security varied systematically in relation to these variables. One might find specific relationships between characteristics of individuals and evoking conditions. For example, an analysis centered on only one category of person, such as females of low social standing, might support a hypothesis if one could relate a particular social threat to a corresponding syndrome.

Textbooks in evolutionary psychiatry like Stevens and Price (1996) and McGuire and Troisi (1998) summarize research findings pertaining to psychiatric disorders that are based on evolutionary theory and discuss important clinical applications of the theory. They draw on and add to earlier studies in the evolutionary biology of psychiatry, the study of anxiety disorders with an evolutionary emphasis, and paleobiology such as Wenegrat (1984, 1990), Marks (1987), K. G. Bailey (1987), Gardner (1982, 1988, 1995), and Nesse (1984, 1990).

A unitary perspective is most useful in the pursuit of therapeutic goals. The quintessential clinical reality of many syndromes of psychopathology can be equated with principles of evolutionary biology. A successful treatment depends on an understanding of the mutually reinforcing character of specific proximal causes/mechanisms within the organism that produce the syndrome, and with ultimately conditioned functional/adaptive concerns about how threats to safety and security perturb the mechanisms in question.

Functional/adaptive questions do not just pertain to hypothetical events that might have taken place in the EEA. Most if not all ecological crises that selected for psychopathology in the first place are analogous to events in contemporary settings in both industrial and nonindustrial societies. Moreover, contemporary social environments must have deviated far enough from the EEA as to strain adaptive programs of behavior significantly, rendering some maladaptive. Here, the concepts of ancestralization—how inherited cognitive traits shape and condition current environments—and of environmental mismatch—differences and similarities between current and past environments with respect to specific psychological algorithms—are appropriate (Crawford 1998a, 1998b). Many current environments are enough like those of the evolutionary past to activate several mechanisms of the generic psychopathology defense system, or perhaps other response systems that share operational components with it. Some of the response patterns may still be on target, with different degrees of efficacy, whereas others may simply be off the mark, yet their putative functions and goals need to be appreciated nonetheless. In short, a functional/adaptive perspective on psychopathology seeks to understand its origins with reference to ancestral environments, but its rationale applies to contemporary behavior problems because the context of human behavior, coping, and adaptation are universally applicable.

Disadvantages of the Hard Evolutionary Conception

Of the many weaknesses of a unitary, that is, functional/adaptive and causal/mechanist, perspective regarding psychopathology, I will mention only a few. The idea of psychopathology as a threat response system seems counterintuitive, since it views psychopathology in positive terms, that is, as ultimately desirable, whereas our cultural and clinical tradition sees it in negative terms. This raises logical and empirical problems.

There seem to be two ready criticisms of the hard evolutionary conception. First, it seems unrealistic or nonsensical to equate causal mechanisms and processes in the central nervous system, that is, the body, with functional, adaptive ones situated in mental, interpretive operations, where psychological influences reside. One cannot possibly make a point-by-point integration of molecular, neurobiological events with interpretations of events in the social environment. Thus, it is impossible to synthesize causal/mechanistic and functional/adaptive questions. This criticism is weakened if the hard unitary view seeks mere correspondence between causal/mechanistic compared to functional/adaptive events and processes, acknowledging the limitations of reductionism, on the one hand, and instead allowing for emergent phenomena and levels of scientific domains and discourse, on the other. Moreover, virtually all evolutionary biologists address activities of organisms and in fact repeatedly seek to integrate causal/mechanistic questions with functional/adaptive ones (see Curio 1994).

A second criticism would stipulate that the hard evolutionary conception simply works up old material in new form. For example, central tenets of social and cultural psychiatry involve relating aspects of the environment that have functional, adaptive significance, such as social stress and object loss, to the unfolding and course of any number of varieties of psychopathology. A socially and psychologically dynamic psychiatry, it would be asserted, has always involved looking at the environment of the individual in functional, adaptive terms and examining its role on mechanisms within the individual that regulate and shape social and psychological behavior. One cannot deny that functional and adaptively significant causal explanations have analogues in the history of human populations, although their relevance to contemporary psychopathology might be questioned. This criticism is weakened by the new and penetrating insights that have been proffered about psychopathology by clinicians immersed in the tenets of evolutionary biology (K. G. Bailey 1987; McGuire and Troisi 1998; Nesse 1990; Nesse and Lloyd 1992; Stevens and Price 1996; Wenegrat 1984, 1990) and by a failure to appreciate the unique implications of examining behavior generally and psychopathology in particular from the standpoint of biological functions and goals shaped during phylogeny.

There is also the problem posed by the character and identity of the response or defense system itself, which is the hallmark of the hard evolutionary conception. Can psychopathology constitute a general, adaptive response system against threat, or is it better to regard merely some varieties of psychopathology as consisting of a set of such responses, each aimed at solving a different class of problems? In either case, an array of adaptive problems are at issue and algorithms and mechanisms designed to solve them would need to have many subcomponents and levels, whose functions would overlap with respect to stimuli that elicit them and information processing measures that operate them. The threat response system formulation suggests that a particular ecological problem may have any number of responses, yet psychopathologies are not equally likely responses to a situation, nor are all individuals equally vulnerable to all varieties of them.

The way psychopathological syndromes respond to threats suggests a degree of selectivity, willfulness, and premeditation that is not accounted for in ordinary conceptions of psychopathology. Consider how the regulator of the response system monitors the responses of group mates to the syndrome and modulates its operation accordingly. Any functional analysis of behavior, including that of a syndrome of psychopathology, that is conducted in terms of decision and optimality theory runs into this problem of whether or not responses are conscious. One could argue that responses need not be consciously chosen but rather reflect the operation of innate routines and mechanisms, and indeed some innate defense mechanisms are self-deceiving (see Alexander 1987a, 1987b, 1990; Nesse and Lloyd 1992). However, to argue that the course and duration of psychopathological syndromes depend on responses of group mates that require modulation is inconsistent with the idea that psychopathology is an unconscious response mechanism.

It is obvious that each domain of reference of psychopathology as a defense system—that is, functional/adaptive vs. causal/mechanistic—poses conceptual and empirical problems and that mastering both sets of them would be difficult. Pursuing issues of mechanism immediately suggested by empirical data may in some instances be facilitated by neglecting functional ones that could distract and delay inquiry. And vice versa: Making sense of how a syndrome as a whole communicates and solves adaptive problems may suffice, given immediate purposes or concerns, and taking into account the functional significance of specific, microprocess responses may be superfluous and distracting. An indeterminate proportion of organism responses that the causal/mechanist identifies may represent secondary, ancillary, random, or mechanical responses that simply have no communicative function or have no bearing on the evoking cause. Some of the pitfalls of the old psychosomatic medicine need to be recalled: Many internal physiological responses and reactions that a mechanistic, cause-oriented person identifies may be totally devoid of functional significance.

A functional/adaptive point of view ultimately requires determining what kind of and the degree of risk posed by the (or a particular feature of) social ecology, and this gives rise to many complications that might prove very difficult to overcome. For example, to assess risk one should know at least two things: (1) the danger and threat to security of an ecological or social conflict crisis in the behavioral environment, and (2) how the organism weighs this danger in light of its perceived capacities and strengths. (In their analysis of the social competition hypothesis of depression, Price et al. [1994; see also Stevens and Price 1996] use the concept of resource-holding potential [RHP] to designate how an individual evaluates threat in a dominance hierarchy in light of resources.) To determine and weigh each of these measures may involve a large number of issues, some of which may not be ascertainable because of individual differences in bias, history, or biography, and this applies to inferences about the EEA or to analogues in the current environment.

Some varieties of psychopathology simply do not lend themselves to a functional/adaptive interpretation as constituting a response or defense system to perceived threats in the behavioral environment. This clearly applies to varieties of psychopathology stemming from defects in the apparatus, lesions and infections (Nesse and Williams 1994). It applies to disorders that are linked to single gene effects and most likely represent deleterious mutations, such as Huntington disease and phenylketonuria. A functional/adaptive view of psychopathology by definition is linked to processes of natural and sexual selection, but these two disorders reflect the operation of nonadditive genetic factors, that is, dominance variance and epistasis, which are not directly linked to selection (J. M. Bailey 1997, 1998). Moreover, the manifestations and course of these disorders consist of behavior dilapidation with a relatively uniform, inexorable, and downhill course. Autism and at least some varieties of schizophrenia also resemble clear defect psychopathologies, with more or less progressive, downhill courses. The high prevalence of schizophrenia in human populations and its link to cognition and language has raised the possibility that it represents a breakdown of, hence a cost associated with, a mechanism that marked a turning point in human evolution, but the idea that it may constitute a functional, adaptive defense to ecological hazards in the EEA has not been entertained.

Evolutionary View of Psychopathology and Culture

A final limitation of an evolutionary conception of psychopathology is that it excludes aspects of culture generally and symbolic culture in particular. To view evoking circumstances, elicited mechanisms, and response patterns of behavior, that is, psychopathology, in terms of the logic of evolutionary theory is to concentrate largely on economistic questions of trade-offs between benefits and costs as these pertain to survival and reproduction—that is, topics dominated by individualistic, utilitarian concerns and ultimately implicating genes. The power of biological theory lies in its reduction to matters involving constraints, environment, and fitness as these make sense in light of natural and sexual selection. Aspects of symbols and their meanings as emphasized by cultural anthropologists appear not crucial. Culture in fact has been discussed analytically from a biological point of view (Tooby and Cosmides 1992). While different conceptions of culture viewed in relation to evolution have been identified and discussed, the general tendency has been to regard domain-specific behavior algorithms as dominant factors, shared traditions, and conventions of behavior somewhat significant in modifying how psychological adaptations are played out, and systems of symbols, including myths and rituals, of secondary or minor relevance.

In the line of evolution leading to anatomically modern, language-proficient humans, aspects of symbolic culture certainly become prominent. A distinguishing feature of the transition to the Upper Paleolithic Era is the proliferation of evidence of artifacts, cave painting, decoration of tools used for subsistence, and patterns of arrangement of dwellings, all of which imply aspects of language, culture, and artistic creativity (Chase and Dibble 1987, 1992; Donald 1991; Gamble 1993; Pfeiffer 1982). It is very likely that in ensembles of psychopathology, aspects of meanings, myths, rituals, and narratives about cosmology played important roles in explaining manifestations and attempts at healing. Moreover, there are good reasons for viewing the creative explosion of culture as having a more particularistic, slower, gradual emergence (Foley 1987, 1995a, 1995b), although this is a much contested issue (Gibson and Ingold 1993; Knight 1991, 1996; Mellars 1996a, 1996b, 1996c; Mellars and Gibson 1996). Even during earlier phases of evolution, in other words, symbolic culture would have played an important role in psychopathology.

It is likely that varieties of psychopathology had a significant content and meaning in relation to group traditions and values even before the creative explosion or revolution was singled out as characteristic of the Upper Paleolithic Era. Given that a symbolic niche is one in which interpretive and valuational considerations play important roles, one must suppose that this applied to incidents of psychopathology. What this implies is that in tracing the origins of psychopathology during the phase of human evolution leading from protoculture to culture, one would have to allow for the influence of symbols and their meanings in explaining the cause, significance, and outcome of psychopathology. While one need not exclude purely economic, utilitarian questions involving trade-offs as calibrated using optimality theory, some allowances would have to be made for aspects of meaning and interpretation as formulated by cultural anthropologists. A strict evolutionary conception of psychopathology tends to exclude, or at least minimize, these considerations, and this can be regarded as a limitation.

A concomitant of the emergence of language, culture, and cognition during biological evolution is the creation of symbolic content and meaning in the social spaces of hominid communities (Deacon 1997; Donald 1991; Mithen 1999).

Evolutionary residuals consist of underlying mechanisms, adaptations, algorithms, and/or structural systems designed in EEA to meet biological imperatives, for example, survival and reproduction. They are subject to perturbations, deficiencies, and defects. The result is manifestations of psychopathology registered in impairments in areas surrounding motivational and functional goals. Anxiety, as an example, consists of paralyzing symptoms of fear, dread, and avoidance/withdrawal. Depression, on the other hand, is associated with despondency and loss of morale, motivation, and energy. And a variety of psychopathology, such as somatoform disorders, produces numerous bodily complaints and preoccupations that can hinder the pursuit of basic subsistence tasks and in theory undermine social relationships. Of course, from an evolutionary standpoint, all of these disorders and their symptoms and costs need to be examined critically. For example, they could represent elements of functional strategies designed to obtain advantages or, at least, to make the best of a bad situation, in which case the costs are more apparent than real. At any rate, during phases of human biological evolution, each of these varieties of psychopathology was configured and enacted in terms of behaviors that entailed functional deficiencies and were played out in terms of evolutionary residuals, as adumbrated by evolutionary psychiatrists (McGuire and Troisi 1998; Stevens and Price 1996; Wenegrat 1984, 1990).

The manifestations of psychopathology, however, do not just passively and mechanically undermine behavior and adaptation. Chimpanzees have some awareness of pain, sickness, and suffering in group mates (Fábrega 1997). The behavior of an amygdalectomized free-ranging adult rhesus monkey gives rise to abnormal contextual behaviors, and group mates notice this, for they aggress at the operated animal and appear to ban it from the group (Dicks et al. 1968; Kling et al. 1970). Similarly, monkeys also appear to react negatively to manifestations of psychopathology that involve infant abuse (Troisi et al. 1982). Such forms of awareness among group mates of the behavior of victims of psychopathology obviously expand during phases of human evolution. Consequently, a content and meaning superstructure must be presumed to have actually permeated earlier varieties of psychopathology in hominid communities in light of evolving traits related to language, culture, and cognition. What an individual was anxious about, whether it be nonexistent but probable predators, sexually predatory behavior from males in the case of females, and conflicts of interest with a dominant and stronger member of the group, in the case of a male, would constitute the social reality of the psychopathology in question. Depression manifestations would have a corresponding content and meaning; for example, feelings associated with a loss of status or of repeated failure to reach success in mating or subsistence. The sickness picture of a somatoform condition would likewise be played out in an idiom of what physical ineptitude meant in the economy of life of the individual; for example, a failure to engage in subsistence and a heightened need for dependence on kin.

Expansion of content and meaning during evolution implies a more elaborate form of consciousness and sense of self, a greater incorporation of a sense of historical and group identity, cosmological awareness, and even a sense of spirituality (Boyer 1994; Damasio 1999; Mithen 1999; Rappaport 1999). Content and meaning of this type come to permeate the impairments in the realization of evolutionary residuals of behavior that must be presumed to constitute the fundamental idiom of psychopathology. To say this is merely to emphasize that origins of psychopathology involve continued interference and deficiency in areas related to fitness but in a context of meaning and value that becomes more complex as language, culture, and cognition expand the symbolic capacities of hominids.

- Evolutionary psychiatrists agree that psychopathology is rooted in the evolutionary residuals of behavior programmed in ancient environments but disagree on the emphasis they give to pure breakdowns of functions compared to environmental mismatches.
- While psychopathology suggests something negative, the evolutionary interpretation of psychopathology suggests positive aspects that should be examined critically.
- If one follows the conventional wisdom that ultimate, functional biological questions should be integrated with proximal, causal mechanistic ones, then psychopathology might constitute an actual adaptation.
- While it may not be realistic that varieties of psychopathology constitute true psychological adaptations, it is reasonable that they are part of an organism's general response defense system and can promote better coping with environmental hazards.

- Psychopathology as defense can be equated with an adult organism's attempt to solve stressors in the short run. It also fits an organism's attempt to bring into play long-term developmental behavior strategies so as to solve structural hardships in the ecology.
- While psychopathology conforms to the idea of disorder as a harmful breakdown of a natural function (based on evolutionary theory), its obvious connection to intercurrent social and cultural factors limits the overall value of the harmful dysfunction thesis.
- Although not without limitations, looking at psychopathology in evolutionary terms from a scientific, practical, and even therapeutic standpoint has many advantages.

Part II Psychopathology during Human Biological Evolution

Chapter 7

Searching for Psychopathology in Nonhuman Primates

To propose that some aspects or types of psychopathology have a basis in the evolutionary past of Homo sapiens is to raise two questions. First: How far back into the phylogenetic past can one plausibly hypothesize about and offer constructions of episodes of behavior that meet conditions for psychopathology? Second: What might such episodes have consisted of? In other words, what were the manifestations of the earliest varieties of psychopathology, and what social consequences did they produce? In light of the psychological and social character of psychopathology and knowledge of evolutionary biology, the first question points in the direction of animals who bear an affinity to the human species. To answer the second question, one should examine the observations and insights of ethologists who study the natural behavior of higher apes and the experiments of comparative psychologists on nonhuman primates designed to study the social and psychological characteristics of their behavior.

This chapter reviews the work of selected primatologists and ethologists whose observations implicitly raise the question of prehuman varieties of psychopathology. To be sure, these researchers, with a few exceptions that will become evident, shy away from using a concept like psychopathology as it is ordinarily understood with respect to humans. They either avoid the term altogether, place it within quotation marks, or appear to assume that their observations and experiments reflect mechanical disturbances of behavior that have little direct relevance to naturally occurring forms of psychopathology. By and large, they view disturbances, anomalies, and/or maladaptive changes of behavior pretty much exclusively from within the context of ethology and evolutionary theory. This approach is a legacy of the contrasting frameworks of clinical psychiatry and evolutionary biology.

With a formulation that allows us to begin to reconcile these perspectives, we may examine this work with the aim of delineating a model of psychopathology among higher primates. A rough outline of disturbances of social and psychological behavior among contemporary nonhuman primates will provide a useful frame of reference for conceptualizing the origins of psychopathology in human beings.

Studying Behavior in Nonhuman Primates

The higher primates are living representatives of prehuman species and the best available approximations of the adaptive behavior of our hominid ancestors. These include Old World monkeys and the chimpanzees that have been the objects of research studies such as Cheney and Seyfarth (1990), Strum (1987), Kummer (1995), Goodall (1986), Nishida (1990), Wrangham et al. (1994), and de Waal (1989, 1996). Zihlman (1997) provides an up-to-date, succinct review of recent field studies of primates and also of their comparative anatomy, molecular biology, paleontology, paleoanthropology, and genetics. Zihlman argues that chimpanzees, in particular Pan Paniscus (although Pan Troglodytes also share many characteristics), offer the best living model of the last common ancestor between apes and humans, a separation said to have taken place around 5 to 6 million years ago. However, many nonhuman primates, but especially members of Pan (and presumably the earliest hominids), are said to show, with some exceptions, somewhat related forms of anatomy, life history characteristics, cognition, and behavior. Hence, descriptions of abnormal behavior or psychopathology among nonhuman primates provide a rough approximation of the kinds of behavioral disturbances that might have been manifested in the early hominid line.

These organisms live in groups with different types of social organization depending on their species, size, and ecological setting (Dunbar 1988; Wrangham 1979). Relationships among group members are complex and involve dominance hierarchies, suites conveying threat and intimidation, alliances among kin and nonkin members, grooming, foraging and hunting excursions, sexuality and mating, and parental caring, to name but a few. Members communicate by means of facial displays, actions, gestures, and vocalizations. Groups survive by learning to avoid and ward off predators, developing effective foraging and hunting strategies, mating and caring for the young, and maintaining orderly relations among group members.

With respect to contemporary nonhuman primates, psychopathology can mean one or both of two things. First, and from the standpoint of the organism itself, psychopathology can refer to how individuals behave, or fail to behave, such that they are unable to carry out ordinary routines of living. More specifically, they fail by not providing for their nutritional needs, by not learning to avoid predators, failing to mate, being unable to care for offspring, in the case of females in particular, and being unable to compete for and obtain needed resources. Thus, more or less adapted organisms must fulfill certain behavioral desiderata (Broom and Johnson 1993): they must acquire information about the fauna, flora, and physical characteristics of the environment. They must learn the signals and calls that the group has evolved to point to important issues, to carry out basic survival and reproductive tasks, to adopt behavioral routines so as to make sense of the behaviors of conspecifics, and to learn routines of behavior that further their interests and do not thwart or endanger them. A deficiency or abnormality in communicating and interpreting behaviors crucial to basic subsistence and reproduction fails the individual's requirement of survival and reproduction; hence it is by definition maladaptive, causing stress and reducing fitness; this could be called an individualistic, functional criterion of a psychopathology.

Second, psychopathology can be located in an analytically separate, group-centered frame of reference (not necessarily empirically unrelated to the first). In this instance, the proper term might be *social behavioral pathology*, for it involves the effects of a potential abnormality of behavior on conspecifics. The primary organism's abnormality, in other words, can thwart, endanger, threaten group mates, or contravene the integrity of the group as a whole. Most individual psychopathologies are likely to be overlooked or ignored, but if they pose a serious hindrance to the group's existence and stability, the individual will be opposed, banished, excluded, or eliminated. In this instance, the group sets the standards for what constitutes a psychopathology, since through the actions of selected others and a presumed and implicit consensus, it deprives an individual of the necessary support, resources, and social protection entailed in group membership. The group's responses curtail an organism's ability to exist and reproduce, and in this way establishes what is maladaptive and pathological. Finding a complementarity between individual and group affairs with respect to psychopathology is consistent with a tenet of evolutionary theory that links individual fitness issues with issues involving the evolution of social systems (Wrangham 1979).

In many respects, the criteria pertaining to an organism-centered perspective on psychopathology complement the group-centered criteria. Thus, logically and empirically, maladaptive behavior of an organism—for example, excessive or inappropriate aggression—often thwarts the group's integrity. Conversely, strategies and plans required for group survival place constraints and stressors on vulnerable individuals, producing behavior that is deficient, destabilizing, noxious, and/or destructive, leading to exclusion or physical retaliation, even to banishment or elimination. Clearly, the organism-centered and the group-centered criteria for psychopathology are not always complementary. Organisms showing maladaptive behavior can in theory be protected and cared for by altruistic group members, at a cost to themselves and to the group's needs and pursuits. Since biological evolution is ordinarily thought of as involving harsh competition, selfish pursuits, and self-interest, altruism in general is theoretically problematic— especially in the form described here and among groups of nonhuman primates. In short, only in certain types of groups and among selected segments of them, such as kin and reciprocally interactive nonkin, is there likely to be protection and caring of members displaying maladaptive behavior, that is, psychopathology. One can readily appreciate that with respect to psychopathology such concepts as altruism, caring, and social tolerance constitute emergent phenomena that will be encountered somewhere in the line of evolution leading to modern humans and will entail distinctive forms of cognition and social morality (see de Waal 1996; Nitecki and Nitecki 1993). These concepts and the emergence of psychopathology itself as a social reality of hominid groups bear a close relationship to the theory of sickness and healing. (See also Fábrega 1974, 1975, 1997.)

The Fabric of Behavior Where Psychopathology Manifests

A basic assumption of this book is that psychopathology constitutes a complex ensemble of psychological, social, and physiological behavior and that its characteristics were manifest in hominids during the biological phases of human evolution. Thus, cognition, emotion, social communication, arousal, motivation, and basic physiological parameters serving vital internal biological functions— for example, sleep, appetite, activity patterns, sexuality, and sense of bodily integrity—and external ecological ones—for example, reproduction, fitness, and survival—form the material out of which varieties of psychopathology were produced and shaped during human biological evolution. The assumption about the holistic character of psychopathology of hominids is but a hypothetical projection to human phylogeny of the generalization present in contemporary clinical theory about human psychiatric disorders.

It can be assumed that characteristics of human psychopathology probably antedated the pongid/hominid split of around 5 million years ago and constituted potential traits among the very earliest hominids. Presently, it is the organization and rationale of social behavior among nonhuman primates that is the focus. It is important to emphasize at the outset that the stage onto which a possible variety of nonhuman primate psychopathology is projected, in other words, the setting or space where it is configured and enacted, is intricately and complexly organized from a social behavioral and communicational standpoint.

The work of Frans de Waal (1989, 1996) on the theme of aggression and hostility can be used to illustrate the organization of behavior among nonhuman primates. He shows that peacekeeping and reconciliation are themes that describe how conflicts of power and breakdowns of social order are handled. Rather than merely constituting a drive or instinct that simply gets released and discharged once a particular threshold of provocation is physically exceeded, aggression and hostility should be viewed as the product of social decision making. How aggression is *managed* among nonhuman primates involves hostile displays and acts between combatants, into which are figured considerations about the value placed on social relationships and the status of group activity and dynamics.

Observations and experiments summarized by de Waal (1995, 1996) and formulated in terms of his Relational Model indicate that hostility is associated with empathic understanding and responses among noncombatants. Aggression and resort to violence are not always just the result of status conflicts or mating rivalries that have produced bursts of negative emotions and that have exceeded controls or gotten out of hand. Instead, they in some respects are unwanted outcomes, and costly ones at that, since they can undermine important social relationships among combatants, and their allies, that preexisted the conflicts. Combatants show reconciliation behavior following outright hostility and violence. Efforts have usually been made by conspecifics who were bystanders to forestall the expression of overt hostility among would-be combatants. In the event this cannot be accomplished, they then frequently console the victims.

Careful observation of behavior discloses that hostility and violence are associated with high arousal and anxiety on the part of both combatants and, one can presume, also among bystanders. This is manifest in their actions and bodily mannerisms. Measures of various behaviors indicate that hostile episodes are preceded and followed by activities designed to reduce arousal and anxiety. The alleviation of distress and what can be described as empathy, in short, are integral features of how aggression and hostility are configured and played out among chimpanzees and monkeys.

Species differences in social behavior pertinent to the expression and communication of hostility—and one can assume also for other traits pertinent to the expression of psychopathology—exist between chimpanzees and monkeys, and among members of the genus Macaca for monkeys. These differences pertain to group cohesion, strictness of the dominance hierarchy, social tolerance, rates of affiliative behavior, and conciliatory tendencies. Moreover, experiments have shown that social behaviors that mediate reconciliation, and perhaps consolation and other aspects of empathy, are not just fixed, innate, or hard wired since they can be learned. This was indicated by the fact that juveniles of species of comparatively hostile rhesus monkeys show greater measures of reconciliation behavior during a six-week period of co-housing with juveniles of the more sociable, less aggressive stump-tailed macaques. The latter appeared to act as tutors for the more aggressive and volatile rhesus. Differences between reconciliation tendencies of experimental compared to control rhesus persisted following the period of co-housing.

It should be recognized that de Waal's formulation applies only to aggression and hostility and not to psychopathology per se. Moreover, it pertains to but one or two rubrics of behavior that one might wish to equate with psychopathology; namely, aggressive dyscontrol and anxiety. However, one can assume that many other manifestations of psychopathology are configured and enacted in a socially complex yet structured way. In other words, the signs and symptoms of a variety of psychopathology do not just spill out or mechanically disrupt and intrude into the group as extraneous, foreign objects. Like the aggression, hostility, and anxiety discussed by de Waal, all manifestations of psychopathology are likely to be textured into the complex social fabric of group life among nonhuman primates and inside their psychology as well. Complex and organized patterns of cognition, emotion, and social communication are aspects of behavior that describe many activities and concerns of nonhuman primates, and any or all of which can serve as a manifestation of psychopathology. And to complete the picture, the various physiological parameters that underlie behavior—for example, sleep, appetite, activity level—can safely be assumed to also show variation in relation to instances of psychopathology, although these are generally not as well studied, or reported on, by ethologists and field primatologists.

The Behavior of Baboons as a Platform for the Study of Nonhuman Psychopathology

In his analysis of the social behavior of baboons living in captivity and in feral conditions, Hans Kummer (1995) provides material that can be used to hypothesize about the possible character and implications of psychopathologies in nonhuman primates. I will use this as a platform from which to discuss this topic in prehuman groups. Kummer's monograph contains a number of interesting observations that bear on the subject of the origins of psychopathology. Kummer provides several illustrations of how social context and ecology affect innate capacities and tendencies so as to produce unusual, aberrant behaviors and accompanying social responses. He explains the behavior of baboons as an interplay between genetically based behavioral mechanisms and adaptations, on the one hand, and the contingencies and pressures of the environment, on the other. In his estimation, the collision between these two sets of factors determines the form and content of anomalous behaviors. He tends to interpret the latter largely from the standpoint of evolutionary theory, taking into account ecological and contextual factors. However, on occasion Kummer goes so far as to use concepts consistent with psychopathology as defined here.

Working initially with baboon groups housed in a Zurich zoo, Kummer made close observations on the nighttime behavior of his subjects. He noted the sleep movements and stirrings of four females whose rank in the group was well established. During eight nights he counted "how often each of the four females stirred while she was presumably sleeping, as her eyes were closed" (33) and showed that high rank was correlated with fewer stirrings and movements. His comments and interpretations are interesting: "On several occasions I saw Nacha, and only Nacha [the lowest-ranking female], while closing her eyes and apparently asleep, raise her brows—the threat expression, a signal she hardly ever had a chance to give in her daytime life [because of her low rank]. A disturbance of socially oppressed animals so profound that it extends into their nighttime sleep has not, to my knowledge, been studied in detail. I would be skeptical about it if other researchers had not since then described still more astonishing psychosomatic effects of social position among primates" (33). Kummer goes on to quote literature describing how rank level influences a range of behaviors and details of reproductive physiology.

In the unnatural environment of the zoo, with food regularly provided and predators nonexistent, Kummer found that social behavior and relations became more inventive and seemed to luxuriate. He suggests that the synchronization of estrus cycles among fertile female baboons may constitute an adaptation that by satiating the harem leader and official male partner of the group's females might give the females opportunities to seek impregnation by other males. In the condition of the zoo, however, the genital swelling of females served to attract fellow females. He describes how the various females responded to each other sexually: "In the wild we never saw such explicit sexual activity between females; . . . [they] did not merely show interest in a feminine sexual stimulus, then, but actually reacted with male copulatory movements, although their behavior toward males was entirely feminine. Male baboons also mount one another with copulatory movements. In contrast to female hamadryad homosexual behavior, such behavior occurs among males even in the wild, where it had the important function of appeasement in situations that require coordinated behavior. We never observed intromission or orgasm between individuals of the same sex, and I know of no case in which a primate rejected heterosexual behavior in favor of homosexual" (36). In the seemingly furtive sexual escapades of some females, Kummer observed attempts at social concealment during mounting by the dominant harem leader. Moreover, when a low-ranking male and his mistress appeared to catch the eye of the harem leader (the owner of the female), unusual and seemingly pathological, or at least self-destructive, behavior resulted: "If Pasha's [the leader of the group] . . . gaze turned on [the female and her lower-ranking lover], they fell terrified into each other's arms and clawed their hands into each other's hair again and again, kecking [a staccato sound signifying fear], with their faces turned toward the patriarch. Often the young female would actually leap onto her lover's back, and he would flee with the precious load in great agitation, as though he were trying to save her from a leopard; . . . the young pair would dash *toward* Pasha for salvation and embrace in front of him clutching at each other with great urgency, while he usually looked mildly on. Then, separately and somewhat more calmly, they returned to their love nest" (36–37).

Kummer offers an interpretation of this strange, seemingly self-defeating and anomalous behavior. In discussing how in wild communities of baboons males herd their females by threatening and resorting to violence, he makes the following observations:

> If the female dares to wander far away or is actually near other males, he rushes to her and gives her a neck bite. . . . And now something remarkable happens. Instead of running away from the attacker, like any "sensible" animal, the female presses herself screaming against the ground or turns straight into the jaws of the attacking male. After the bite, she follows close behind him, making a quacking sound, and begins to groom him hectically as soon as he sits down. This behavior illustrates the paradoxical flight of the fearful zoo baboons to their menacing, Pasha, the only one with whom they found refuge and comfort: it functions to bond the female to her male. In the zoo, this urge was absurdly taken over by the followers and juvenile males as well. . . . That the neck bite is a particular punishment for a female that has wandered too far away was not noticeable in the zoo, where confinement made straying impossible. (101)

In this instance, then, strange, anomalous, seemingly maladaptive or pathological behaviors were produced by the interplay of inherited adaptations—herding behavior, neck bites—and environmental conditions created by confinement in the zoo, where there were no good opportunities to hide and pursue amorous adventures. And, interestingly, both adulterers participated in the behavioral scenario described. In other words, the lovers both acted out the more or less innate tendency of the female to bond to the harem leader and acknowledge his ownership, given his status and behavior, particularly in response to his signal actions. All of this was played out in a way that appeared maladaptive: upon being caught with a female, the cuckolding male carried her to her owner and his boss, thereby seeming to advertise their errant sexual behavior. In actuality, they were following herding instincts and communicating appeasement and fidelity and seeking punishment—and, perhaps, attempting to distract the harem leader from their amorous pursuits.

Another example of how innate tendencies can be affected by social circumstances to produce unusual and seemingly maladaptive behavior is illustrated in Kummer's account of a female harem chief named Vecchia. The situation was created when a younger and stronger male named Ulysses became superior to the older harem leader, Pasha. "Even though Ulysses looked like a Pasha, he wasn't one, and his new, elderly alpha female Vecchia took advantage of that" (53). Vecchia, who had previously occupied a relatively low rank, began to commandeer other females of the group. She meted out punishment to Ofe, who made a play for Ulysses. "Vecchia . . . charged at Ofe, who once had been her superior in rank, bit her in the neck, and shook her, as only a very angry male would normally do with a female. This was clearly the behavior of a [male] harem leader. Ulysses did not intervene, then or later. When Ofe fled to him after being bitten by Vecchia, he ran away" (53). On other occasions, Vecchia would supplement punishments handed out by Ulysses, the titular leader and chief. When another female, Sora, transferred to Ulysses, the latter showed comparatively little interest:

> At first Vecchia was satisfied with barring Sora's attempts to flee to Ulysses during quarrels. But I soon noticed that during the rest periods around noon, Sora continually groomed Vecchia and never Ulysses, as would have been customary. . . . If Sora still wanted to approach Ulysses, Vecchia casually stepped into her way and blocked all of Sora's attempts to go around her. . . . Sora progressively became Vecchia's female rather than Ulysses'. . . . One day Sora tried to present her [genital] swelling to Ulysses but, to be on the safe side, did it halfway in Vecchia's direction. Ulysses responded and copulated with Sora. Immediately afterward she hurried around behind Ulysses's back as usual. However, Vecchia was already sitting there and demanded for herself the grooming that was the male's due. This now became a regular occurrence. A little later Vecchia took over the final element of the male's role: she mounted Sora and bit her gently in the back, as males sometimes do while mating. . . . In her everyday life, she had virtually changed sex; she had taken the place of the ineffectual harem leader in everything but reproduction. (53–54)

This unusual, seemingly deviant and pathological behavior situation was created by the interaction of inherited sexual tendencies and clashes between personalities and the requirements of well-established roles. Apparently, Ulysses's passivity and Vecchia's assertiveness were compensated for by the adoption of dominant, masculine behaviors on the part of Vecchia. The pathology of this individual (Vecchia) and situation, however, is calibrated mainly in the minds of the observers. Relations between Sora, Ulysses, and Vecchia continued to be social and functional, for it is not known whether any of the participants lost any status or access to food, not to say inclusive fitness, as a result of an arrangement that complemented but did not supplant natural reproductive strategies.

Kummer's analysis is illuminating:

> Vecchia had not become chief of the harem because of any superiority to Ulysses in rank or strength. She had simply learned to do what Ulysses—incomprehensible to me—left undone, and had estimated correctly the extent to which he would let her play leader in his place. She made her own way to the top, from low-ranking harem female, disdained and chased by all, to executive officer of the family. It is remarkable that this position, otherwise exclusively the province of males, was one to which a female would aspire. I was even more surprised that her years of being persecuted had not extinguished either the abilities or the self-confidence she needed in her leading role. Perhaps a feeling for one's identity can only be destroyed if one has an identity to begin with. No such thing [self-identity] has been documented in primates below the intelligence level of the chimpanzees. (54)

In support of the previous formulation, Kummer cites research to the effect that when all-female groups are set up experimentally, the females transfer to the highest-ranking female, who then plays a role ordinarily played by the male harem leader. "Even without a male, then, female hamadryads form copies of one-male families. Their first contribution to marriage is fleeing to the aggressor and the second is the tendency to be concerned almost exclusively with the highest-ranking member of the group" (125). Kummer's conjectures on the topic of aberrant and anomalous behaviors are worth quoting further: "In captivity, where predators and hunger are not problems, forms of behavior can develop that are never seen in the wild but nevertheless are not pathological. On the contrary, they are richer manifestations of a natural disposition within the innate reaction norm; they are not 'abnormal' but supernormal" (132).

The influence of the environment in shaping highly adaptive and not just anomalous behavior suites is illustrated by what Kummer termed "protected threatening" situations. "The tactic of protected threatening, so highly perfected in the Zurich Zoo, existed only in elementary components in the natural environment" (134). What Kummer refers to as protected threatening involves a suite of behaviors consisting of a female placing herself between a female rival opponent and the male harem leader. In this position, she would face her antagonist and simultaneously threaten her while presenting her own behind to the male leader in an act of sexual conciliation and deference. This would induce the male to attack the female's opponent:

> At the sleeping cliffs of Erer and Saudi Arabia, we never saw this perfect combination of threatening, ensuring one's own safety, inciting an ally,

and blocking an opponent's way. Usually an adult female merely pressed herself against her male, screaming; at most she placed herself in front of him and screamed at her opponent.... These primitive stages were equivalent to the behavior of youngsters barely two years old in the zoo group. The protected threatening in the Zurich Zoo was the combination and perfection of behavioral elements that occurred only in infantile forms in the wild. The ritual is so tactically perfect and effective that any observer would be likely to regard it as a fully developed, adaptive product of evolution and assume as a matter of course that it was part of the natural behavior of the species. Nevertheless, it is a product of [the unnatural living conditions of zoos made possible by] civilization. (134)

Kummer here shows with some irony how genetically based tendencies and programs relating to normal behavior pertaining to social rank and sexuality can be shaped by ecological circumstances to produce a highly organized and effective strategy for handling naturally occurring conflicts. The irony is that the unnatural, civilized environment of the zoo was what created this seemingly natural and adaptive behavioral scenario.

Note that both this adaptive sequence of behavior and previous pathological ones are in some ways similar: all more or less reflect the creative integration in the special environment of the zoo of behavioral tendencies found under natural conditions. Clearly, it is the observer's interpretation of creative solutions to contextual problems that leads to the qualification of normal and adaptive as compared to abnormal and pathological.

A further irony devolves when one takes into account that unnatural, civilized environments of zoos have in several instances been described as contributing to the production of highly pathological and maladaptive behavior. Kummer reviews the situation taking place in the Cologne zoo, which involved the aggressive mistreatment of infants and females by subadult males. This resulted in the killing of infants, and in one instance, at least, a male killed three of his own offspring. Thus, the behaviors are maladaptive even when judged from the standpoint of inclusive fitness theory, as well as social organization and ordinary social conventions. (Compare this with similar behavior of langur populations of the Dharwar of India as reported by Hrdy 1979.) Under natural conditions, male hamadryad baboons usually show maternal behavior toward infants, and in the Cologne zoo the behaviors of the males in question actually started out this way, only to become increasingly aggressive and harmful. Kummer suggests, "Infants appear not only to attract tender responses, but they evidently also stimulate responses that may harm them" (139). Unusual circumstances taking place in the community of the Cologne zoo are held to have contributed to the baboons' maladaptive, pathological behavior.

What Kummer is pointing out is that social environmental conditions in a population can alter and shape the natural, genetically based programs of baboons and other higher primates, including, of course, man. Unusual behavior could not be the result of genetic programs if the conditions under which it occurs never endured long enough during the evolution of the species to have been selected for. Conditions in zoos are simply too unnatural and civilized to have occurred under natural conditions; hence behavior observed, especially maladaptive behavior, cannot be the direct expression of genetic programs.

Kummer points out that the provisioning of food and protection from predators in a zoo can to some extent liberate or release organisms from their inherited tendencies, in the process contributing to the production of either adaptive or maladaptive behavior. When food is scarce and predators prominent, as in natural environments, the harsh conditions of social life can shut off social behavior programs. As he puts it: "Harsh conditions are evidently the icehouse of social behavior" (136). In contrast, the conditions of zoos allow genetically ingrained social programs to luxuriate in a genetically uninformed, blind manner, and this can lead to behaviors that one could place at a different end point on a prosperity scale. In other words, in some instances it can lead to seemingly adaptive behavior, and in others to maladaptive, pathological behavior, depending on conditions.

Kummer's analysis is worth quoting in full:

> Conditions of luxury allow free play to develop by emancipating the organism from its genetic programs.... The strategies of an animal can be evaluated by two different criteria. The first is the *survival value for the genes:* using strategy A, how well can the organism in the given environment preserve its own life, produce and raise children, and take care of relatives that also have genes for strategy A? Survival value is the criterion for evolution.... The second criterion is the *gratification value for the individual.* What strategy will the animal prefer to another if it has the choice? In other words: What forms of behavior and which of their immediate effects act as rewards and are the end and goal of a search?... The second criterion is the gratification value for the individual. (139; emphasis in original)

Evolutionary theory informs us that under natural conditions survival and gratification are linked in the production of behavioral sequences. The genetic systems of organisms have been selected so as to provide satisfaction from behaviors that have survival value, and this continues to work well until unnatural conditions are created: "The particular alienation from the environment experienced by zoo animals takes the form of giving them more spare time and spare energy than their species ever had when living free. Under these conditions higher animals begin to play with their gratification system, the word 'play' being meant here in its broadest sense [in the zoo].... The evolved gratification systems developed their own dynamics and went their own way, regardless of its survival value" (141).

Kummer's observations of elderly baboons as they lose their social position in the group and their ready access to females parallel those of Goodall (1986) and are instructive for what they suggest about social stress and psychopathologies: "The sudden loss of their females caused a change in the three clan elders at Cone Rock. Until then they had been the most imposing males in the band, robed in their massive mantles [of hair]. Within a few weeks after they had been deposed, their faces turned dark and their mantles became sparse; they lost weight and looked old. Admiral's shoulders were soon almost bald, and the skin of his lower abdomen hung in folds. . . . It must have been the profound experience of defeat and loss that changed the males in this way. . . . Testosterone causes the male sexual characteristic. It was precisely these sexual features of the adult male that the three oldsters had lost: the reddish face and the light-colored, dense mantle" (237). This reaction certainly constitutes a form of grief and mourning; in this instance, to the passing away of the organism's status and prestige and its associated benefits. Reactions of this type have ordinarily been judged as natural, given the circumstances; yet they can be equated with behavioral developments that can have maladaptive consequences.

Kummer goes on to indicate that old age and its social biologic consequences can have both adaptive and maladaptive results. "Presumably some deposed family leaders leave the clan where they had spent their whole life and wander alone, at an advanced age, from band to band. Why they do this is a mystery" (238). "Neither of the two males [who had been deposed] tried to acquire new females—for example, by becoming followers again. Neither would look at his former spouse, even when they were resting only a few steps away; . . . [they] now spent time with their children as never before—mainly with their small sons, and not as protectors but in a leisurely way that involved much mutual grooming. . . . Not every deposed family leader becomes a true exile, a miserable drifter from band to band" (238–39).

Along the same lines, Kummer reports that in one instance deposed males fought against a rival band alongside junior males who earlier had taken over their own females: "The net effect was that they defended their own former females for those males who owned them now. The victims fought for the perpetrators. . . . Once again, we see something that confounds human empathy. The deposed leaders exacted no moralistic revenge for the suffering, so obviously severe, that the young males had inflicted on them—just as the latter had no moral compunction about inflicting it in the first place. Biologically, for the old males to fight in aid of the younger made more sense than to stand aside spitefully, as many a man would have done in the old males' position, because some of the younger males were their relatives" (239–40).

Other unusual behaviors on the part of older group members were noted. In some instances, for example, older, deposed males functioned much as protectors, keeping watch over group members. In other instances, deposed males were observed taking chances and placing themselves in dangerous situations. "Old females will also live more dangerously;.... peripheral females, usually older, are just as willing to explore and take risks as the old males" (241).

Kummer's interpretations are stimulating and instructive: "Evolution has invented a thousand variations on ways to care for the young. Care for the old is unknown; ... evolution has not produced any kind of behavioral *system* that serves exclusively to help the old.... The old animal at the end of its reproductive phase is released to some extent from the guidance of its genes and so is relatively free. Genes have contributed to its behavior as a child and to its reproductive strategies; now, in old age, many of these are unemployed, and there are hardly any specific genes for behavior in the final stage of life" (242). Along the same lines: "The behavior of the aging baboons can give us something to think about. Some of them never overcome their loss and release; they become miserable wanderers. But others remain, are more courageous than their younger companions, go further into the face of danger, and take on greater discomfort. This is just the opposite of retirement in a literal sense, as withdrawal into a rather feeble self-indulgence. They have retired only from the compulsion to succeed" (243).

That striking personality differences exist among members of natural troops of baboons and that some of these raise the question of psychopathology in the minds of primatologists is also illustrated in Strum (1987). Her account of social behavior among baboons and, in particular, features of the ranking system among females, as well as friendship patterns among males and females, came as a result of careful observations over a number of years of the life routines of baboons in Kenya. One example is David, a male unusual in that he did not emigrate from his natal group, as is typical for adolescent males of the groups Strum studied. A failure to "separate and individuate" created special problems for David, and while his unusual behavior could be interpreted as a result of this, his unusual temperament and personality may have operated as a reason for his dependency on his group in the first place. Strum writes:

> It is difficult to tell whether baboons are neurotic in the same way humans are. David, however, came closest to what one might imagine a neurotic baboon would be like. He took failure badly; when females rebuffed him, he would lurk behind bushes and rocks. . . . Gradually, almost imperceptibly, David distanced himself from his family. When a crisis developed, he was slow to arrive on the scene and when he did get there he defended his mother and sisters only half-heartedly, often leaving before the crisis was settled. Then he began to ignore his family's

calls for help completely. Instead of sitting with them during the day, grooming and being groomed, he kept away. Whether this was deliberate or accidental I never knew. By now he was sleeping alone. I came to realize that if I had just begun to study the troop, it would be hard for me to tell that David was a member of [this troop].... There were no [social] clues that would help identify his mother or sisters. (112)

In the case of David, a measure of social expertise and success developed over time. Strum's description of a female named Thea provides another illustration of this theme of personality idiosyncrasy as compared to possible natural forms of psychopathology among baboons:

> Thea was, in fact, a bitch. Her status in the troop was second only to her mother's, and she used it tyrannically; she was unprovokedly aggressive, intimidating other females in situations where Peggy [her mother] would have calmly quelled the whole matter with a rebuking glance or approached and waited for what she wanted. Moreover, Thea was always poking her nose into other people's business. Whenever females or juveniles were involved in a tiff, Thea would be there in seconds, adding her weight sometimes to one side, sometimes to the other, frequently almost schizophrenically switching sides unpredictably. She often managed to *prevent* the quarreling individuals from settling the argument.... When the situation got this far, Peggy would occasionally come to Thea's support, her added weight resolving the issue on the spot and cutting short Thea's muddling influence. Most of the troop gave Thea a wide berth, so these situations occurred less frequently than they might have, but at times she could stir up such a to-do that some of the adult males would intervene and she would find herself the new underdog. Here ... Peggy would lend her support to her daughter, usually tipping the balance in the direction of a face-saving retreat on Thea's part.... Her lack of calm and her generally high level of aggression sometimes extended to her own children: certainly she seemed to have less time for her daughters than did other mothers. (40–41)

Aberrant Behavior in Chimpanzees

The idea of pathology, like that of disease, is based on cultural notions of what is normal or healthy. Stated succinctly, if one intends to stipulate that a form or segment of behavior of a particular individual is pathological, then one must show that it deviates from the norm of behavior pertaining to the group or species to which the individual belongs. But even if an appropriate methodology were found with which to carry this out, we would still be plagued with the problem of specifying what environmental context should frame the normal or healthy. On general grounds, and particularly with respect to more evolved complex groups that display behavioral plasticity like the baboons discussed by Kummer, it is difficult to arrive at a norm or standard of behavior precisely because ecology and environment play such influential roles in producing behavior, which varies in relation to local conditions. Kummer's formulation encourages one to pursue the question of naturally occurring nonhuman psychopathologies.

As indicated earlier, the literature of primate ethology makes few inferences about possible forms of psychopathology based on interpretations of social and psychological behavior. This reflects the different points of view of clinical and biological observers. The dominant approach in ethology is to explain behavior—however pathological viewed anthropomorphically—using tenets derivable from evolutionary theory. Infanticide among chimpanzees constitutes a case in point. This type of behavior has been the object of considerable discussion and controversy, and a number of theories have been invoked to explain it, including in particular sexual selection (Hrdy 1974, 1979). Crowding and social pathology are frequently considered as possibilities, but they are rarely invoked as significant or likely. I will give primary attention to chimpanzee infanticide associated with cannibalism. For examples of infanticide in other species and factors considered relevant from the standpoint of biological theory, usually as a reproductive strategy of males, see Hrdy (1974, 1979), Struhsaker and Leland (1985), and Watts (1989).

Goodall (1977) reported two types of fatal attacks on infant chimpanzees at Gombe: Infants of stranger females seized by adult males of the Kasakela community, that is, intercommunity killings, and infants within the Kasakela community killed by females of the same community, that is, intracommunity killings. One female chimpanzee, Passion, and her daughter killed infants on three separate occasions, two of which involved the infant's mother. In all instances, violent and brutal attacks were directed at the mothers of the victims, but these attacks desisted once the mothers acquiesced or were overcome and failed to resist: "One striking aspect of the infant killings of the Passion family is that, on each occasion, the attack on the mother seemed to be related solely to the acquiring of the infant for food, and once they had established their claim over their prey they made no further aggressive attacks on the mothers" (269). Another striking feature of the reported intragroup infanticide/cannibalism is that, apparently, while eating their prey, the perpetrators were observed to show "none of the bizarre behaviors observed during male cannibalism" (267).

Some strange behaviors were recorded nonetheless, and these involved the perpetrators' interactions with others: "Once again, the family shared the body of their victim. On this occasion a number of other individuals appeared during the course of the meat eating. An adult female approached and stared as they fed, then picked up a tiny scrap of meat from a branch. She sniffed it, then threw it away and wiped her fingers" (268). The third case of intragroup infanticide/ cannibalism involved Passion, her adolescent daughter, Pom, who was an integral partner in the three attacks, and Melissa, the mother of the infant: "Melissa did not leave, but stayed with the family until nightfall, watching as they fed on her infant. Once she picked up some intestines which had been laid over a branch: she sniffed them and then carefully put them back. Once Gremlin cautiously reached for the hand of her dead sibling and sniffed it, then let it drop. Twice, as Melissa moved close to Passion, with submissive gestures, Passion responded by 'reassurance' . . . once embracing Melissa and once touching her hand" (269).

The intercommunity cases of infanticide recorded by Goodall are notable in a number of respects. First, the perpetrators were males. Since the females were outsiders and could have been in the process of incorporation into the group in which the perpetrators were members (female transfer is common among chimpanzees), this could have been a factor: it worked to the reproductive advantage of the males since following the infanticide the mother will soon resume estrus and is in a potentially fertile condition. Goodall indicates that this is not likely since an earlier mother who had been victimized in the same way did not successfully integrate into the group that included the perpetrators. She also mentions population crowding and/or regulation as possible reasons for the infanticide, but she cannot be certain (see Power 1991). The second point of interest is that in two of the instances, the victims were cannibalized. Third, bizarre behavior was observed on the part of the perpetrators during and following the attack. In two instances, the infants died during the attacks:

> After the infant had been killed, Humphrey and Mike [two of the perpetrators] fed on the flesh of one thigh for about 10 min. Then Humphrey climbed down and, as he walked, smashed the infant's body time and again onto the ground. The others followed and gathered around Humphrey when he sat, watching as he and Mike ate. Soon Humphrey moved off, flailing the corpse so that the head smashed against the ground. Once more he sat and fed, sharing with Mike, but after 20 min. he stopped. Mike continued to eat a detached leg while Humphrey investigated and played with the body. His behavior was bizarre: he pounded on the corpse with his fists, pressed down hard on the chest so that air was forced audibly through the lungs, poked his thumb repeatedly into an armpit causing the arm to jerk, peered into the face and briefly groomed the body. (Goodall 1977, 263)

There followed several attacks on the lifeless body by Humphrey and a different participant in the incident named Satan: "The last chimpanzee to carry the body was Satan. He kept it for about one and a half hours, during which time he was mostly very gentle with it. He took it into a nest and spent some time grooming it, especially the face. During this time he nibbled at very little of the flesh. He dragged it behind him as he traveled, wedged it in a tree whilst he fed on fruits, and left it on the bank of a stream while he drank. He then jumped the stream and began to move away but suddenly stopped, looked back at the corpse, and hurried back to get it. About 6 h. after the capture of the infant, Satan arrived in the feeding area dragging the remains of the body behind him, then throwing it forward, picking it up and repeating these actions" (262–264).

Goodall records similar bizarre behaviors that took place during the other attacks:

> [A perpetrator] seized the body and began to display with it, dashing it repeatedly against the ground and tree trunks. Next he sat down and pounded on its head with his fists, time and again. He pushed his hand into the thoracic cavity, withdrew and sniffed it, then wiped his hand on the ground. Then he abandoned the body. It was picked up by an adolescent male (Goblin) who slapped at it a few times and then left it lying on the ground. A young female approached, sniffed the body, and briefly groomed it. . . . [Regarding still a separate incident:] One of them held [the corpse of the infant victim] upside down from his foot and patted it. The other dangled it in front of him, hung it over a branch, and played with it. . . . Other males present reached out and touched the body from time to time . . . but it is not clear whether this was begging, or curiosity. (264)

Goodall describes a final instance of intergroup infanticide that did not involve cannibalizing the corpse, but it also contains several examples of strange and bizarre behavior. In this instance, a female stranger from another group who was carrying her infant in the ventral position was chased up a tree by Kasakela males, some of whom followed her there, barking, displaying, and attacking her.

> The Kasakela chimpanzees then quieted: most of them sat in the tree with the stranger, some staring at her, some actually feeding. Presently the stranger approached one of the males, presented and, twice, extended her hand to him submissively. She received no response. After this it seemed that she tried to leave. . . . Immediately some of the males climbed down and followed her along the ground, looking up, whilst others followed through the branches . . . 30 min. After she was first attacked she again approached one of the males (Satan) and presented to him, twice reaching to touch him. Satan actively moved from these contacts (the second time he picked a large handful of leaves and scrubbed his leg where her hand had rested). Immediately after this she was attacked again. . . . During the violence her infant fell: the mother instantly dropped down after her, but adult male Jomeo, who had been down below, had already seized the infant. . . . Finally, as the mother

escaped (bleeding profusely) Jomeo raced up a tree with the infant. Figan followed and, at once, took possession of the infant. He began to leap through the branches in a charging display, smashing the infant against the branches and trunk as he did so. He leapt to the ground and continued to flail his victim against the rocks and he ran. Finally, after charging for some 40–50 m, he flung her from him. On this occasion the infant was not eaten. Satan approached, picked her up, groomed her, and then gently put her down. She was then "rescued" by a 4-year-old male . . . who carried her for over an hour, constantly supporting her since she was too badly wounded, and probably too shocked, to cling unaided. . . . [Satan then took the infant once again and] carried her, holding her while he fed, supporting her as he traveled, grooming her from time to time while he rested. (265–266)

The infant was subsequently "rescued" by other male chimpanzees and eventually left in the bushes. The wounds on the infant were severe and she died later that night.

Goodall is not able to explain the bizarre quality of the behavior that she so richly records. The aggressiveness and violence of the behavior she suggests can be linked to the general nature of aggressive behavior in these and other nonhuman primates, something that has since been the object of considerable interest (see Wrangham and Peterson 1996). As indicated, she suggests but is not convinced that overcrowding and population control may be proximate factors. In some respects, the behavior recorded by Goodall brings to mind the ritualized behaviors and even the "fixed action patterns" described by classical ethologists (Thorpe 1956; Tinbergen 1952) and emphasizes an individualistic approach to aggression and violence, an approach that currently seems less dominant in primatology (de Waal 1996). If one were to anthropomorphize liberally (but see Kennedy 1992), it could be said that the character of the routines that embody the infanticidal attacks in some ways suggests the release or lack of control of impulses and/or the repetitive enactment of programs of behavior, for they resemble the systematic, frenzied, and sustained violence perpetrated in human sexual homicides, as well as other types of homicides where the perpetrators seem driven consciously or unconsciously to injure, hurt, despoil, and reestablish and reemphasize their dominance and control in a rageful manner. The possibility that the behavior released is actually dissociated or reflects an altered state of consciousness cannot be ruled out.

There appeared to be something psychological if not symbolic to the behaviors of infanticide described by Goodall. She writes: "The adult males, both at Gombe and at Budongo, showed behaviors which differed from that seen during normal meat eating. For one thing only a few of the individuals present actually took part in the eating of flesh. . . . For another, very abnormal patterns were observed" (279). (The patterns described in the previous quotations.) This suggests that the perpetrators appear to be exulting while engaging in repeated symbolic displays of aggression and violence, in addition to actually perpetrating such acts on dead victims. Symbolic elements are evident not only in how the violent attacks are enacted but also in the admixture of seemingly competing impulses or programs of behavior. The rescuing behavior in the last case of intergroup infanticide and the grooming, patting, gentle stroking behavior that is part of all the routines appear to be anomalies to Goodall and other researchers, who see such behavior as bizarre. Goodall attempts to make partial sense of the "rescuing" elements of the third case: "One might speculate that the sex of the infant was significant here: male Japanese monkeys and hamadryad baboons direct a greater amount of "paternal" protectiveness to female infants than towards male infants. . . . However, since we only know the sex (male) of one of the two stranger infants who were actually killed, such speculation is premature" (271–272). Reasoning anthropomorphically in ways that can certainly be challenged (Kennedy 1992), this type of rescuing, grooming, almost tender behavior seems to compete with or undo the aggression and rage. All of the unusual or bizarre qualities of the infanticidal behavior (to make a larger and not only anthropomorphic but also a clinical/forensic leap) resemble the violence that is observed in human cases when, very often, those perpetrating acts of violence appear to be dissociated and in an altered state of consciousness.

Goodall was not able to satisfactorily explain cannibalism as a form of primate behavior. She does suggest that it may constitute a more widespread and problematic subject: "But the fact that cannibalism can also occur in the gorilla and the Red tail, species which have not yet been observed to hunt mammals for food despite intensive studies, suggest that the issue may be more complex in the primates than in the carnivores [where it also occurs]. That gorillas, chimpanzees and man all share cannibalistic tendencies is a fact which will be of great interest to students of behavior, but cannot be meaningfully discussed until more incidents (particularly in the gorilla) have been observed" (1977, 281).

A case report such as Goodall's and other observations of researchers conducting naturalistic field studies of nonhuman primates bring into focus questions of abnormal or psychopathological behavior. This material is interesting because the cases reported differ from laboratory studies of nonhuman psychopathology, where the goal is to actually model, through laboratory manufacture, human varieties of psychopathology. Naturally occurring forms of aberrant social and psychological behavior pose obvious quandaries, for, as already indicated, researchers appear reluctant to apply the concept of abnormality or pathology.

Suomi (1978) postulates that the absence of a mother early in life tends to induce later abusive mothering. This hypothesis has been supported in two independent reports of a study conducted by Japanese ethologists. For example,

Hasegawa and Hiraiwa (1980) indicate, "Primiparous mothers are usually rough and awkward in their handling of infants, and it is characteristic of the orphaned primiparous mothers that their infants die from being mishandled by their mothers" (155); and Hiraiwa (1981) writes: "The infant mortality rate among the primiparous, orphaned mothers was 45.5% (5/11), which was significantly higher than that of among the multiparous, non-orphaned others" (319). Thus, it is well established that orphaned primiparous mothers display aggressive, incompetent infant-handling behavior, eventually losing about half of their infants during the first year, despite showing no apparent abnormality in their overall social behavior.

As illustrated in my analysis of Goodall (1977), infanticide and cannibalism are other categories of potentially abnormal or pathological behavior that have been reported to occur under more or less natural conditions and which researchers have sometimes found difficult to explain. Usually, primate infanticide is viewed as a strategy to attain high individual reproductive success; for example, intergroup male infanticide is said to be aimed at removing a future competitor because a chimpanzee group constitutes a patrilineal social unit, although other explanations have been proffered, such as social pathology, and population regulation (see Hrdy 1974, 1979). Takahata (1985) reports a case of male chimpanzees killing and eating a male newborn infant in the Mahale National Park in Tanzania. In this case of intragroup male infanticide, chimpanzees killed infants of the same unit-group, possibly their own offspring. This was explained as possibly being due to the males' false notion that the infant in question had been sired by a male of another group because its mother had been absent from the group for several months before parturition.

In another case report by Norikoshi (1982) pertaining to observations in Kasoje Forest of western Tanzania, adult males also killed and ate an infant of their own group. This case of infanticide/cannibalism was associated with unusual responses by group mates:

> The cannibalism observed at Kasoje resembled more ordinary meat eating. However, it differs in that the males spent considerable time eating the infant without showing much excitement, also in that other individuals present were apparently little interested in the eating. One could have assumed that cannibalism would be more inhibited when eating an infant of one's own community than when eating a strange infant. The exact reverse was the phenomenon which was actually observed and remains entirely unaccounted for [that is, the behavior of the mother was considered unusual].... First, her apparent lack of emotional distress and concern at the cannibalism of *Humbe* [the 2.5-month-old cannibalized infant]; and second, her intimate association with two adult males that she observed eating and/or handling *Humbe's*

body ... [the behaviors of the mother] were perhaps related to the fact that the individuals who were eating *Humbe* were high-ranking males and absolutely dominant over her. Her apparent lack of distress and concern might also have been related to her general performance as a mother ... [since she] could have been considered a bad mother. We often observed her dangling her infant by its leg high up in trees. We also often observed her newborn infant left unattended. (72–73)

The behavior of the mother suggests a state of dissociation. This difficult-to-explain episode, which could represent a naturally occurring case of psychopathology in chimpanzees, and possibly analogous or homologous with respect to cases/responses of hominids pertaining to conjectured forms of behavioral abnormality, also suggests interpretation and labeling by group mates. Norikoshi writes:

I have already mentioned that the infant was eaten very slowly in this cannibalism and that strong inhibition was exercised over it. In relation to this inhibition, I will mention the fact that *Kajugi*, the first-ranking male, often attacked the individuals who were eating the infant or approaching it, and got them to leave the scene of cannibalism. It is open to argument whether the behavior of *Kajugi* aimed at prohibiting cannibalism or at obtaining the flesh. The next day, he had the pelt of the infant with him. The result, however, is that this behavior did help in keeping cannibalism from spreading, to say the least. The following is known as a similar case: When a cannibal female took an infant of another female, a high ranking male lent his help in giving back the infant to its mother [cp. Goodall 1977]. There is no question that cannibalism is not a taboo among chimpanzees, but it is almost certain that the behavior of the first ranking male worked as a social mechanism to restrict cannibalism. (Norikoshi 1982, 73)

A Naturally Occurring Animal Variety of Psychopathology?

Evidence that supports the validity of naturally occurring varieties of psychopathology among human ancestors is provided by a series of studies conducted by Alfonso Troisi and collaborators (Troisi et al. 1982, 1983, 1984, 1989, 1991). Among captive primates, infant abuse is not uncommon (Caine and Reite 1983; Nadler 1980). A mother's abusive behavior toward her infant is a relatively common sequela of a prior separation experience (or trauma) that deprived her of contact with her own mother (Suomi 1978). However, such abusive behavior on the part of a mother is associated with generally impaired maternal behavior and other social deficits as well. The coexistence of abuse behavior and seemingly adequate prior maternal care on the part of female monkeys is thus unusual; and the lack of known trauma in the case under consideration further compels attention.

The initial report was of Okame, a Japanese macaque *(Macaca fuscata fuscata)*, who was born and lived in the wild and later brought to live in a large enclosure of a Rome zoo when she was approximately four years old. Okame's first male infant lived only two months and evidence of the corpse suggested infant abuse. A rigid protocol of behavior measurement was used to analyze the behavior of Okame following the birth of her daughter one year later.

Abusive behavior patterns (ABP) were defined as consisting of "any mother's behavior potentially or actually causing physical injury to the infant" (Troisi et al. 1982, 452). Ten categories of behavior were identified and carefully recorded in a controlled fashion. Okame showed no evidence of abuse and a "completely adequate" maternal behavior profile during her infant's first four weeks. When they later appeared, ABPs were usually preceded by infant's attempts of separation/emancipation, emotional disturbance of the infant, and/or difficulties of mother-infant communication. In contrast to abusive behavior of socially deprived mothers of other studies who generally showed this type of behavior in response to the infant's attempts to establish physical contact, Okame did not. Furthermore, there was much to suggest that Okame was actually maternally caring and responsive. For example, "every abuse bout was followed by close contact intimacy." In comparison to nonabusing controls, Okame showed high infant contact scores and did not display avoidance of contact with her infant and was judged by the researchers as not a rejecting mother. Okame was actually said to be protective of her daughter while still abusing her.

In a separate study Troisi and D'Amato (1983) pursued the question of "diagnosis" of Okame's behavior. They pointed out that it did not qualify as morphologically abnormal—that is, distorted, strange, idiosyncratic behaviors not found in normal animals—and suggested the applicability of the category of contextually abnormal; that is, behavior of the animal's otherwise normal repertoire but performed in the wrong context (Erwin and Deni 1979; Goosen 1981). However, a careful analysis of the categories making up ABP showed that Okame's behavior could not be readily classified as aggressive. In their own words, the researchers stated: "all the ABP's can be found in categories of normal behavior . . . most ABP's are not allocated to the aggression category [and] most ABPs can be found in the play category" (Troisi and D'Amato 1983, 169).

The researchers suggested that while abusive behavior is, in an evolutionary or ultimate sense, "maladaptive and thus functionless," its short-term functions, more properly construed by them as consequences, are actual physical injury to the infant. Because aggressive behavior, as defined by Hinde (1974), implies a measure of intentionality—that is, directed at causing physical injury to another individual and not accidental injury—Troisi and D'Amato rightly showed restraint in classifying Okame's behavior as aggressive. What was striking was that Okame handled her daughter during ABPs as an inanimate object. It was not due to her failure to learn specific motor skills of maternal behavior. The researchers speculate "that a latent condition of abnormal emotionality temporarily disorganizes those complex motor patterns which allow normal monkey mothers to cope with anxiety-triggering situations involving their infants" (Troisi and D'Amato 1983, 172).

The validity of the claim for naturally occurring varieties of psychopathology among human ancestors is strengthened by Troisi and D'Amato's (1984) further discussion of their research. They emphasized once again that Okame alternated between violent abuse and attentive maternal care, scoring highest on maternal warmth, protectiveness, and possessiveness, and lowest on maternal rejection. Their conclusions are worth quoting in some detail: "The results from this case are not consistent with the learning defect explanation of monkey infant abuse—abusive mothering would be caused by a failure of the early environment to provide experiences critical for basic-perceptual motor development of infant-rearing skills. Clearly, Okame possessed the complete repertoire of maternal abilities and displayed it appropriately for the most part of the time spent in interacting with her infant. Rather, the intense anxiety she showed in the relationship with her infant indicated that she suffered from an emotional disorder. In our opinion, this emotional disorder can be well described by referring to Bowlby's model of anxious attachment" (Troisi and D'Amato 1984, 107).

Troisi and D'Amato's observations regarding the response of group mates to the behavior of Okame are interesting. They report that the group was clearly affected by the maltreatment. "Most of the monkeys left the proximity of Okame when she was abusing her infant. The alpha male remained and watched closely and displayed obvious conflictual behavior (yawning, scratching, threatening the observer). On two occasions we witnessed the alpha male and some females threatening Okame, who was abusing her infant. However, there was not actual correction of abuse" (Troisi and D'Amato 1983). On balance, their observations suggest that the form taken by Okame's psychopathology had a measure of "content and meaning": it was noticed, responded to emotionally, and produced negative effects on group integration, although no efforts were made to calm or redirect Okame.

Troisi and D'Amato's observations are certainly consistent with postulates about social cognition of monkeys. The latter are not expected to show pedagogy or deliberate attempts to teach or correct group mates, forms of behavior that are explained as cultural learning, the exact origins of which are in dispute (Tomasello et al. 1993). Moreover, the observations confirm the point discussed earlier regarding how unlikely it is that monkeys, or even chimpanzees, for that matter, would show meaningful, coordinated, let alone positive, responses to those afflicted with psychopathology. Here, the responses of group mates appear to have mainly consisted of muted, negative sanctions of Okame. The reaction of group members to manifestations of psychopathology constitutes an important topic in any formulation about the origins of human psychopathology.

Troisi and collaborators (1989) have gone on to study cases of abusive behavior of a mother who was born in a stable social group of macaques living in a large enclosure of the Rome zoo. The behavior of the mother's own behavior as an infant had been observed under controlled conditions. The case involved the behavior of Chiocchiola, who had been abandoned by her own mother but adopted by Chiarella, who had given birth to her own daughter, Chacma, two days previously. For all intents and purposes, Chiocchiola was well cared for by Chiarella and so was, of course, Chacma. There was nothing to suggest that Chiocchiola had been abused, traumatized, or unduly affected by her mother's abandonment. While emphasizing that abandonment should not be equated with maternal separation, the researchers cautioned against drawing firm conclusions about its effects based purely on observation data. Nonetheless, consistent with findings pertaining to Okame, the abusive mother was extremely possessive toward her own infant and infrequently tended to reject him. Here and elsewhere Troisi and collaborators discuss the complex question of etiology of a variety of psychopathology consisting of anxiety manifesting in abusive behavior toward infants, raising the question of developmental experiences interfering with the learning of maternal behavior, intercurrent disease, social stress in the group, as well as individual, genetic differences in vulnerability.

In a later study Troisi and D'Amato (1991) treated these varieties of macaque psychopathology with an anxiolytic agent used in human cases of anxiety. An improvement of abusive behavior and of other indices of anxiety was noted. The temporal profile of administration and behavior change, along with changes in dosage schedule, proved illuminating: "Diazepam treatment was associated with a consistent decrease in maternal anxiety, as measured by the indices of maternal warmth and protectiveness [and] reduced the abusive mother's inhibition in rejecting and punishing her infant . . . This finding supports the view that infant abuse is a distinct behavior pattern due to an abnormal emotional condition rather than an exaggeration of normal maternal aggression" (Troisi and D'Amato 1991, 572).

Evidence that supports the view that the behaviors described earlier constitute a more or less naturally occurring variety of psychopathology among nonhuman primates—that is, not laboratory manufactured—is provided by a series of controlled studies performed by Dario Maestripieri (1998, 1999). An early study (Maestripieri 1994) established that abusive behavior occurred in "strict temporal association with intense psychosocial stress" (48). Maestripieri (1998) also investigated whether abuse of infants among group-living rhesus macaque mothers was due to an adaptive reduction of parental expenditure on infants. Another study (1999) established that infant harassment was not an accidental result of infant handling or the result of inexperience but could be a consequence of competition among lactating mothers. Besides showing that abusive females scored higher than controls on measures demonstrating that they were highly controlling mothers—for example, higher protectiveness and rejection—they also received similar or higher scores on measures of parental expenditure in the offspring. This was in contrast to what would have been expected of the adaptive hypothesis, namely, a reduction in expenditure. This supported the view that infant abuse is a form of behavioral pathology.

Are There Naturally Occurring Varieties of Psychopathology in Chimpanzees?

Goodall's more general observations (1977, 1986, 1988, 1990) pertaining to the social behavior of chimpanzees have a bearing on the question of psychopathology in nonhuman primates and, by extension, on the origins and significance of human psychopathology. Goodall's observations on the behaviors of the chimpanzees of Gombe disclose examples of what can loosely be termed psychopathology. I have already noted the brutal infanticides followed by cannibalism that were perpetuated by Passion and her daughter, Pom. Goodall's views on Passion are instructive: "She was a loner . . . had no close female companions, and on those occasions when she was in a group with adult males her relationship with them was typically uneasy and tense. She was a cold mother, intolerant and brusque and she seldom played with her infant, particularly during the first two years. And Pom, being the first surviving child, had no sibling to play with during the long hours when she and her mother were on their own. She had a difficult time during her early months, and she became an anxious and clinging child, always fearful that her mother would go off and leave her behind" (Goodall 1990, 33). Parenthetically, if one assumes that Passion's anxious, stressed nature also prevailed during her pregnancy with Pom, then Goodall's observations of Pom are natural evidence of the consequences of psychopathology and stress on mothers for the behavior of their offspring.

Pom's early experiences with her mother may have been significant from a behavioral standpoint. Pom's own child was killed in an accident when the child fell from a tree on a very windy day, an incident that Goodall in part attributes to Pom's lack of sensitivity and caring. "It would be unfair to blame Pom entirely for the accident, to accuse her of negligence. It could have happened to any infant. Yet I cannot imagine Fifi losing a child in this way. For Fifi, like Flo before her, like all really attentive chimpanzee mothers, is alert to potential danger" (41). Goodall compares the cautious behavior of mothers in similar circumstances.

The case of Passion and Pom is interesting in another respect. This pair's involvement in attacking and eating the infants of group members was described above, behavior that Goodall views as bizarre and pathological, clearly out of the range of normal: "Whereas the killing of infants by the adult males . . . seemed to be *consequences of the attacks on their mothers* (similar equally brutal attacks on stranger females were seen on many other occasions), Passion and Pom attacked the mothers of their victims *only in order to acquire the infants as meat*. Once they had possession of the babies, no further aggression was directed toward the mothers" (1986, 284; emphasis in original). Here and from her other publications, one can infer that Goodall (1971, 1977, 1986, 1988, 1990) views this and other types of behavior as highly deviant and pathological. She explains that most instances of infanticide in nonhuman primates are due to the acts of males directed at the offspring of competitors during the appropriation of females and, at any rate, these instances of infanticide are not followed by eating the infants. The mothers of the victims continued to react with fear at the approach of Passion but apparently could not communicate to others the basis of this fear. Many of the females of this group were fearful and avoidant of her for the rest of her life.

Further support for the claim of a diagnosis of psychopathology of Passion is afforded by Buirski and Plutchik (1991), who developed a chimpanzee version of the emotions profile index (EPI) that was originally used to study the emotional distribution of human adults. Actually, a baboon version has also been developed and used successfully (Buirski et al. 1973): "The baboon study demonstrated that independent raters could use the rating system with high reliability. It was also found that those baboons that were rated as being dominant were least social, most aggressive, and were groomed most by other baboons" (Buirski and Plutchik 1991, 208). Returning now to the matter of Passion's personality and behavior, the following quote is relevant: "*Passion* was judged to be considerably more aggressive than the other females . . . more depressed . . . and more distrustful. . . . She was considerably less trustful . . . less timid . . . less controlled . . . and less gregarious. . . . The overall impression of *Passion* was that of a disturbed, isolated, aggressive individual who would be considered in human terms to exhibit a paranoid spectrum of traits. This image of Passion which emerged from the EPI-Chimpanzee Form rating scales is especially interesting in the light of her past history and subsequent behavior" (208). This report provides a detailed summary of Passion's deviance and abnormality as a mother. That the personality of chimpanzees can be rated by observers familiar with their behavior has been reported by J. E. King and Figueredo (1997), J. E. King (in press), and Lilienfeld et al. (1999). Individual differences in the behavior of chimpanzees appear to be well appreciated by the chimpanzees themselves, as the observations of de Waal (1989, 1996) suggest.

Goodall describes a number of chimpanzees whose behavior also appears deviant and relatively maladaptive if not pathological. The case of Flo and her son Flint is a case in point: "Flint was still dependent on his mother, riding on her back and sharing her nest at night, when he was eight years old. And by this time Flo herself was to some extent dependent on her son. If they came to a fork in the trail and each took a different direction, Flo was as likely to whimper, turn, and follow Flint, as he was to give in to her. The final anomaly was Flint's extreme depression after Flo's death, followed by his illness and death three and a half weeks later" (Goodall 1986, 204).

The case of Jomeo provides an interesting example of a chimpanzee observed over a long period of time and whose behavior has been singled out by Goodall. He is unusual because despite having been physically large and endowed with strength he somehow failed to make full use of his natural talents and to establish himself in a successful way in the social system of his group: "From adolescence onward, [Jomeo] was almost entirely lacking in social ambition" (Goodall 1990, 151). He failed to establish himself successfully in the social hierarchy. His unusual social history is instructive:

> We know nothing of his childhood, for he was already a young adolescent when first I met him. . . . In most respects he was a perfectly normal adolescent, but he did have one idiosyncrasy. When he came to camp with one or more of the big males, Jomeo, like any other youngster, was seldom able to get a share of bananas. And so, like the other adolescent males, he quite often arrived by himself—which meant that we could hand him his very own bananas. This was when the odd behavior showed itself—the moment he set eyes on the fruits he began to scream. Not just a few small screams of irrepressible excitement—which would have been quite understandable—but loudly, and for a couple of minutes at a time. Naturally, any chimps who happened to be nearby rushed to camp to see what was going on—and helped themselves to Jomeo's bananas. For at least six months he behaved in this peculiar fashion. And then, quite suddenly, the screaming stopped. (151–52)

Jomeo's early adolescence seemed normal enough. His initial expressions of aggression and assertiveness appeared "vigorous, impressive and audacious": "It seemed then that Jomeo was firmly established on the ladder that would lead, ultimately, to a high position in the dominance hierarchy. But then something happened. One day . . . just a few months after his successful confrontation. . . . Jomeo limped into camp covered with deep wounds. The worst was a great gash across the sole of his right foot which took weeks to heal and which left the toes permanently curled under. We shall never know who or what attacked Jomeo, but whatever it was that happened, it seemed to affect his whole subsequent career. His blustering displays towards the community females, even the lower-ranking ones, abruptly ended" (152).

This incident appeared to have affected Jomeo strongly, for thereafter he was easily intimidated. Following an aggressive interaction with an infant and his mother, Jomeo, "this time, in marked contrast to his performance the year before, he fled before her and, screaming in fright, took refuge up a palm tree. When she began to climb after him, Jomeo, screaming even louder, leapt to another tree, tumbled to the ground, and raced, helter-skelter, away.... By that time, Jomeo had become the heaviest male at Gombe, and his chicken-hearted behavior made him the laughing stock of his human observers.... The frequency with which a male displays is, of course, an important factor in determining his position in the male hierarchy. Jomeo's frequency had dropped to almost nil after the horrible injury to his foot six years earlier" (153).

Goodall goes on to detail Jomeo's slow maturation and social integration in the group: in several instances be behaved like a "clown" and during which he suffered social "indignities," behaving in a relatively incompetent way: "It was the same when it came to hunting: at first Jomeo, though dead keen, usually bungled the job.... As time went on there were other reports of Jomeo losing his prey to higher-ranking males" (155). A picture of public failures but private, furtive "successes" is described. "But all the time, despite his new accomplishments—his unchallenged authority over the females, his improved display techniques, and his increasing skill in hunting—Jomeo continued to be plagued by countless small indignities" (155–156).

Jomeo's actual courting behavior with females is also relevant. According to Goodall: "Jomeo's scores are the lowest in all measures of consorting behavior. He is thought to have taken females on fifteen consort ships only, during as many years. Only five of these included the periovulatory period [when females were in estrus] of the female concerned. Seven were with females already pregnant with the offspring of other males. It should also be mentioned that Jomeo's performance in the promiscuous group situation was equally poor; his frequency of copulation was usually low. Thus it would appear that Jomeo has been unable to compete successfully with other males in any reproductive sphere" (1986, 476).

Despite these failures, Jomeo eventually became a respected senior member of the group and a trusted friend; however, he seemed always to occupy a subservient position in social relations. Goodall writes: "I have wondered so often about Jomeo's fascinating character, his strange lack of any sort of dominance drive. If he had not been wounded as an adolescent, would he have gone on to become a high ranking male? Probably not for, after all, his brother Sherry showed the same inability to cope with adversity. Was this a genetic, inherited trait? While this is possible, I suppose, it seems far more likely that it stemmed from the personality, the child-raising techniques, of their mother Vodka ... [who was shy and a] very asocial female, spending most of her time wandering, with her family only, in peripheral parts of the range" (1990, 159–160).

COMMENT

Many types of behavior by nonhuman primates occurring under natural conditions have been described as bizarre, aberrant, and hence deviant. In many instances, earlier traumas and developmental problems are used to explain the recorded abnormal behavior. This mode of reasoning is understandable, given what we know about the laboratory manufacture of animal models of human psychopathology. In many instances, the abnormal behavior is not seen as adaptive in any way.

Behavior that raises the question of whether there could be natural varieties of psychopathology primarily involve acts of aggression and hostility. This could be a prepared mode in which behavior breaks down; in other words, because aggression and resorting to violence is a component of many natural, normal situations involving survival and reproduction, a final common pathway for many adaptations, it may constitute a vulnerable link in what otherwise is orderly, adaptive, and controlled aggressive behavior. When a vulnerable individual is subjected to proximal instances of stress and strain resulting from group conflicts or ecological challenges, its information-processing systems are perturbed. The individual is strained and overwhelmed, as it were. The result is a passive, mechanical breakdown of behavior.

Infanticide by males presents an ambiguous and complex situation with respect to the possibility of natural psychopathology. Looked at in ultimate terms of conditions prevailing in the EEA, such behavior appears to be a well-established reproductive strategy among the males of some species (Borries 1997; Hrdy 1979; Struhsaker and Leland 1985). Indeed, even in current environments, infanticide has been considered adaptive. Hence, in these species, usual varieties of male infanticidal behavior cannot easily be ascribed to psychopathology if certain conditions exist that are conducive to this strategy; for example, if the males are recent immigrants into the group, if the infants they kill were probably not sired by them, and if the male sees a good probability of impregnating the female.

Male infanticide among chimpanzees is more ambiguous. First of all, it does not appear to be common or standard behavior in this species. Furthermore, when male infanticidal behavior has been observed, researchers have not obtained empirical data to prove that it constitutes a facultative adaptation. Yet ethologists certainly entertain a sexual selection explanation for this behavior in male chimpanzees. Like good biologists and evolutionary psychologists, ethologists seem to assume that such behavior, which appears to constitute part of an adaptive reproductive strategy in langurs, has to be ruled out as an explanation of infanticide among chimpanzees. They assume that a potential for this type of behavior is at least part of an innate facultative adaptation that is resorted to unconsciously, or through released or dissociated behavior, when certain conditions, like those of the EEA, where the adaptation was presumably designed, warrant it. Researchers do not necessarily see a male chimpanzee's infanticidal behavior as homologous with that of its common ancestor with the langurs, for instance, but imply that the program or mechanism in question is somehow hardwired by natural selection and activated when certain releasing environmental conditions come into play. Presumably, if conditions like those prevailing in the EEA of chimpanzees were in fact different from those found today—not likely to be the case, despite Power's claims (1991, see below)—male infanticide would occur more frequently and in a pattern that made sense in light of the theory of sexual selection. Alternatively, conditions in the EEA of chimpanzees might have led to the establishment of behavior programs that competed with, nullified, or appropriated (exapted) that of a phylogenetically older male infanticidal reproductive strategy, like the one seen in monkeys.

Something resembling male infanticide has clearly been described for humans. Using the logic of evolutionary theory, Daly and Wilson (1988) have been able to explain patterns of homicide of children perpetrated by parents. In some instances, males driven by rage and jealousy resort to the murder of children whom they did not sire. Daly and Wilson offer compelling explanations for a type of human aberrant behavior (homicide) by relying on a strict logic drawn from evolutionary biology. In humans, male infanticide, and homicide more generally, have often been explained using a clinical idiom in light of associated social conditions of poverty, deprivation, and psychopathology in the individual and the family. Daly and Wilson imply, however, not that such behavior in humans is adaptive but that a program or mechanism that is consistent with a designed reproductive strategy appears to be brought into play when, again, vulnerable individuals are strained and stressed by ecological conditions. That male infanticide, or some forms of homicide in general, can have an evolutionary explanation—that is, it was adaptive in the EEA and is still triggered by prevailing circumstances—yet also constitutes a psychopathology is a clear example of the quandaries raised by the differences between an evolutionary and a clinical framework.

The logic of evolutionary theory suggests that modern humans have a naturally designed evolutionary residual or program whereby infanticidal behavior can be activated or released when prevailing ecological conditions resemble those of the EEA, conditions that had once threatened survival and especially reproductive success. In other words, biological or cultural mechanisms of behavior that under ordinary conditions constrain male infanticide can be contravened in vulnerable individuals if certain ecological conditions are met. This view is consistent with the formulation of Bailey (1987) regarding phylogenetic regression. Be that as it may, one can say that at the ape end of the human evolutionary continuum, male infanticide conforms to a sexually selected program of behavior and that moral, clinical considerations are simply unwarranted because of anthropomorphism. Toward the modern human end of that continuum, on the other hand, such behavior, when released or expressed, perhaps in a dissociated mental state, gives rise to contrasting, opposed interpretations. In this instance, the behavior is usually explained as a manifestation of psychopathology, although it still may be consistent with that of a naturally designed reproductive strategy. This implies that during some phase of human evolution between these poles of the continuum, the paradox of male infanticidal behavior would have been more obvious.

The Possible Role of Social Ecology and Social Organization on Psychopathology in Chimpanzees

Even when evolutionary residuals are more in evidence, if not clearly dominant, the behavior of primates, in this case chimpanzees, has been explained clinically. Margaret Power (1991) proposes that environmental changes introduced by humans are the cause of behaviors that have been erroneously ascribed to species traits, constituting instead induced forms of psychopathology. As implied earlier, the social and behavioral ecology of a group has a strong influence on its members' behavior. A later chapter will discuss the importance of this topic in paleoanthropology with respect to human evolution. Essentially, one's interpretation of the origins and evolution of psychopathology must be conducted with a steady focus on changing ecological conditions—as well as, of course, correlated changes in culture and cognition. In a fundamental sense, a group's social ecology sets basic constraints and creates the fundamental conditions that shape the social and psychological behavior of its members, and all of this changes in accordance with the process of natural selection and human evolution more generally. I have already stressed the importance of concentrating on the social ecology when interpreting behavior of nonhuman primates in reviewing Kummer's studies. This section presents a similar but more critical emphasis on the role of social environment, this time on chimpanzees.

Power (1991) provides an interesting perspective on what one can think of as psychopathology in chimpanzees. "The Egalitarians—Humans and Chimpanzees" is an extended analysis and interpretation of the literature pertaining to communities of chimpanzees and contemporary groups of hunter-gatherers. Power contends that such chimpanzees, very much like human foragers, have evolved an immediate return foraging system characterized by a mutual dependence system of social organization. It is characterized by open groups that range over familiar, undefended territories that actually overlap and that feature fluid interchanging leader-follower relations that are generally beneficial and have few costs. The groups are nonhierarchical, nonaggressive, highly egalitarian and characterized by indirect competition which "lacks the stress and tension of direct competition" (16).

Power summarizes and interprets the effects on chimpanzee individual and group behavior of changes in the system of food provisioning that took place in Gombe and Mahale during the 1960s. She reasons that chimpanzees became frustrated at the imposition of human control over the food supply, and the effect was a major change in their behavior: "Obtaining food is considered to be the most basic of physiological needs. Interference with the ability to do so can be expected to be more deeply disturbing than overcrowding or even restricting access to mates ... the natural form of food competition of chimpanzees is *indirect*, the separate simultaneous seeking of the same resources. ... [The new] situation—being aware that a desired food is present but not obtainable except at unpredictable intervals and through direct competition—is abnormal. ... Deliberate irregularity in feeding time forces the animals to remain tense and vigilant throughout the whole indeterminate waiting period, increasing the stress level of the whole experience" (33).

In discussing the behavior of chimpanzees before and after the introduction of changes in food provisioning, Power presents a controversial interpretation that challenges contemporary views on the naturalistic behaviors of chimpanzees. She ascribes many of the accepted features of chimpanzee social organization and behavior to the altered, unnatural social circumstances imposed on these chimpanzees by food provisioning and its consequences. The common view of chimpanzee communities stresses their competitiveness, proneness to intragroup aggression, and at times murderous territoriality. This view has been forged out of empirical observations strongly influenced by data drawn from Mahale-Gombe communities and by extrapolations from evolutionary theory. Power ascribes this view to observations on behavior following the altered, unnatural social circumstances imposed on these chimpanzees. She presents a much more sanguine, egalitarian, positive picture of the natural behavior of chimpanzees than many other researchers.

Power's controversial generalizations are somewhat weakened by the wave of recent studies of naturalistic communities in the Tai and Kibale forests, which are somewhat consistent with the model drawn from Gombe-Mahale, which points to aggression, competition, and greater territoriality. Moreover, her argument is heavily influenced by group and species selectionist ideas, which are generally not taken seriously by evolutionary biologists today. However, some of her comparisons and interpretations may be valid. Moreover, one does not have to agree with all of her conclusions to find value in her interpretation of social organization and behavior pathologies in chimpanzees. She presents an interesting perspective on features of what I view as psychopathology among nonhuman primates.

Power discusses everyday patrolling and scouting activities among chimpanzees, the nature of social organization, aspects of territoriality and competition, the conduct of peer relations, parent-offspring relations, mating behavior, alliance formations, and forms of intragroup and intergroup aggression and violence, including infanticide and cannibalism. She notes various forms of behavioral pathology, including scapegoating, sexual hostility and violence, and displaced anger and violence.

A dominant theme in Power's analysis involves social aggression and sexual behavior. She essentially challenges generalizations and interpretation of normal social relations in these areas: "It is the excessive apparently unprovoked attacks on the females and young that raise the first suspicions that this aggression-based dominance hierarchy is not the normal form of organization of chimpanzees, and that the alpha Gombe animal is a despot, using his power oppressively, rather than serving the group as a protective leader" (76). Here, Power is using a group selectionist basis to infer behavioral pathology among leaders, which is certainly suspect, but her point about excessive aggression being potentially maladaptive cannot be faulted: "The fact that these over aggressive animals could act in this way suggests a breakdown of the adapted control mechanisms (i.e., the structure of attention and application of sanctions) which . . . control deviants in an undisturbed group. It seems possible that their aggressive behavior toward the vulnerable females and young was frustration-induced scapegoating, and that these domineering males were simply aggressive, *uncontrolled* despots" (77). Although her interpretations are heavily anthropomorphic, they point to behaviors that one could infer are pathological viewed in terms of social organization and cooperation.

Power is equally concerned about the effects of social aggression on infant and adolescent chimpanzees: "Unlike the gentle, affectionate (generalized) caretaking response from each and every member of the wild society, the infant and juvenile chimpanzees at Gombe (post 1965) experience an uneven and often unpredictable amount of tolerance—ranging from affectionate acceptance, to simple impatience, to rough aggressive play, to the threatening or murderous attack. The young, like the adult females and other venerable individuals, are often scapegoats for powerful males' frustration" (88–89).

Power indicates strongly that one form that behavioral pathologies can take involves physical injury including violent death of infants: "The causes of the infant deaths are not known, but . . . [referring to Goodall] female chimpanzees are quick to leap out of the way during the 'charging displays' of the males or other violent social activities. [Goodall] suggests that one cause of infant mortality may be injuries caused by falling from the mother. When a mother must run from the displaying males the situation is potentially very dangerous for a newborn offspring . . . as the very young chimpanzee infant is not able to grip securely to its mother's hair" (94).

Power claims that the altered circumstances of aggression that pose a danger to females and infants has changed the behavior of mothers, rendering them much more possessive and protective, essentially shielding the infants from social interactions with others that in natural, more normal groups would be adaptive for the infants. In this instance, then, social disorganization and psychopathology involving aggression are costly, since they produce maternal behavior that could hinder the development of prosocial behavior of infants.

Power takes issue with traditional views of natural behavior as indicated by the number of offspring produced. Her views on bad mothering and fathering—directly and indirectly produced by the altered social organization and heightened aggression/danger—constitute forms of psychopathology that can have long-term consequences on offspring and the group: "Adaptation is for behaviors that lead to reproductive success—not the number of times an individual mates, but the number of their offspring, raised to adulthood, that can pass their genes to a next generation. When a primate society is exposed to any kind of stress, the progeny are particularly vulnerable. Under stress, parental treatment of children is affected [negatively]. . . . When the young grow up, they tend to have their own parental behavior affected negatively and so the original stress can be passed on to a third or fourth generation" (95).

In discussing the reported infanticide and cannibalism at Mahale of recent years, Power counters the interpretation of contemporary researchers that such behaviors are not pathological, since they do not involve bizarre actions toward the corpses of the slain infants. Instead, she questions the normality of the behavior of mothers who showed tolerance toward adult male cannibals by letting their own children play with them, and she finds pathology in the cannibalism itself, since it cannot be said to promote inclusive fitness of the males concerned, as such behavior is claimed to do in other primate species: "The victimized Mahale mothers had both groomed and been groomed by, and also copulated with, the infant-killing males for several years before their infants were killed—and apparently also afterwards. . . . This whole sequence of behavior seems to me bizarre: that is, extraordinary, and involving striking incongruities, maladaptive and pathological. While the physical environment may have changed little, the social and emotive climate of the provisioned chimpanzee of Mahale seems as changed as that at Gombe, and essentially in the same negative fashion" (102–103).

In a similar vein, Power discusses a number of additional adverse consequences of changes in the system of provisioning at Gombe and Mahale. These include such things as persecution of vulnerable co-members, prolonged dependency and delayed maturation of infants and juveniles, negative aspects of weaning behavior involving abnormal behaviors and signs of depression. She describes changes in patterns of male retention in the group and female migration, with adverse social consequences. In addition, there are recorded abnormalities in the experiences and apprenticeship of adolescence. Changes in group organization, changes in foraging behavior, curtailment of the fission-fusion mode of group composition, competition with neighboring groups, and restriction of intergroup relations are described as affecting diet, changing activity levels, altering opportunities for relaxed play, and diminishing opportunities for the release of tension—all of which adversely affect variety and quantity of food, short-term and long-term health, including a needed spread of genetic material among larger breeding populations.

Consider Power's observations about how pathology and abnormality in behavior can result from changes in the social organization and social ecology of food provisioning:

> There is much irrationality in the recent direct competition among the Gombe and Mahale males. Their aggressive competition to gain dominance or alpha rank is irrational, in that [it does not yield selective advantage].... The competition is stressful, costly in terms of energy, and dominance rank does not correlate with copulation frequencies. The aggressive attempts of the males to deny other male group members sexual access to estrous females are irrational in that this too is stressful, and costly to all involved. Also usually futile ... in that, while the possessive male is engaged in driving off one competitor, other males grasp the opportunity to mate with "his" female. Recent attempts of the Gombe and Mahale males to patrol and maintain a closed home range are irrational in that a more efficient use of the type of habitat (patchy distribution of food) in which they live would be maintenance of open groups and foraging ranges.... There is considerable evidence that suggest that the Gombe—and perhaps the Mahale—groups are moving towards breakdown as functioning societies. (143–144)

Power culls some of the literature involving chimpanzee behavior and presents a controversial interpretation. She challenges the general consensus about the behavior of chimpanzees, as viewed from within standard and conventional evolutionary theory and judges much of their normal, adaptive behavior as maladaptive and pathological. She ascribes this negative behavior to intervention by humans and alteration of ordinary routines, particularly for getting food. Her argument can be faulted on a number of grounds (Kemper 1992; McGrew 1991;

White 1993; but see Cohen 1993). However, Power highlights interesting examples of at least psychopathology-like behaviors in nonhuman groups and provides a basis for qualifying such behaviors as strictly pathological.

Questions about how best to make sense of behavior in primates and humans, the quandaries and ambiguities of a clinical compared to an evolutionary point of view were reviewed earlier. In general, environmental changes causing strain and stress should be assumed to release behavior that may or may not necessarily have had an adaptive basis in the EEA. Furthermore, Power's formulation for the effects of social stress on infants—for example, as a consequence of maternal harassment, distraction, and outright attacks—are well taken and offer a compelling picture of the subtle effects across generations of circumstances of social disorganization and conflict. Studies by Suomi (1978, 1996) and Schneider (1992a, 1992b), involving rhesus monkeys offer some support for Power's conjectures regarding the nature of psychopathology in chimpanzees and perhaps its genesis in communities of prehominid and hominid groups.

The Laboratory Manufacture of Nonhuman Psychopathology

The first part of this chapter reviewed the possible naturalistic patterns of social and psychological behavior in nonhuman primates that resemble human psychopathologies. In fact, these behaviors are sometimes formulated by ethologists in terms of psychopathology. Some of these nonhuman primates, such as chimpanzees, are the surviving relatives of the last common ancestor of Homo sapiens and the pongid line, and all are high on the evolutionary line. The fact that they can display under feral conditions behavior resembling human psychopathology also supports the validity of claiming that psychopathology had its origins during human biological evolution. This chapter presents an additional body of evidence to support this general claim. This involves the actual construction of animal models of psychopathology, some involving nonhuman primates.

The quest for animal models of psychopathology has a long history in experimental psychology and psychiatry. It is the basis of an important tradition of contemporary clinical studies of psychopathology (McKinney and Bunney 1969). Of course, the research does not necessarily involve higher animal forms. Jeffrey Gray's monograph (1987) on fear and anxiety, as an example, discusses many experiments conducted on rodents; and observations on the behavior of rats and mice up to baboons subjected to various experimental procedures have been used to help elucidate human psychiatric disturbances (Hofer 1984, 1987, 1995; Sapolsky 1987, 1990a, 1990b, 1998).

Many observations of abnormal behavior of monkeys and chimpanzees are the result of inducing chemical-physiological lesions and neuroanatomic ones involving actual surgical ablation of local areas of the brain. With respect to the former, studying the effects of chronic methamphetamine intoxication on cats, rats, monkeys, and chimpanzees is a classic attempt to mimic paranoid and schizophrenic human pathologies. (See Ellinwood and Duarte-Escalante 1972; Ellison and Eison 1983; Kraemer 1985, 1986.) Experiments conducted by Friedman (1994) on five maternally reared, socially experienced, and behaviorally normal bonnet macaque *(M. Radiata)* males involving the effects of administering oral yohimbine are another example. This agent, which produces a syndrome of behavior termed acute endogenous distress, was shown to virtually halt activities requiring sustained attention and perceptual motor control, but not activities on similar but easier tasks. This suggests that the interference was definitely cognitive and did not depend on sheer motor disturbances. This chemical lesion study offers a good model of pathological anxiety in humans and is thus another example of the manufacture of psychopathology. Animal models of depression implicating the locus ceruleus have been reviewed by Weiss and Simson (1986).

Behavioral disturbances produced by neurosurgical ablation or chemical lesions generally involve induced physiological changes that are unlikely to be replicated under natural circumstances. Nevertheless, because they alter basic neuroanatomical structures and/or neurophysiological mechanisms, they offer a window through which one may observe the likely parameters of psychopathology in nonhuman primates—and hominids, by extension. In some instances, observations made during these studies offer tantalizing clues to how manufactured varieties of psychopathology might fare in natural settings. For example, lesions of the amygdala, orbital frontal cortex, and other limbic structures produce disturbances correlated with the animal's loss of social standing (Myers 1972; Raleigh and Steklis 1981) and with a failure to participate in or rejoin the group (Kling and Steklis 1976). Fuster (1980) reviews the behavioral effects of lesioning the prefrontal cortical system in primates. He summarizes the effects of surgical ablation and reversible lesions on sensory discrimination, motility patterns, performance on delays tasks, as well as instinctual and emotional behavior. In many instances, chemical lesions produce equally compromising deficits in the social behavior of animals (Redmond et al. 1971).

Keep in mind that the manufactured varieties of psychopathology produced in the laboratory are associated with a range of interferences in adaptive behavior: changes in motor/locomotion, foraging, feeding, self-grooming, and avoiding predators. The disturbances that are observed, then, could be a result of physical weakness due to inanition, dehydration, or intercurrent infection. In short, here one sees the apparently natural overlap between psychopathology and general sickness.

Just as there is scientific evidence of behavioral deficits or forms of psychopathology among nonhuman primates produced by direct lesions or interferences of cerebral mechanisms, much research in comparative psychology and psychiatry involves the behavioral effects of functional or developmental-experiential trauma (behavioral lesions) inflicted on chimpanzees, monkeys, and other animals. This variety of manufacture may not directly destroy or impair the functioning of the brain but produces equally devastating effects on social behavior. Studies of the effects of maternal separation and social isolation of infant monkeys and chimpanzees on their subsequent behavior are examples of this tradition (see Harlow and Mears 1979; Harlow and Novak 1973). A wide variety of disturbances in social behavior, parenting, and sexuality have been described. These results underscore the critical influence of social relationships and affectional-social communications more generally during development on the ontogenesis of normal, adaptive behavior.

Research on the behavioral consequences of social deprivation reinforces the importance of social intelligence in human evolution and its potential link to psychopathology (Humphrey 1976). On the one hand, the ultimate or evolutionary origins of psychopathology stem from the delicate integration established in the environment of evolutionary adaptedness between emotional bonding and communication and the construction of socially intelligent behavior routines. This integration is established by natural selection not only during infant development but also throughout the life span. On the other hand, if one explains psychopathology as caused by proximate, contemporary factors, one must consider specific developmental experiences that perturb the establishment of this same delicate integration of emotional bonding and communication required for the acquisition of socially intelligent, normal behavior.

In a fundamental sense, then, the theory of human evolution, like a host of other narratives of transformation, is the story of the important influence between emotional bonding, attachment, and communication and the construction of socially intelligent behavior (Atkinson and Zucker 1997; Carter, Lederhendler and Kirkpatrick 1997; Serban and Kling 1976). The origins and significance of psychopathology may be explained in light of events taking place in that story. Kraemer (1988, 1992) covers the effects of social perturbations during development on neurobiological mechanisms and processes that can culminate eventually in such pathological behavioral routines as fight/flight, searching/exploring, protest, and despair. Kraemer's cascade hypothesis about the mechanism of the induced changes in behavioral and brain-related routines during development clearly is consistent with the ideas just presented.

Characteristics of Grossly Abnormal Behavior in Nonhuman Primates

Goosen (1981) classifies breakdowns in the behavior of monkeys produced and studied under laboratory conditions involving social deprivation and isolation.

The conditions of these experiments are drastic and highly unnatural and lead to the manufacture of rather gross forms of psychopathology. Goosen classifies these as (1) stereotyped locomotion and gross rhythmic activity, such as walking in loops, somersaulting, twirling, rocking, hopping, and (2) self-directed and "bizarre" behavior, such as self-aggression, self-holding, and digit-sucking. He explains the first type of behavior: "Stereotyped locomotion largely results from seeking specific kinds of social contact (which however are not available) while avoiding other (available) forms. The evolution of this behavior might be related to the fact that, under normal conditions (when animals are not caged), walking is likely to bring the individual to a place where the form of social contact sought is found or where the animal's motivations are otherwise changed. Under deprived conditions, then, the individual's environment hardly changes, so that some animals apparently persistently continue seeking" (704). On the other hand, Goosen explains self-directed and bizarre behaviors as follows: "A number of abnormal activities can be regarded as idiosyncratically distorted ways of maintaining certain forms of social communication when there is no partner present. As such, one might say that some individuals 'interacted' with a self-invoked 'ghost-partner.' In other words, it might be that the abnormal activities were shaped by exteroceptive feedback from the animal's own actions insofar as the feedback resembled the normal social stimuli of which the animal had been deprived" (709).

Integrating work in primatology with concerns of veterinary science and comparative medicine, Mitchell and Clarke (1984) provide a useful summary of the development of social behavior of primates. They discuss a range of issues, such as prenatal, developmental, pubertal changes, and endocrine factors, based on research on feral populations and observations and experiments on primates in free-ranging but confined settings and in captive situations. They offer the following formulation of abnormal behavior of primates: "Use of the term 'abnormal' behavior assumes some knowledge of the concept of 'normal' behavior. Normality is best illustrated in feral conditions, and abnormal behavior is most often seen in artificial or captive conditions. Abnormality can be qualitative or quantitative in character" (42). To illustrate abnormality, they offer these examples: "Prolonged self-clasping, crouching, eye poking or shielding, and bizarre postures and movements are some of the many abnormal self-directed movements that may be seen in primates, particularly in captive primates" (42). Mitchell and Clarke continue: "The varieties of abnormal behavior as a result of impoverished captive rearing are enormous in the great apes. Rocking, self-sucking, wrist biting, and a large number of bizarre behaviors are seen in impoverished gorillas. Rocking, swaying, twirling, pivoting, digit sucking, lip contortions, head banging, eye poking , and great numbers of idiosyncratic and bizarre movements also occur in deprived chimpanzees. Such behavior declines to some extent when the chimpanzees are housed with wild-born individuals or in more free-ranging situations. . . . Much of the abnormal self-directed behavior of chimpanzees may be performed as a result of boredom, tension, or frustration rather than as a result of early social isolation" (43). Many other categories of abnormal behavior are included, some already cited by Goosen (1981), such as repetitive-stereotypies, and abnormalities of feeding, motor activity, sexual behavior, aggression, and parental behavior.

Capitanio (1986) presents a more elaborate classification scheme of abnormal behavior: (1) self-directed behaviors—self-clasping, rock/sway movements, self-orality and aggression; (2) repetitive motor stereotypies—pacing, twirling, somersaulting; (3) appetitive actions—ecoprophagy, polydipsia, hypo- and hyperphagy; (4) physical actions—hyper- and hypoactivity, frantic searching, withdrawal, reduction in explorative activities, suppression of social play; (5) abnormalities in parental behavior—neglect, abuse, infanticide; (6) abnormal forms and frequencies of sexual behavior—failure to achieve species-typical copulatory positions, bizarre autoerotic activities, failure to brace during intercourse; and (7) abnormal forms and frequencies of agonistic behavior—hyperaggressiveness and hyperfearfulness out of proportion to social stimuli.

The latter three domains of pathology are interpreted by Capitanio (1986) as "disorders of coordination and communication": "The basic motor patterns are present—the animals appear to be unable to coordinate the behaviors in a normal sequence; . . . the ability to coordinate complex sequences of behaviors probably occurs relatively early in life, and social play is probably an important mediator for the development of such skills. . . . Isolation rearing results in a maturational delay of affective behavior . . . [suggesting] . . . a strongly canalized 'program' exists that allows for the sequential development of affiliation, fear, and aggression. Isolation may produce its effects partially by delaying the occurrence of this sequence . . . and partially by producing deficits in an animal's ability to modulate its agonism by appropriately 'reading' and responding to another's signals" (441).

The principles proposed by Capitanio (1986) and summarized by Mitchell and Clarke (1984) are worth enumerating since they provide a useful background involving the use of nonhuman primate data with which to conjure up formulations of psychopathology in the hominid line. Mitchell and Clarke summarize pathology in primates as follows:

> 1. Pathologic behavior is not limited to animals whose early experiences include social deprivation. Significantly altering a primate's environment from that seen in the feral state can produce bizarre behavior. . . . Abnormal behaviors occasionally occur spontaneously, even in the wild. 2. The more *severe* the early social privation (e.g., closed chamber versus wire cage), the greater and the more obvious the

behavioral pathology. 3. The *longer* the period of privation, the more severe and persistent the abnormalities. 4. The *earlier* the period of privation, the more severe the deficits. 5. Some abnormalities can be reversed. Maternal care among abuser mothers becomes more appropriate with subsequent offspring. Increasing cage size (providing activity for bored primates) decreases stereotyped repetitive whole-body movements. "Infant therapists" have been used successfully on isolate-reared Rhesus monkeys. 6. Males of many species, especially species that are sexually dimorphic physically, seem to be more susceptible to the effects of privation. 7. The number of different abnormal and often stereotyped behaviors increases from monkeys to apes to humans, as well as from New World monkeys to Old World monkeys. Few reports exist showing sexual and agonistic pathology among prosimians or New World monkeys, although there are many reports of inadequate parenting in these species. 8. Most of the data and principles on [sic] abnormal behavior have been derived from work on Old World monkeys, and mostly on Rhesus monkeys. Caution should be taken in generalizing to less well-known species. (47–48)

Scientists have been able to manufacture psychopathologies by strongly manipulating an organism's normal access to social stimulation from its mother and conspecifics during development. Many of the psychopathologies are persistent, and the timing and extent of separation or isolation during various developmental periods correlates roughly with their severity and duration. In the short term, such psychopathologies are termed separation-related responses and the maternal or social deprivation syndrome. They are analyzed and interpreted as a protest response, involving agitation and hyperactivity, and a despair response. Both express the organism's behavioral reaction to being deprived of biologically necessary communication and stimulation from maternal, nurturing figures and conspecifics. Kraemer (1988) reviews the neurobiology of this syndrome of abnormal, pathological behavior in primates (manufactured forms of psychopathology). Coplan et al. (1992) studies the relationship between developmental and functional disturbances of behavior caused by social deprivation and behavior disturbances caused by so-called chemical lesions, that is, treatment with yohimbine. Higley, Linnoila, and Suomi (1994) review the general rationale for studies of this type.

Claims of Psychopathology in Nonhuman Primates

All of the preceding seemingly abnormal forms of primate behavior involve gross and highly visible syndromes of behavior. While they are linked socially to established interactions, social relations, and social structural considerations pertaining to the group as a whole, they are nonetheless patent deviations. They illustrate a qualitative dimension with respect to psychopathology. I will turn now to potentially more subtle forms of psychopathology and to the measurement of quantitative deviations. Like other varieties of psychopathology, the subtle, quantitative varieties can also be produced under laboratory conditions. However, they involve procedures less drastic than those of outright manufacture. Moreover, observations are made in settings with conspecifics that involve social interactions, making the behaviors seem more natural.

An example of this approach to the question of possible psychopathology among primates is illustrated in a study of rhesus monkeys born and raised in normal settings. Suomi (1978, 1996) examines heritable differences in behavioral reactivity and impulsiveness that have distinct consequences on development. The study reveals that subgroups of monkeys reared in wild conditions are highly reactive and impulsive. Individual monkeys so characterized show differences in exploratory behavior and susceptibility to stress and separation. Because they are highly impulsive, they are unable to moderate their behavior, which makes them socially inept and incompetent. Because some are shunned and become solitary, their ability to survive in the wild is severely compromised and they eventually perish. Experiments conducted on rhesus monkeys clearly show that these deviations in behavior are characteristic of individuals not maternally raised.

Actual experiments in which the environments of females are manipulated while pregnant or caring for their newborns and infants involve an interesting rationale. The infants are subsequently tested to assess their neuromotor coordination, exploratory behavior, temperament, pattern of interactions with group mates, and neurobiological response patterns, such as their production of stress hormones under different conditions. Researchers then compared quantitative ratings of experimental and control infants—those not exposed to the stress inherent in challenging the mother. The abnormality of infants of stressed mothers is thus demonstrated by the quantitative results of group differences on a variety of parameters analogous to those used in human studies. Since psychopathology is equated with differences in adaptive, coping, and/or socially compromised behavior—for example, impulsivity, social restraint, boldness—it is measured against the performance of group mates. When the ratings are quantified, they usually reveal statistically significant differences between the two groups. For example, infants or juveniles whose mothers were stressed, either behaviorally or hormonally, show activity profiles that reflect adaptive or maladaptive coping.

One experiment involved infant monkeys whose mothers were subjected to variable demands on feeding. In essence, experimenters made it difficult for mothers to obtain food regularly and forced them to work for it. As a result, the mothers were stressed and less able to provide enough food for their infants.

(Note that this environment is not unlike that encountered in the wild, where access to food is uncertain.) Such alterations in the social environment have been shown to have effects on the behavior of infants that are manifested late in life, well after the period of environmental perturbation. It is the presumed effect of the perturbed mother-infant bond that contributes to the stress that later produces abnormal behavior, and all of this correlated with consistent changes in neuroendocrine responses. Similarly, peer-reared infants who lose control over their environment demonstrate short-term changes in their behavioral responses to novel environments, as well as long-term differences in their behavior when administered an anxiogenic compound.

A logical extension of this body of research is to examine the effects of stress on pregnant females and then to test the effects on their newborns, infants, and juveniles. Mary L. Schneider (1992a, 1992b) and collaborators (Clarke et al. 1994; Clarke and Schneider 1993; Schneider and Coe 1993; Schneider, Coe and Lubach 1992), illustrate this tradition. Mild stress during pregnancy, such as being subjected to controlled noise for a ten-minute period five days a week for fifty-five days, contributed to lower birth weights, delayed self-feeding, increased distractibility, and less motor maturity in newborns. At six months, prenatally stressed infants demonstrated more disturbance in behavior and lower levels of exploratory motor behavior; half of them showed an abnormal response—falling asleep—during testing. At eighteen months, juveniles who were subjected to stressful conditions showed more abnormal social behavior and less normal social behavior than those in the control group. In other experiments, pregnant females were subjected to hormonal activation as opposed to actual stressful circumstances, and the hormonal responsiveness of their offspring was tested when they reached the age of 14.5 to 16 months. In all instances, the experiments produced abnormalities in the infants that could be expected from previous studies. A final procedure involved subjecting prenatally stressed eight-month-old infants whose mothers had been exposed to the regular application of noise and not the hormonal activation, to social separation and tests of their social behavior and brain biogenic amine levels. Mothers subjected to chronic unpredictable stress produced offspring who spent more time clinging to their surrogates, exploring, eating, and drinking. They also tended to cling and seek contact with their cage mates more than others, who showed more locomotion and social play with cage mates. Long-lasting effects on noradrenergic and dopaminergic activity were also demonstrated.

Research involving the recording of naturally occurring individual differences in coping and adaptive behavior and, more directly, the effect of prenatal stress constitutes a new chapter in the quest for animal models of psychopathology. Rather than manufacturing models or versions of psychiatric disturbances per se, the experiments produce individuals whose social coping skills are nonetheless compromised, if not impaired, and whose behavior is analogous to varieties of psychopathology. Such impairments have long-lasting effects. The versions of impaired adaptation produced can be likened to the kinds of stresses feral mothers and infants are regularly exposed to in the social environments of wild populations. These experiments enable us to conjure up the role that psychopathology, as reflected in impaired social behaviors and associated neurobiological response patterns, might play in the wild. This raises the question of what role such varieties of psychopathology may have played during the evolution of our nonhuman primate ancestors.

Experiments that study the social behavior of rhesus monkeys imply that behavior resembling psychopathology and/or behavioral maladaptation are necessarily entangled with how individuals ordinarily function, perform, and adapt to their social circumstances. What one may otherwise regard as individual differences in the more or less normal aspects of rhesus monkey social behavior can be explained in terms of concepts pertaining to the study of psychopathology. Behavioral pathology and, indeed, relative behavioral health or competence is another way of measuring differences in the temperament and social behavior of individual rhesus monkeys.

Studying the analogues of human forms of psychopathology in a wild community of primates, which may resemble that of very early hominids, based on quantitative differences, could account for individual proclivities in affiliative behavior, aggressiveness, exploratory activity, success at mating, and no doubt the ability to cope with environmental hazards and predators. This means that psychopathology examined in these more subtle quantitative terms is rooted in behavioral normality and competence, and vice versa, and that the two cannot be differentiated without great difficulty. Outliers at the extremes of a continuum of social behavior might possibly be labeled abnormal or pathological, but within the continuum pathology is inextricable from normality.

This quandary about the relativity of psychopathology when behavior is examined from a quantitative standpoint also applies to a more qualitative approach. What an observer may term pathological with respect to a qualitatively striking alteration of behavior can be understood as natural, if not normal, given a change in ecological conditions (Kummer 1995).

- That psychopathology has an evolutionary basis suggests not only that it was found in hominid groups but may have also been a feature of the last common ancestor. This raises the question of its existence in the higher primates, our earliest living ancestors.
- Hans Kummer's observations on the behavior of baboons under natural and captive conditions led him to raise questions about animal analogues if not varieties of psychopathology.

- Jane Goodall studied chimpanzees in natural settings and got to know them well, and her assessments of the behaviors and personalities of some of them pointedly suggested abnormalities, even different varieties of psychopathology.
- Alfonso Troisi's observations and experiments of monkeys living under comfortable conditions of captivity uncovered abnormal and aberrant behaviors that were peculiar and specific to the individual, also suggesting natural varieties of psychopathology.
- Experiments involving selective brain lesions, toxic chemical agents, and interferences of normal development and attachment have provided animal models of psychiatric disorders indicating that higher primates are vulnerable to psychopathology.
- It is reasonable to conclude that even before the pongid/hominid split, individuals were not only vulnerable to varieties of psychopathology but under stressful circumstances came to manifest them, supporting the idea of psychopathology in the hominid line.

Chapter 8

Responses to Psychopathology in Nonhuman Primates

ORIENTATIONS OF BEHAVIOR researchers interested in evolutionary biology have varied in their approach to the question of psychopathology in animal and particularly nonhuman primate societies. Among some there appears to be a general reluctance to using the concept, perhaps because its special connection to clinical and cultural concerns endows psychopathology with special human values such as rendering a claim of such conditions applicable to animals inappropriate and anthropomorphic. On the other hand, some researchers interested in better understanding human varieties of psychopathology for theoretical and especially pragmatic, treatment-related reasons embrace the idea of laboratory-produced models and, more important, that of naturally occurring varieties of psychopathology.

This chapter analyzes in further detail themes and questions that become apparent when the concept of psychopathology as applied to nonhuman communities is examined critically. The principal focus is on the consequences and social responses of group mates to manifestations of psychopathology in a member or victim. Some of the issues discussed in earlier chapters from a general, theoretical standpoint—for example, differences between cultural/clinical engaged compared to evolutionary/detached viewpoints—are given more focused attention and are examined from a somewhat different standpoint.

Social Spaces Emplacing Psychopathology in Animal Communities

In visualizing psychopathology in the social behavioral space of hominid communities, two separate but interrelated factors need emphasis. These are, first, its behavioral manifestations, the effects it produces on the victim, and, second, the social responses it evokes from group mates. The first factor has received primary attention in the preceding chapter. Here emphasis is given to the second factor. In order to elucidate this topic, it will be necessary to give some further attention to characteristics of psychopathology because the two factors, manifestations and responses, are not only connected but interactive.

The idea of evolutionary residuals has been used to describe and explain aspects of human behavior that are part of the phylogenetic history of primates generally and hominids in particular. It designates the mechanisms, algorithms, and/or behavior systems and structures that were ultimately caused, naturally designed in EEA, and are responsible for behavior routines. The purpose of the mechanisms is the realization of biological goals of survival, reproduction, and associated social strategies of behavior involving kin and nonkin. How this is accomplished through behavior is explained by drawing on central concepts and traditions in evolutionary biology; for example, optimality and life history theory.

In mammals and nonhuman primates, evolutionary residuals constitute complex psychobiological response tendencies and patterns. They are determinative of social and psychological behavior. In primate species that are highly social, biological goals—termed biosocial for this reason—are structured in relationships that members have with group mates that pertain to such things as foraging, establishment and maintenance of rank, avoidance of and protection from predators, and mating and parenting. During later phases of human biological evolution, with the emergence of language, cognition, and culture, evolutionary residuals continue to operate as important determinants of social and psychological behavior. In this context, the degree of their influence on behavior, in other words, whether, and if so, how, under what conditions, and to what extent, cultural factors influence evolutionary residuals—for example, override, neutralize, oppose, complement—is a matter of considerable controversy and contestation (Alexander 1987a, 1987b, 1989; Durham 1991; Lumsden and Wilson 1981; E. O. Wilson 1975).

Structures of behavior geared to biosocial goals may be conceptualized as innate drives or instincts, terms no longer used in a technical sense. Essentially, the respective mechanisms are responsive to social signals that originate in the external world and impinge on and influence the biological state of the individual. When activated from a neuropsychological and neurophysiological standpoint, structures of behavior shift information processing to more or less automatic, stereotypic, and segregated forms of functioning (McGuire and Troisi 1987; Troisi and McGuire 1998). Besides influencing internal states toward optimal, regulated or atypical, dysregulated modes of functioning, the mechanisms also propel animals in their activities to reach biological goals, resulting in something like the mindless chains of behavior described by Richards (1987, 1989a, 1989b).

In highly social and socially intelligent primates, evolutionary imperatives are enacted in complex relationships with group mates. These involve conflicts of interest, competition for rank and status, territoriality, mating and parenting, and the like. The relationships are mediated through social displays and communicative signals that are part of the reproductive and survival strategies of the species. A strategy here may be conceptualized as a rule that specifies for the individual something like "in the event of stimulus S, do or implement behavior B." Social strategies are more complex since the activities, feelings, and intentions of group mates constitute the relevant stimuli, and the goal of a social display and strategy may be to influence such psychological states.

Animals are tightly locked into the imperatives of specific biosocial goals, as these emanate from stimuli and signals from the environment and are carried out in complex social relationships. The social fabric of a group essentially comprises communicative signals, displays, activities, and exchanges and social relationships in the service of biosocial goals. This fabric of behavior is the medium in terms of which varieties of psychopathology are configured and enacted. Given such patterns of social behavior, one can ask how they might be affected by and/or come to embody manifestations of psychopathology and how group mates are likely to respond in such a context. A basic question is whether the individual can survive on its own and, if so, in what ways the psychopathology will impact on and alter ongoing social activities that make up normal, everyday routines in the group.

Given this setting of primate, hominid behavior, many factors need to be taken into consideration in visualizing a possible scenario of psychopathology. First, of course, are the various types of normal, biologically determined exchanges that make up behavior seen in relation to biosocial goals. They include relations pertaining to social rank, motivation, aggression, reciprocation, and mating and reproduction as classically set forth by Hinde (1974). Communicative signals and exchanges involving vocalization, facial display, and bodily posture in terms of which an individual relates to its group mates constitute the template or blueprint of normality in the group. It is with respect to this pattern of behavior that any deviation or abnormality such as a psychopathology might be calibrated. This applies to nonhuman primates and was a feature of hominid behavior during phases of human biological evolution.

In visualizing the relationship between normal behavior and psychopathology, two useful concepts to keep in mind are that of a defect and that of handicap (Berkson 1970, 1973, 1974, 1976, 1977; Nesse and Williams 1994). A *defect* refers to an abnormality of some sort, and not merely in a statistical sense as unusual or rare. Rather, it implies a characteristic of appearance, morphology, function, or behavior that can potentially interfere with an individual's behavior and its attempts to adapt to its circumstances. A defect has an obvious relationship to the idea of harmful dysfunction. A *handicap* refers to whether an attribute of the individual, including in particular a defect, actually does prove a disadvantage. Thus, the latter concept brings into play the environment of the individual, specifically, the extent to which it influences functioning such as to socially select against the individual as a consequence of that attribute or defect. The environment, of course, has a physical component and a social one, that is, involving group mates, each of which may differentially select against the individual. In short, to claim that an individual has a handicap is to make an ecological statement that brings into the equation of adaptation the individual and its whole environment. A defect need not prove a handicap, and whether or not it does depends on its character in relation to the individual's behaviors with respect to the physical and social environment. The latter differ across populations of a species and, within any one population, as a function of any number of possible considerations. The effects that a defect has, then, should not be regarded as necessarily a fixed, static attribute but a dynamic one depending on the individual's overall environment. With respect to the focus pursued here, this means that characteristics of the physical habitat, the group, and the actual social consequences of and responses to an individual's psychopathology need to be kept in mind in assessing its biological consequences. Furthermore, in theory, manifestations of any one of the many possible varieties of psychopathology need to be given attention.

In addition to the architecture of normal behavior and the social effects of possible defects, there are other factors that need to be considered in conceptualizing animal varieties of psychopathology. These include characteristics of the ecology—pressure from predators, availability of food sources; characteristics of the group—its size, density, and stability; characteristics of the victim of psychopathology—age, sex, rank, kinship status; and characteristics of the psychopathology itself. All of these factors can be expected to affect the enactment and consequences of a psychopathology in predictable ways. For example, a harsh environment will impose more stringent requirements on the victim, infants will resort to strategies and display signals different from those of adults, males will be affected by different considerations than females, as will also individuals of different rank and kinship status.

To pursue the analysis further, it is useful to give attention to manifestations or varieties of psychopathology in light of the discussion of defects and handicaps. A clinician interviewing a potential patient may see and feel such things as anxiety, depression, antisocial tendencies, impaired cognition, suspicious paranoid behavior, and undue somatization. Indeed, the patient may report associated symptoms, and family members of the victim of psychopathology may have already noted characteristics of the disorder. But in an animal community there exist no real reports of symptoms; and group mates are unlikely to have a conceptual category such as psychopathology on the basis of which they may spot signs of a potential victim.

To diagnose a variety of psychopathology, then, one must take into consideration its role in the social economy of behavior of the individual; for example, the relative adaptiveness, or degree of handicap, resulting from its manifestations. Obviously the phase of evolution marked by the emergence of language, cognition and culture of the animal constitutes a determinative factor. Elsewhere I have given consideration to aspects of sickness and healing that bear on the question of how chimpanzees in particular are likely to show and manifest conditions of disease, injury, and pathology (Fábrega 1974, 1975, 1997). Some of the same generalizations would apply to monkeys, although classic studies of how they see their social world contain little information on this topic (Cheney and Seyfarth 1990; but see the work of Troisi and Maestripieri). From a psychobiological standpoint, to the extent that a variety of psychopathology matches conditions of sickness it can draw on social routines, strategies, and displays that serve to communicate suffering, debility, and disability and thereby elicit a measure of support and toleration in the light of conditions discussed earlier.

Responses of Monkeys to Group Mates with Defects

One way of approaching an analysis of the manifestations of and responses to psychopathology is to conceptualize it as a defect (Nesse and Williams 1994). Gershon Berkson (1970, 1973, 1974, 1976, 1977) has discussed the question of behavior of and social responses to animals with defects. One of his areas of interest was on the use of artificially produced lesions in monkeys as a basis for clarifying what characteristics of an animal's behavior and morphology its environment selects for. He pointed out that environments permit rather dramatic variability in structure and behavior and that by means of selected lesions of animals, the defects of which can be observed and inferred, it is possible to describe the extent to which various environments select for particular defects by following the animal's adjustment through time. Studying how the animal fared would lead to a clarification of the characteristics of the individual that are or are not necessary for survival at a particular time and place.

Berkson conducted a series of studies using macaque monkeys that involved observing the behavior of and responses to infants who had been surgically operated in the eye. The operation produced a lesion in the pupils such as to create scar tissue that produced visual defects. Studies conducted on caged animals established that a searching optical nystagmus was prevalent and the animal groped for food on the ground. Otherwise, it appeared normally responsive to stimuli carried in other sense modalities. In the natural environment the infant's locomotion was noted to be clumsy and cautious, and it almost never jumped.

Infant monkeys were first studied in the heavy forest of the natural but complex environment of a sixty-acre island in Thailand. The habitat in question was potentially selecting for the visual defect since a large monitor lizard lived there and it featured long, arid periods during which no fresh water was available, requiring the animals to actively search. A few of the animals managed to survive for up to seven months. They disappeared under ambiguous circumstances during a period when no freshwater sources were available. However, the thickness of the forest precluded a determination of the circumstances, particularly whether predation by the lizard was a factor. Similarly, the adult females were breeding at the time, and a reduction of maternal care may have been a contributing factor. Another series of studies was conducted in an island environment southwest of Puerto Rico that had comparatively few predators and ample food and water. Here the animals survived a considerably longer time and thus permitted more observation.

Survival of the infant in the wild was a function of the special attention provided not only by their mothers but by other members of the group as well. Berkson reported the following: "mother monkeys gave compensatory care to blind infants by retrieving and carrying them at an age when control infants stayed with their group by themselves. In that sense, the monkeys responded to the blind infants as they would have to younger animals" (Berkson 1974, 235). Berkson, then, explained the responses of mothers and group members more generally in terms of patterns of adult-infant relationships. Distress signals and other communicative behaviors of normal infants resembled those emitted by animals with lesions. Such signals ordinarily elicited care and help. Berkson conjectured that a balance between the strength of such distress signals and the strength of reciprocal social behavior tendencies on the part of group mates—for example, maternal and general supportive behavior tendencies of adults—determined how animals would actually fare in the wild with a defect. "Undoubtedly, the general level of aggression within the group, interspecific differences in tolerance of deviation, familiarity of the defective individual, and also the character and severity of his defect are other factors important in his status" (Berkson 1974, 236). When mothers withdrew from the group to consort with males they would sometimes take their blind babies with them but often left them in the mangrove. "Sighted babies might follow their mothers but the blind ones could not do so and stayed in the forest or on open ground, often calling hoarsely. The blind babies were never left completely alone, however. It is remarkable that there was always another animal of the group near them. In addition, two individuals regularly stayed with the blind infants during this time, retrieved them when they were in trouble, and carried them if the group moved (Berkson 1974, 242).

Researchers have pointed out that fractures and lesions that injure body parts are not infrequently observed in animals that otherwise manage to hold on in their group. Others have reported that in some instances members of communities are able to maintain their rank despite a physical disability (reviewed in Berkson 1974, 1977). In these instances of survival with functionality there is essentially no sharp break in demeanor and attitude, no clearly profiled deviance or abnormality of the individual in question, no major disruption in patterns of social behavior, and consequently little apparent cost incurred in area of social relations by the injured animal. The behavior of blind infants provides a lucid example of a special defect and disability, for it only minimally impacts on the structure and organization of behavior. Berkson points out that the symptoms of a monkey with a visual lesion consisted of subtle modifications of routines of social behavior ordinarily normal for members of the species at that age and that these must have been sufficient to elicit toleration and support. His reports indicated that there was little tendency for the group to exclude the handicapped so long as they were socially responsive. He indicated that auditory and tactual modes maintained social affinity, but there was no indication that olfactory discrimination between animals was particularly sensitive.

Berkson reasoned that a similar visual defect in a mature animal would probably result in behavior having a more dramatic impact on the individual and on the group: awkwardness, social ineptness, and a gross impairment in implementing adult social strategies, not to say basic functional routines related to survival in the habitat. However, results would still depend on the severity of the blindness, the rank of the victim, its system of alliances, and characteristics of the group at the time. The behavior could still not be described as abnormal or deviant in a morphological sense, although, to the extent social signals could not be identified correctly from a functional standpoint leading to inappropriate behavior, a contextual abnormality could be said to prevail (Erwin and Deni 1979; Goosen 1981).

The effects of a variety of psychopathology on the individual and the response it elicits from group mates in particular may thus be comprehended from the standpoint of what its manifestations would add to or subtract from the normal routines and strategies of behavior of the individual given its circumstances. Normal routines and strategies involve the complex displays and signals that in a socially communicative way function as evolutionary residuals to achieve biosocial goals. The developmental phase of the individual would be a critical factor, as would other personal characteristics noted above. Bilateral lesions of the amygdala, an important subcortical structure involved in emotion and the regulation of social behavior as well as memory, produces gross abnormality in social behavior. Animals subjected to this lesion do not survive very long, although one did. Observation of amygdalectomized monkeys showed changes in the social behavior of group mates, mainly attack and avoidance. However, reports did not provide details—for example, the signals and displays—or allow clarification of how and why group mates responded in the way they did (Dicks et al. 1969; Kling et al. 1970).

De Waal and collaborators (1995) reported on the behavior and responses of group mates to a defect/handicap produced by autosomal trisomy in a community of macaque monkeys. They describe at length the behaviors of an infant named Azalea, whose condition qualified as a handicap and as a naturally occurring variety of psychopathology. Azalea lived thirty months, and during her first eighteen months the rate of grooming she received was similar to that of her peers. However, following this, the scores for grooming she received substantially exceeded those of her peers. As Azalea increased in age, her mother's approach rate toward her began to exceed that of all other mother-daughter pairs, a possible consequence of her mother's sense of awareness of Azalea's dependence. "Azalea was born into the highest ranking matriline, which offered strong protection against potential aggression. Azalea's mother never showed signs of rejecting her, although she also was not overly interested in her daughter [at least initially, see above]. Azalea's 1- and 2-year older sisters, on the other hand, did pay extra attention to her; they carried Azalea around well beyond the normal age for such sisterly care and protected her against other monkeys. If others did things to Azalea to which any other rhesus infant would object, such as plucking out hair during grooming, a sister would often interrupt the activity even though Azalea herself had not uttered the slightest protest" (de Waal et al. 1995, 381).

In the latter part of Azalea's life, during a time when cohorts played actively with and aggressed against each other, peers seemed to lose interest in her as all measures of behavior fell below average. "At this age, negative interaction sharply increased among the normal peers themselves, probably reflecting the beginning of the process of rank establishment. Azalea seemed to be left out of this process" (de Waal ct al. 1995, 382). Azalea showed a lack of understanding of the dominance hierarchy that was not attributable to knowledge of her family's place in the group.

The most striking finding of this study, then, was that a monkey who has a variety of psychopathology featured by mental retardation and who is born into a highly structured society of an even aggressive species can nevertheless be fully accepted, tolerated, and even integrated. De Waal and collaborators make explicit that until the end of her life they could not detect evidence that suggested even the slightest indication of hostile rejection. A lack of peripheralization was a function of the behavior of her mother and other kin. The report in question (see also that of Fedigan and Fedigan 1977, involving a case of an infant with cerebral palsy) indicates increased protection and high social tolerance. However, what the response of the group would have been to a comparable defect in an older individual is not clear. In line with the rationale outlined earlier, one can assume various responses, depending on type of defect, age-sex grading, group, environment, and the like. In this light, the following comment by de Waal and collaborators is interesting: "Possibly with increasing age Azalea would have become an outsider to her group because of an inability to learn her rank relative to others" (de Waal et al. 1995, 388).

Research reported by Troisi and collaborators involving the anxiety-related behaviors of abusive mothers, another seemingly naturally occurring variety of psychopathology in monkeys, indicates that changes in behavior of group members resulted in relation to the abuse. Here, in contrast to the behavior of blind monkeys, and similar to those whose amygdala had been destroyed, the behavior of the mothers by definition was regarded as abnormal on clinical and evolutionary grounds. However, it proved difficult for Troisi and collaborators to classify the abuse syndrome as aggressive using morphological criteria. In fact, categories of play seemed to better fit the abusive behavior patterns, leading the researchers to question ordinary interpretations of this category they did not pursue. Similarly, while indicating that group responses were discernible, Troisi and collaborators unfortunately did not provide further details.

Dario Maestripieri's studies were directed at examining the character of abusive behavior patterns mothers directed toward their infants. He further clarified and validated inferences of Troisi and collaborators regarding the possibly pathological character of maternal infant abuse:

> infant abuse seems to be a stressful experience for the whole social group. Adult females other than the mother are responsive to the abused infants' cries and often attempt to approach the infants and take them away from their mothers. Abusive mothers, however, usually avoid such approaches. On a few occasions, I also observed interventions from the alpha male (the most dominant male in the group). He would approach the abusive mother and mount her shortly after the abuse. This is a behavior shown by males in tense situations and it may serve to reduce the likelihood of a conflict. It is hard to tell what other individuals understand about infant abuse and whether they experience something close to empathy. A simple explanation not involving complex emotional states or cognitive processes is that other individuals have a general tendency to respond to infant distress calls with nurturing behavior. (Maestripieri, personal communication, May 19, 2000).

Explanations and Implications of Care Giving Responses in Monkeys

This is an appropriate place to bring up the matter of how evolutionary biologists explain the responses of group mates to a victim of psychopathology in monkeys. In general, they draw on two related concepts pertaining to the social behavior of the species. From the standpoint of ultimate causation, the concept of inclusive fitness is used to explain behaviors directed at promoting the well-being of kin; and in the case of behavior of nonkin, the concept of reciprocal altruism. From a purely descriptive standpoint, researchers invoke general attributes of social behavior of the species: members form groups that are highly affiliated socially and their repertoire includes communicative behaviors that promote individual adaptation by means of altruism and mutualism. For example, Berkson emphasized natural maternal behavior elicited by distress calls of infants, and both he and Maestripieri suggest that, in the case of immature and adult members of the group who are not kin, nurturing behavior that is natural in the species is responsible.

Troisi and McGuire (1990) have emphasized a variation of the idea of natural social behaviors of a species: invalid care. They define this as behaviors and circumstances "consisting of changes in feeding rank order, resting behavior and grooming behavior (these all being to the advantage of the sick animal)" (Troisi and McGuire 1990, 973). These types of behaviors and changes are brought into play as responses of group mates to a victim's manifestations of sickness, including, presumably, psychopathology.

Invalid care is a component of altruism and resembles and would probably include general care giving or epimeletic behavior, which I have discussed in a formulation about the evolution of sickness and healing (see Fábrega 1974, 1975, 1997). It has been given a broad interpretation by Troisi and McGuire (1990) in their discussion of the possible role of deception in human somatizing disorders like hypochondriasis and malingering. They first review cases that seemed to involve the dissimulation of disease in animals, one of which was contained in a report by Caine and Reite (1983) on infant abuse among pit-tailed macaques. Caine and Reite pointed out that in their series, most mothers who abused their infants were behaviorally normal. However, a case of hysterical paralysis took place in a mother who was judged as behaviorally deviant: when placed in her group she moved about as though injured; when examined she showed no evidence of pathology; and when housed alone subsequently, the hysterical behavior suggesting injury and suffering disappeared. Troisi and McGuire used this case to suggest that such patterns of sickness behavior may represent evidence of medical deception or disease dissimulation, the purpose of which is to elicit invalid care and thus obtain benefits. This illustrates the close connection mentioned earlier between the manifestations of psychopathology in a victim and responses to this of group mates, considered in relation to other environmental circumstances, of course.

Troisi and McGuire, then, suggest that as a result of an innate or learned capacity to deliver invalid care—in reality, the whole complex of natural care giving behavior discussed above—animals may have been selected to use manifestations of sickness, and may even exaggerate them, as a basis for deceiving group mates and obtaining benefits. An evolutionary escalation of this strategy is the selection of a capacity for self-deception, by means of which animals are better able to deceive group mates into delivering invalid care: If animals believe and feel that they are really sick, then their strategy to obtain favors and benefits is likely to be more behaviorally compelling and, hence, more successful. In short, the idea of possible strategic benefits deriving from manifestations of sickness, including psychopathology, and ideas of other and self-deception in relation to these manifestations, are brought into the equation of animal responses to disease and psychopathology.

The material reviewed in this section and the previous one provides a frame of reference that one can use to evaluate responses to psychopathology in animal communities placed high in the line of evolution. Social responses to the same sign or symptom of psychopathology will depend on the environment in which the group lives. An animal's defect as reflected in behavior will tend to be more or less normal in various environments and group responses may also vary with the environment. Responses to a feature of psychopathology can be a result of its severity, its impact on normal social communication, and the extent to which the environment weighs, distinguishes, and hence selects between the behavior of the victim and that of other group members and modifies tolerance of the group to deviance. If groups are stable and socially affiliated, an individual with a particular variety of psychopathology will survive provided its manifestations do not handicap basic physiological, vegetative functions or sociability. Kinship bonds will be most important in maintaining a victim in a group, but relations with certain other members of the group will also be influential. An individual with a psychopathology will more than likely, but, again, depending on its manifestations, tend to have a subordinate status. However, this will be influenced by how much of a handicap the psychopathology poses, the degree of protection by dominant animals, and the degree to which the animals live in an environment that does not evoke much competition and hardship. Responses of animals to defects and related evidence of disease and psychopathology in a group mate include patterns and strategies of behavior that resemble sickness and healing. The latter may conceivably be part of a complex pattern of deception involving the self as well as group mates.

Studies conducted on monkeys provide a glimpse of how phenomena one can conceptualize as psychopathology manifests and is responded to in feral groups of comparatively distant relatives of Homo sapiens. Five points can be emphasized. First, the studies illustrate that the idiom that manifests a variety of psychopathology in animal groups consists of the seemingly normal social routines that play out evolutionary imperatives in relation to biosocial goals. Second, the diagnosis of an animal variety of psychopathology requires a consideration of how the functions of the biosocial goals and strategies of the animal are thwarted or undermined in the short run. In other words, while the routines of behavior may have been ultimately caused in relation to optimality and life history theory considerations bearing on reproduction and survival, it is how the routines affect adaptiveness in the current environment that is important (Troisi and McGuire 1998). Third, the social responses of group mates, how they perceive and respond to a psychopathology, are crucial because they can enable, by supporting or tolerating, or thwart, by exploiting or attacking and rejecting, the victim's efforts to meet biosocial goals. Fourth, the manifestations of and responses evoked by psychopathology in monkeys provide a snapshot of situations that bring into play evolutionary imperatives and constraints that are highly determinative. These situations can be assumed to constitute part of the phylogenetic package of hominids in the area of psychopathology and responses to victims. During later phases of human evolution, behavior and psychopathology come to acquire greater symbolic content and meaning as a function of the evolution of language, culture, and cognition. A contested question in evolutionary biology and social science is the extent to which evolutionary imperatives remain determinative of social and psychological behavior in human communities. Fifth, analyses of monkey varieties of psychopathology make apparent the differential impact that it can have on the whole life cycle of the individual depending on when it takes place.

How monkeys coped with visual defects, with the disorganizing effects of anxiety associated with parenting, or with social behavior deficits resulting from extirpation of a subcortical structure, is only part of the story of what impact these disturbances have. In the event blind infants reach adulthood, one can assume that the anxiety and distress caused by visual defects during a crucially important stage of development will have important implications for adaptiveness. Should abused infants live through their period of hardship, so will they continue to show psychopathology. It may also be the case that mothers who abuse may come to show sequela of their anxiety in other spheres of social activity during subsequent phases of adulthood. If amygdalectomized monkeys survive their ordeal in the short run, they may be assumed to continue to also show related deficits throughout their lives. Thus, the ontogenetic, developmental, and life cycle phase when a psychopathology takes place has implications not only for how it manifests and is dealt with by group mates in the short run but also its long-term impact on fitness. Finally, psychopathology and responses to victims constitute a situation that resembles sickness and healing and that raise complex questions of deception, counterdeception, and the dissimulation of disease, injury, and pathology.

Social and Psychological Behavior of Chimpanzees

Ethological knowledge about the social and psychological behavior of chimpanzees can provide an additional frame of reference or baseline for describing the changes in how psychopathology was configured and played out during human evolution. Because they are closer to humans, evolutionary residuals and imperatives inherent in their behavior including psychopathology are moderated in some way by evolved mental capacities—for example, involving cognition, social attribution, deception—that are assumed to be a consequence of their placement on the evolutionary ladder at stages removed from that of monkeys. It should be noted here that in their synthesis of primate cognition, Tomasello and Call (1997) emphasize similarities between chimpanzees and monkeys and draw a sharp line between them and human cognition as it manifests during the latter part of the first year of life and beyond. Yet, field studies of chimpanzees in particular consistently suggest the kinds of emotional behavior and personality attributes discussed here.

A measure of self-awareness and a more extended form of consciousness beyond the here and now (Damasio 1999; Donald 1991) is generally thought to describe social relationships among group mates of a community. This would suggest that greater social intelligence, cunning, flexibility, and subtlety would characterize their relationships with group mates. Still, from the individual standpoint, the behavior of chimpanzees varies with respect to obvious demographic variables such as age, stage of life, and gender. Researchers are more willing to explain chimpanzee behavior as a function of temperament, personality, and intelligence, which one may regard as rooted in genetic differences and developmental experiences. Finally, as in the case of monkeys, it also varies with respect to rank and social status, which devolve from the interplay of the preceding factors but also from the composition and history of the group.

Chimpanzees, who are often used as examples for the behavior of the earliest hominids, show psychological characteristics that reflect high levels of cognition and culture (de Waal 1989, 1996; Goodall 1986; Parker et al. 1994; Quiatt and Itani 1994; Russon et al. 1996; Wrangham et al. 1994). In reports of field observations, individual chimpanzees are described as confident, tentative, fearful, impulsive, happy, depressed, angry, antagonistic, and playful. Less frequently used descriptors are worried, unhappy, jealous, distrustful, avoidant, expansive, rejecting, or brooding. While the latter descriptors mainly refer to negative mental states, they also presuppose more complex states of social awareness and experience, indicating a level of mentation, consciousness, and emotional recognition that seems beyond what ordinary chimpanzees may be capable of experiencing.

Various groups of chimpanzees are organized differently, and each may have its own group-specific, historically contingent traditions for foraging, social exchange, leisure activity, and bonding relationships. These cultural traditions appear to reflect characteristics of the environment and are the result of simple trial and error as well as imitative and guided, if not actually pedagogical learning (Tomasello et al. 1993; Wrangham et al. 1994).

Chimpanzees seem well aware of these group relationships and how they can be manipulated in pursuit of their needs. More important, they appear to be capable of intentional behavior and cognizant of the emotional dispositions and intentions of group mates (Byrne 1995; Byrne and Whiten 1988; Whiten and Byrne 1997). For this reason, they are said to show the rudiments of what has been termed a theory of mind: they operate as social creatures with a seeming awareness that others have points of view, desires, intentions, needs, and more generally, states of being. Chimpanzees have been described as though they were capable of realizing that the self and others have different points of view; they may use this awareness to deceive group mates in the service of their own self-interest. As mentioned earlier, this is contested by Tomasello and Call (1997).

The close link between intention and social attribution in relation to behavior raises the question of how these capacities can be expected to influence not only the cause and expression of psychopathology but also how a victim is likely to be perceived by group mates. In the present context, it ties intrinsic features of psychopathology to extrinsic responses to victims, as discussed earlier. In a broader sense, the interconnection between causation, intention, expression, and interpretation of psychopathology in higher primates underscores that the origin of human psychopathology not only involves individual centered events (a capacity or vulnerability to develop different varieties of psychopathology) but also the connected perceptions and responses of group mates in social ecological contexts that vary and that can play influential roles.

Sickness and Healing among Chimpanzees?

Most highly social mammals and especially primates like monkeys and chimpanzees show a passive accommodation to the physical infirmities of group mates (de Waal 1996; Goodall 1986; Kummer 1995). For example, when one is injured, diseased, and walks in a slow and labored way, members of a group of foraging animals may slow their pace to allow the injured group mate to catch up. Some species might share food and in this way to nurse the sick member. We have also seen that several species of monkeys show this behavior but that it is directed toward handicapped infants. Presumably, distress-related signals and stimuli that are normal in the case of nondefective infants, and which elicit appropriate helping behavior from mothers and kin, simply persist longer and possibly are more compelling in the case of infants with defects.

Ethologists emphasize the striking subtlety of communicative behavior among chimpanzees such as to suggest a difference from monkeys. A greater sensitivity to the plight of group mates as well as a greater sense of compassion is posited. In the midst of intragroup hostility, pursuing a prey, or avoiding a predator, it is unlikely that social accommodations by chimpanzees of the type described would be manifest, however. In general, one can assume that when imperatives of survival and reproduction conflict with the social reciprocity and mutualism to help a weaker member of the group, for example, as in competition over scarce resources, food, or mates, that pure self-interest would win out. As Berkson suggested, it would be important to determine the degree to which infants, infirm adults, elderly group members, as well as pregnant female monkeys and chimpanzees are favored under these conditions.

The topic of how animals respond to disease or injury of group mates under different social conditions has not been thoroughly studied (see Fábrega 1974, 1975, 1997). Responses of different varieties of species to defects and handicaps of monkeys were reviewed earlier. These have been described as natural, caregiving behaviors of the species and explained in terms of mechanisms ultimately caused or naturally designed in EEA. Primatologists, especially those who study chimpanzees, have reported on social behavior in the context of disease and injury. As reviewed in detail elsewhere (Fábrega 1974, 1975, 1997), an awareness of others is reflected in chimpanzees' behavior toward those showing disease, injury, and bodily pathology. Most observations have involved how group mates behave toward individuals showing deformity, physical lesions, and handicaps—that is, visible evidence of pathology. Even in individuals with systemic diseases (hidden lesions), such as infection, the flow, subtlety, and organization of behavior is modified; for example, the individual is likely to be slowed down, to vocalize and communicate less, to grimace, to avoid social exchanges, to not aggress, and generally to show less interest and emotional expression in association with approaches by group mates, possibly even in situations involving competition over rank. In these instances, to the extent that group mates do not alter their behavior and especially if they show support and toleration, they could be said to be responding to physical and visible, that is, public, manifestations of a medical problem. Berkson has discussed pertinent literature of adult wild monkeys surviving lesions and fractures; presumably, the individual's social behavior and system of alliance was maintained, and whether rank was affected would have depended on the range of factors reviewed earlier.

Whether the diseased or injured chimpanzee is aware of being in an altered state is not clear. Chimpanzees are said to have some awareness of self and of others (Byrne 1995). Consequently, if an injured chimpanzee cries out in pain, desists from social activity, looks at a bleeding laceration, or limps and shows impairment, one is forced to wonder about its awareness of its altered condition.

One also wonders about the awareness of group mates regarding the injured or diseased individual (see Goodall 1986). Elsewhere, I have suggested that among chimpanzees the natural responses to disease, injury, and pathology, including behaviors one can interpret as indicating and communicating sickness, as well as healing or nursing others, suggest a psychological adaptation designed by natural selection (Fábrega 1997). In actuality, such behavior could be merely reflexive or, alternatively, conditionally learned, either in the case of the individual or group mates.

To suggest that chimpanzees might be capable of a full awareness of sickness and of a need for healing sounds like extravagant anthropomorphism. Yet self-awareness and awareness of others are not all-or-nothing, qualitatively distinct traits, but rather graded, quantitative ones, and we are dealing with levels or degrees of self-awareness, other awareness, and indeed on content and meaning as it pertains to ongoing group activities (Troisi and McGuire 1990). The writings of Damasio (1994, 1999) on consciousness emphasize differences of its scope in relation to its phylogeny: a major difference between core consciousness (awareness of the here and now) and core self, on the one hand, and extended consciousness and an autobiographical sense of self, on the other. Many gradations between these two poles of a presumed continuum are suggested. Consequently, awareness of sickness and healing could also be graded phenomena. Whether or not chimpanzees or monkeys, or both, manifest such awareness is contested.

In the case of a group mate's disease and injury, chimpanzees show a distinctive array of social behavior that can be classified under the general rubric of not only support and toleration but also compassion (de Waal 1996; Goodall 1986). Again, most responses apply to circumstances when the group is not being challenged or threatened by predators and in the absence of intragroup competition, rivalry, and hostilities over rank or mating. Moreover, such behavior occurs most frequently with respect to maternal figures, offspring, siblings, and possibly close friends. In these relatively relaxed circumstances, an awareness of the sickness of others seems to evoke displays of succor, empathy, sympathy, cleaning and licking of wounds, feeding, and a generally caring attitude. In short, chimpanzees—and some types of monkeys, as clearly reported by Berkson and de Waal and collaborators—offer assistance to the physically ill, infirm, and handicapped and come to show tolerance toward them.

Whether or not the behavior of chimpanzees reviewed here qualifies as sickness and healing in more than a purely descriptive sense is contestable. To be sure, it could represent a naturally designed mechanism, adaptation, or algorithm, the purpose of which is to cope with injury, pathology, and disease. An argument for this is presented in Fábrega (1997) and will not be reviewed further here. On the other hand, the pattern of the behavior in question can be explained in terms of general concepts pertaining to sociality. Behavior resembling sickness and healing could simply be a part of a broader mechanism of sociality that includes care giving, protection, toleration, support, and general accommodation to a group mate's infirmity.

Troisi and McGuire (1990) used the idea of invalid care, and their formulation of its consequences and implications overlap in part with the idea that sickness and healing could constitute two sides of a singular, that is, special, naturally designed adaptation (Fábrega 1974, 1975, 1997). As described earlier, Troisi and McGuire point out that disease, injury, and pathology (and by extension, psychopathology) can in some instances naturally elicit invalid care from group mates. In fact, they suggest that a capacity for invalid care is found in higher primates and that it may willfully or reflexively, that is, unconsciously, come into play when individuals are in a position to obtain benefits, advantages otherwise denied them by appearing diseased or injured.

Troisi and McGuire use the case of a chimpanzee described by de Waal (1982), an example of conscious, premeditated action as compared to instinctive and impulsive animal behavior, to formulate an interpretation of the implications of invalid care. This is what was reported by de Waal:

> Yeroen hurts his hand during a fight with Nikkie. Although it is not a deep wound, we originally think that it is troubling him quite a bit, because he is limping. . . . Yeroen limps only when in the vicinity; [he] walks past the sitting Nikkie from a point in front of him to a point behind him and the whole time Yeroen is in Nikkie's field of vision he hobbles pitifully, but once he has passed Nikkie his behaviour changes and he walks normally again. For nearly a week Yeroen's movement is affected in this way whenever he knows Nikkie can see him. . . . Yeroen was playacting. He wanted to make Nikkie believe that he had been badly hurt in their fight . . . he knew that his signals would only have an effect if they were seen; Yeroen kept an eye on Nikkie to see whether he was being watched. He may have learnt from incidents in the past in which he had been seriously wounded that his rival was less hard on him during periods when he was (of necessity) limping. (de Waal 1982, 47–48)

This case could represent merely that the principal actor in question, Yeroen, had learned the usefulness of limping behavior and was reflexively using his recollection of this in a way that appeared premeditated, skillful, and cunning. On the other hand, the actor may have been displaying and making use of its awareness of the general functions of social mutuality; in this particular case, of invalid care behavior. In other words, that he intended and enacted sickness behavior as a way of deriving benefits. It might also be claimed that the actor was manipulating a sickness healing adaptation which, it needs to be emphasized, could operate partly as reflex and partly as conscious. All of this raises the knotty question of self-awareness in the higher primates and whether chimpanzees differ from monkeys (Tomasello and Call 1997). The interpretation of self and other deception in the context of invalid care and dissimulation of disease is one that could also be applied to the notion of a sickness healing adaptation.

Do Chimpanzees Show Sensitivity to Altered States of Being?

One can extend this line of reasoning and claim that members of a chimpanzee, and perhaps monkey, group behaviorally accommodate to those diseased and injured. Moreover, visible manifestations of suffering under relatively relaxed group and environmental conditions may give rise to behaviors that could be called healing (Fábrega 1997). However, chimpanzees act differently not only toward sick and suffering group mates but also those showing a range of other mental states or conditions. They seem to be aware of responding differently not only to those in pain but also to those who display anger, frustration, irritability, fearfulness, and general moodiness. Chimpanzees respond to such behavior by avoiding, rejecting, consoling, contesting, confronting, or supporting their group mates, depending on circumstances and preexisting relationships. Offspring, maternal figures, and siblings are more likely to be accorded special treatment (de Waal 1989, 1996). For these reasons, one can say that chimpanzees show a rudimentary sensitivity to the psychological states and even personalities of others (Byrne 1995).

Researchers can only presume the emotional implications of chimpanzees' behavior by studying differences in social interaction. Such differences in emotional states and sensitivity to the conditions of others are part of the social fabric of the group. Chimpanzees obviously cannot communicate in language, so researchers have no way of knowing their psychological states—what they are thinking about, what they are feeling, what they are responding to—when they are alone or inactive. Only through vocalizations, emotional displays, and motor behavior in response to what has immediately transpired or what is about to ensue with respect to others can one speculate about their psychological state. Consequently, their full level and/or degree of self-awareness and awareness of others cannot be determined. Yet, as Damasio (1999) indicates, consciousness and a sense of identity have a phylogenetic history. Chimpanzees could show more than a core consciousness and core sense of self (see also Donald 1991) and could have entered into the territory of extended consciousness and that of proto or limited variety of autobiographical identity. Hence, one must retain an open mind as to the psychological dispositions and traits of chimpanzees regarding disease, injury, sickness, healing, and psychopathology. However, analysis of this general topic is instructive because it throws into high relief the types of questions raised by the attempt to explore the origins of human psychopathology.

Psychopathology among Chimpanzees

Let us suppose that much of the behavioral and emotional material generally agreed to be psychopathology can be found in nascent, embryonic form in chimpanzees. If one excludes aspects of psychopathology known to clinical phenomenologists or neurobiologists, that is, the technical psychopathology, and concentrates instead on the social behavior and general emotional manifestations of altered mental states that these clinicians have in mind, one is left with real psychologically and socially meaningful psychopathology. Chimpanzees, one would think, can show most of these common, everyday, lay-interpreted forms of psychopathology. Indeed, just as zookeepers and comparative psychologists cognizant of the behavior of chimpanzees can ascribe personality states and even personality structures to them, it should be possible for other observers to diagnose varieties of psychopathology seen in general emotional, behavioral terms. In human beings, we assume a correspondence between the technical psychopathology diagnosed by the expert clinician and the socially, culturally expressed psychopathology identifiable by a layperson. Similarly, one can claim that chimpanzees also exhibit psychopathology as it is generally understood and that it can be identified and diagnosed as such.

The next question is whether chimpanzees themselves might have a social awareness of psychopathology. This question, deceptively simple, brings up a number of complex considerations. Is a chimpanzee who displays what an informed observer would describe as psychopathology aware of being in a special, altered state? If yes, is this state classifiable in positive or negative terms? Psychopathology is a human, culture-bound concept that is not only symbolically complex but also contingent on culture and history. There are no reasonable grounds for expecting that anything like an awareness of an altered state or condition of psychopathology is possible for chimpanzees. Yet how otherwise might a chimpanzee experience this altered state or condition of distress and impaired well-being? And to complete the fuller picture, how do group mates interpret the condition of the afflicted individual or victim? Such questions are not unreasonable if one decides that the power to distinguish self from others is not a qualitative, all-or-nothing capacity but instead entails levels and degrees of mentation.

Chimpanzees, which occupy the closest position to Homo sapiens on the evolutionary tree, are thought to have deviated or evolved less from the last common ancestor than have humans and thus display more behavioral characteristics of the earliest varieties of Homo than Homo sapiens. Consequently, characteristics of their behavior that they share with man may provide the best model of the behavior of the earliest organisms in the line of evolution leading to modern man. Following this argument, the roots of human psychopathology are suggested by the behavior of chimpanzees with inherited vulnerabilities, adverse developmental histories, and intercurrent stressors in their natural activities. The study of chimpanzees provides a new backdrop against which to pursue this question: What is required in the way of cognition and culture for an individual or group of organisms—chimpanzees, the earliest hominids, early varieties of Homo, or fully modern humans—to be aware that a state or condition of self or a group mate is psychopathological?

Conceived in behavioral terms, whatever can be called psychopathology during human evolution is inextricably connected to advancements in the level and degree of self-awareness and awareness of others. In addition are developments in the sense of group identity and of group history acquired over time. In some respects, this can be conceptualized as involving the interposition of content and meaning between, on the one hand, the evolutionary residuals and imperatives described earlier—involving ultimately caused, infrastructures of behavior, mechanisms, algorithms relating to biosocial goals—and on the other, actual social behavior including manifestations of psychopathology in a specified group and social ecology. The earliest hominids must have begun to be aware of self and others, even without language and symbolic understanding, which is generally assumed to have been set in motion considerably earlier (Deacon 1997). Since self-conceptualization and social attribution were probably more elaborately configured among the earliest hominids, this raises the question of to what extent they were aware of changes produced by disease, injury, and general bodily pathology. Could categories such as wellness and sickness have constituted part of the social intelligence of the earliest hominids? If so, how might they have influenced actions toward the self or a group mate? What other categories pertaining to the behavior and dispositions of others might have been emergent in hominids?

To raise questions about self-conceptualization and social attribution brings up how psychopathology might have been psychologically and socially configured and played out in the group. In a diseased or injured individual, a state of psychopathology might have been associated with a more developed sense of being sick and physically compromised and, possibly, of needing (and expecting?) consolation, care, or healing. When marked primarily by differences in social and psychological behavior, not symptoms of disease, psychopathology could have been associated with more developed ways of conceptualizing, feeling, and expressing how that condition was connected to social circumstances. The afflicted individual might experience negative feelings, such as fear, social disarticulation, oppression, ineptitude, and failure to integrate with the group. It is highly unlikely that group mates would regard such an individual as deviant or abnormal purely as a function of the individual's social propensities and feelings.

Along with the emergence of a capacity for language, or perhaps preceding it, came a capacity for symbolization. Modern humans are assumed to have achieved full linguistic competence around 100,000 to 80,000 years ago, a dating that is controversial and much contested. However, comparative studies of brain and language coevolution suggest that a capacity for symbolization and a form of oral communication employing a limited set of lexical units, or morphemes, and grammatical connectives might have started to take shape much earlier. Symbolization and communication by language can be defined semiotically as an ability to think and communicate by means of abstract systems of representation. In other words, early hominids may have started to rely on conceptual categories that pointed to or signified objects and events, and then followed the ability to relate these to one another by means of protogrammatical connectives so as to make protolinguistic sense.

An individual's understanding of group identity and history was obviously enhanced as a consequence of symbolization. Manifestations of psychopathology would undoubtedly have changed as a result of these social and psychological characteristics, since individuals could now begin to better conceptualize differences in the self and others, communicate about them, and respond to intercurrent events and circumstances accordingly. Moreover, individuals could also probably recall past events and future possibilities better and could begin to imagine in a more elaborated way the intentions and dispositions of others. They could sense their integration with or separateness from the group and could form a concept of the group itself. During this phase of evolution, one can assume that hominids and early members of Homo began to conceptualize and experience more elaborated, complex emotional states, such as trust, distrust, entitlement, loneliness, sadness, envy, and jealousy; at the same time, they would learn how to overcome the opposition of group mates and realize their needs. As organisms or individuals acquire an ability to better define themselves and others, and as the group itself becomes better defined, this knowledge should bring with it a better knowledge of how to advance their own welfare and well-being or fitness. This must have had implications for the way psychopathology was configured and played out in the group.

In visualizing the character of psychopathology in chimpanzees and, by extension, varieties of earlier hominids, it is helpful to distinguish among its manifestations. The actual disorders recognized by psychiatric are simply too numerous, although how some of these might be reconceptualized in the language of behavior systems, motivational functions, and biosocial goals has been accomplished by McGuire and Troisi (1998; see also the aspects of social routines and emotions discussed by Stevens and Price 1996 and Nesse 1998). In a general sense, one may draw attention to disorders that add to or are expressed in behavior routines ordinarily perceived as reflecting mutualism, affiliation, and social integration. For examples, signs and symptoms of fear, anxiety, panic, mourning and grief, despondency and social withdrawal, failure to implement normal physiological functions such as alimentation and sleep, and suggesting physical disability and infirmity, as in the so-called somatoform behaviors, are likely to be implemented in routines that draw on strategies, signals, and displays that are part of the individual's normal repertoire. The communicative intent could be overtly disruptive or divisive although also integrative in the sense that it would seek to appease group mates, be attuned to or consistent with appropriate social roles and statuses, and perhaps be designed to elicit support. This would depend on the type of mechanism and social behavior routine affected and the prevailing social circumstances, including conditions in the ecology and group. At most, assuming an individual's ability to retain command of usual communicative displays and strategies, such varieties of psychopathology would at most be contextually abnormal, as mentioned earlier. To this, one can juxtapose manifestations that are grossly abnormal in a morphological sense; stereotypes of pacing, grimacing, clasping, and self-mutilation. These constitute behavioral pathology in a morphological sense. Finally, there are varieties of psychopathology that are frankly oppositional and antisocial, if not totally in intent and conception, at least in terms of the extent to which they are perturbed by disease, lesions, or functional disturbances during development. These varieties would manifest through aggression, hostility, suspicion, deception, social cheating or failure to reciprocate, disregard for personal space and needs of others, and/or outright rejection of signals of group mates that signal prosocial intents and dispositions.

Comment

Basic questions that need to be clarified if one wishes to understand the origins and evolution of human psychopathology are these: When, how, and why did psychopathology come to be differentiated as a distinctive social and or psychological state or condition? This question can be posed from the perspective of an informed, external observer, what anthropologists term an etic frame of reference, from which a diagnosis can be proffered. As we have seen, it involves untangling the clinical and evolutionary points of view.

The second question is an emic one, applied to the perspective of earlier varieties of humans themselves. When, how, and why did the earliest hominids and varieties of Homo or humanoid individuals come to recognize states or conditions of psychopathology in themselves and in group mates? What forms did psychopathology take during the evolution from protoculture to culture? How were psychopathologies configured, enacted, and interpreted?

A third question is implicit in the preceding ones: What properties of mentation and culture were necessary before psychopathology could assume a recognizable form, either from an etic or an emic standpoint? This question entails the potentially pivotal role of verbal language and a capacity for symbolization in how human varieties of psychopathology came to be configured and played out in social and cultural arenas.

A fourth question is this: Could psychopathology have served some function during human biological, that is, genic, evolution? In other words, could psychopathology as a trait manifested by individuals with a certain degree of culture and cognition have passed through the rigors of a variety of ancestral social and ecological conditions? If so, has it not somehow figured in the process of natural selection? One should entertain the possibility that most varieties of psychopathology would have been actively selected against. However, under some conditions and with respect to some manifestations, psychopathology might have served some adaptive purpose or function, and actually have been selected for, during human biological evolution.

The final question pertains to cultural evolution, the changes in ways of thinking, feeling, and social behaving that have transpired since the emergence of modern humans, sometime just preceding the Upper Paleolithic Era (roughly 80,000 years ago). This question involves a consideration of how psychiatric problems, mental illnesses, have come to be understood and handled in relation to society and culture. Stated succinctly, when and how did content and meaning relating to consciousness, self-identity, group history, and even spiritual/cosmological considerations as embodied in rituals and beliefs come to color manifestations of psychopathology (Boyer 1994; Goodenough 1990; Guthrie 1993; Rappaport 1999)? Questions such as this one have occupied anthropologists, sociologists, and social historians of medicine for many decades, although scholars have rarely conceptualized the topic in terms of human biological evolution. They have been dealt with in terms of cultural evolution, especially with respect to the history of psychiatric diseases and transformation in Anglo-European societies in recent centuries (Fábrega 1974, 1975, 1997). The topic of the cultural evolution of psychopathology involves a consideration of how factors such as agriculturalism, tribalization, state formation, urbanization, social complexity, political economic transformations, secularization, the rise of science, and modernization in general have shaped the character and meaning of psychopathology.

- If higher primates are vulnerable to psychopathologies and manifest them under natural conditions, then this raises the question of how group mates understand and respond to the signs and symptoms of psychopathology.
- How a tightly organized, socially integrated community of nonhuman primates would handle an instance of psychopathology in a member brings to the fore the animal/human, nature/culture question and the associated dilemma of anthropomorphism.

- An occurrence of psychopathology could divide and antagonize group mates or prompt them to exploit the victim; but if a measure of filiality and fellowship had developed, then toleration and even support and caring might constitute responses among group mates.
- Human primates have been described as tolerating and supporting group mates who show physical and in some instances behavior defects, especially in kin and infants, suggesting that some psychopathologies would elicit similar responses in natural settings.
- The nature of the signs and symptoms of psychopathology, conditions prevailing in the group, and circumstances in the ecology influence how group mates are likely to respond to sick, defective, and/or behaviorally compromised victims.
- Descriptions by ethologists emphasize that chimpanzees in particular are prone to showing a great deal of social awareness, tolerance, and even compassion toward compromised group mates and monkeys appear to share the same cognitive and behavior capacities.
- Responses of group mates to victims of disease, physical defects, and psychopathology are explained as part of the package of social, altruistic, and nurturing behavior, but the responses raise the question of a possible sickness/healing adaptation.
- Responses to psychopathology are more complex than responses to disease and physical defects. The latter communicate suffering and helplessness and elicit compassion, but the former may communicate antagonism and opposition, which are socially divisive.
- At the very least, responses of nonhuman primates to psychopathology suggest that the hominids who evolved from them showed similar kinds of behaviors, although these were socially more variegated given their comparably advanced cognition and culture.

Chapter 9

The Setting of Psychopathology during Evolution

THE ANTHROPOMORPHIC tendency to equate psychopathology purely with human thinking, feelings, and related mental sorts of things ignores the necessarily indispensable role of the physical environment in shaping how early hominids associated with one another, formed groups, and subsisted. In turn, these features of the ecology played integral roles in building the social relations in which psychopathology occurred. Thus, while psychopathology connotes disorders of awareness, cognition, emotion, and attendant behaviors, such aspects of disordered psychology manifested themselves in distinct social and behavioral environments and within specific types of social relationships.

During human evolution, myriad external factors constrained and presented opportunities and risks for the aggregation of individuals into groups. Hence external conditions need to be taken into account in calibrating what kinds of psychopathology prevailed during evolution and in assessing their implications. This requires a consideration of comparative social ecological factors, as well as internal ones involving psychology and culture, for it is in the conjunction of both that behavior and perturbations of it as psychopathology become realities. These two dimensions of psychopathology—external and internal—are equally relevant to any formulation of its evolutionary origins, functions, and significance.

Studying Human Origins

Comparing the social and behavioral ecology of hominids during various phases of evolution is a relatively new area of research. In many respects, it is a result of the knowledge deriving from primate field studies and an outcome of methods developed in the newer archeology and paleoanthropology. Researchers are now adding to the work of Sherwood Washburn (1961), Lewis Binford (1985), and Glynn Isaac (1989). The studies of Wrangham (1979, 1986) seek to connect inclusive fitness theory postulates to the evolution of social systems of hominids, in the context of limited resources and tradeoffs affecting the sexes and mating. Mellars and Gibson (1996), Standen and Foley (1989), Smith and Winterhalder (1992), and Steele and Shennan (1996) illustrate this newer trend in paleoarcheology and paleoanthropology. Insights drawn from studies in paleontology and archeology point to more concrete interpretations of early human behavior, but this approach necessarily leaves out topics such as group size, composition, structure, stability, or intergroup relations, and of course details of social and psychological behavior. Examining these phenomena in relation to changing paleoecologies may afford a better understanding of the social and psychological behavior of hominids, and consequently of their possible manifestations of psychopathology.

A large number of scholars have sought to explain the influence of social and behavioral ecology on human evolution. Some have attempted, on the basis of ecological factors such as abundance and dispersion of food, or the presence and formidability of predators, to infer the social organization of the last common ancestor (LCA) shared by the African apes and the line leading to humans (Ghiglieri 1987; Wrangham 1986). Then there is the task of developing and using a model of the social organization and behavior of that group to explain the special character of hominid adaptation and changes in social organization since the time of the LCA (Tooby and DeVore 1987).

Valerius Geist (1978) and Glynn Isaac (see Isaac 1989) offer rich, evocative pictures of the subject. More recently, the work of Richard G. Klein (1999), Clive Gamble (1986, 1993), Robin Dunbar (1988, 1996a, 1996b), and particularly Robert Foley (1987, 1995a) synthesize knowledge about the evolution and behavioral ecology of primate and hominid forms in an attempt to define common patterns in the evolution of varieties of Homo and anatomically modern humans. To varying degrees, these scholars use comparative biology, paleoecology, life history strategy, and evolutionary biology to interpret the patterns and processes that help explain the evolution of anatomically modern Homo sapiens.

Foley (1984a, 1984b, 1987, 1989, 1991, 1995a, 1995b, 1996) provides a useful framework. This is epitomized by his four Cs: the *conditions* under which competition among species and individuals takes place in a particular situation and time, roughly, the prevailing environmental and phylogenetic context; the *causes* or selective pressures that favored specific characteristics of individuals; the *constraints* that set limits and boundaries and in particular the cost-benefit relationships that made the selection of certain traits possible; and, finally, the *consequences* of the interactions of these factors, which would involve the selected, emergent properties of the phenotype, structural or behavioral, that were adaptive, given all the preceding factors. Although Isaac's (1989) home base model of food-sharing behavior by protohuman hominids has been much criticized, it is useful nonetheless for visualizing aspects of behavior and psychopathology during human evolution. Dunbar's formulation (1996a, 1996b) of the role of group size on hominid behavior patterns, including the importance of social grooming with respect to the possible evolution of language, provides an interesting point of reference for attempting to formulate earlier contexts and forms of psychopathology in the human line.

These writings and others like them furnish an ecologically sensitive and rich portrait of normal, adaptive behaviors of hominid forms leading to Homo sapiens. They (1) provide descriptions of adaptive syndromes of ancestral humanoid forms that (2) include evolved social behavioral strategies that are (3) situated in ecological settings featured by distinctive types of food resources and zoological species, including those competing with early hominid forms and those operating as predators. These social behavioral strategies encompass (4) characteristics of how groups were organized as social systems in the habitat. The social system of an ancestral species, then, represents its adaptive strategy or syndrome in relation to the environment. Hence, patterns of adaptive behavior of ancestral hominids allow one to delineate the context of normal behavior and, by implication, how and why certain behaviors could be conceived as abnormal, maladaptive, or pathological.

To provide an orienting context, I will briefly outline the major landmarks in human evolution, starting with prehominids—in particular, the last common ancestor of the pongid, especially chimpanzee, and hominid line. (See Moore 1996, in McGrew, Marchant and Nishida 1996.) The last common ancestor lived in closed social networks. Relations between groups could be hostile, although females might be exchanged between them. Strangers were attacked by the males of the group. Males and females were strongly differentiated, and there were no monogamous bonding relationships. The earliest hominids, the Australopithecines, probably increased their terrestriality, walked upright, and developed postcanine megadonty.

As evolution progressed, early hominids increased their consumption of meat and began to hunt prey larger than themselves. Infants became increasingly helpless, requiring greater nurturing by mothers or maternal surrogates, which increased the male's investment in infant care. Estrus and ovulation were no longer visible in females, and females became sexually receptive for longer periods. The male sexual organ increased in size, and body hair began to disappear. Labor began to be divided along sexual lines, and large coalitions of males began to develop. New traits that emerged in the species Homo include larger brains; the controlled use of fire; the development of vocal communication and eventually a protolanguage or true language, that is, full linguistic capability; the manufacture of stone tools and most probably the use of wood and hides, though evidence of the last two is not preserved; loss of megadonty; less sexual dimorphism; an adolescent growth spurt, most notably in females; and better ways of procuring, transporting, and storing food. Kinship networks developed and participated in the selection of mates.

Social Ecology of Hominids

Attempting to explain the changes in behavior that took place during hominid evolution is hazardous and theoretically complex (see Tooby and DeVore 1986). A referential model, for example, uses chimpanzees to exemplify the social behavior of the last common ancestor. A strategic model, as another example, uses tenets of evolutionary theory as a basis for comparison and analysis. (See Kinzey 1987.) Ghiglieri (1987) compares differences in the posited behavioral suite of the last common ancestor based on models derived by including, rather than excluding, gorillas in the set made up of chimpanzees, bonobos, and humans (see also Wrangham 1986). Maryanski (1993) analyzes social characteristics of primates in an effort to infer elementary forms of organization in the first, protohuman societies. She concentrates on mating characteristics, patterns of transfer in and out of groups in relation to age and sex, and patterns of sexual bonding, and she emphasizes that chimpanzee-like social relations most likely evolved into stronger male-female bonds in which the male was dominant. Moore (1996) also uses chimpanzees as a referential model for the LCA and hominids.

The social organization of hominids during evolution has been conceptualized as a pattern of behaviors structured by the group's arrangement in space and over time (Foley 1995a, 1996; Foley and Lee 1989; Ghiglieri 1987; Lee 1992, 1994; Rodseth et al. 1991; Standen and Foley 1989; Wrangham 1986). Research in paleoanthropology, paleogeography, paleoecology, and primate ethology allow us to model the probable habitat and overall environment of a particular species or subspecies, as well as its social state and group characteristics. A posited social state is derived from relations between individuals based on kinship and sexual status. An assumed pattern of association among members of an ancestral group is used to model ecological conditions, constraints on change, and other pressures. Researchers then speculate about patterns of association and possible future social states on the basis of the model. Human societies may seem to be extremely variable and complex, but they actually involve a relatively small range of social states that can be compared to those of nonhuman primates. Models of social states allow us to chart the possible locations of various hominid species and subspecies and the kinds of shifts that might have occurred during evolution (Foley and Lee 1989; Rodseth et al. 1991). Maryanski (1993) also describes characteristic patterns of social organization, ranging from primate antecedents to the last common ancestor, and then to hominids proper.

SOCIOECOLOGICAL SETTING OF HOMINID ORIGINS

During the Late Miocene Era in Africa, the time and location of the earliest hominids, forests became smaller, climates grew drier, woodland regions became more widespread, and the habitat grew more diverse. As food became scarce as a result of reduced rainfall and increased seasonal change, females were forced to forage over wide areas and, associated with this change, males had to form kinship coalitions to defend these areas. Conditions therefore did not favor the association of females in kinship groups, and the common ancestor of humans, chimpanzees, and gorillas may have lived in groups consisting of alliances of males, with females dispersed (Rodseth et al. 1991).

THE EARLIEST HOMINIDS

Information about the earliest hominids is confined to Africa and involves the Australopithecines and early varieties of the Homo lineage. Unlike earlier species, the earliest hominids walked upright. Bipedal locomotion led eventually to morphological changes in the teeth, a larger brain, and a range of new activities such as hunting for food. Some evidence suggests that early hominids may have used bone implements for digging subterranean plant foods in rocky soils (Klein 1999). There is ample evidence of stone artifact production, classified as the Oldowan Industrial Complex. These artifacts are generally viewed as crudely made and remarkably uniform across time and space. There is controversy about who exactly made the Oldowan tools, since several varieties of early hominids, Australopithecines, early ancestors of Homo erectus, or some other Homo varieties, were contemporaneous. In some locations, evidence indicates that Oldowan "pieces were used for scraping or sawing wood . . . cutting meat, and . . . cutting grass stems or reeds" (Klein 1999, 168). Replication experiments have suggested that the earliest hominids may have preferentially used the right hand and right arm while manufacturing stone artifacts, raising the strong possibility of an associated cerebral lateralization of functions and suggesting incipient humanlike forms of cognition and communication (Holloway 1981; Toth 1985). Evidence about the construction of settlement sites and the use of fire is ambiguous. The wear and polish on stone flakes and the cut marks on bones found at different sites indicate that the Oldowan people relied on animal food, but to what extent they consumed scavenged carcasses as compared to actual prey is not clear.

The social organization of the first hominids is believed to have resembled that of chimpanzees. Consequently, protohominids are held to have formed only temporary, relatively promiscuous sexual consortships, with exclusive sexual bonding not coming until much later. Compared to earlier groups, Australopithecines probably occupied more open, savanna-type habitats that included large patches of grassland and bushland. The threat of predators and scarce resources favored the formation of larger groups. Australopithecines are believed to have lived in large mixed-sex groups, with males connected by kinship constituting the main social linkage. The pressure on females to forage over large areas for food, together with the need for protection against predators, led to the formation of stable associations of specific males or alliances of males. Larger and less structured groups would have weakened male kin bonds. Moreover, in larger groups not joined by kinship, males would have been more concerned about defending their females, resulting in a haremlike structure that still included related males. The result was an aggregation of reproductive units with unrelated females in stable association with a single male and including the kin of both sexes.

The Australopithecines began to walk upright about 5 million years ago, as the climate grew colder, drier, and more seasonal. With hands free for carrying tools, they foraged over larger ranges in search of food in newly opened woodland and savanna territories adjacent to rainforests. Predators, including hyenas, leopards, and lions, posed a constant danger, so that in addition to new food procurement strategies, the earliest hominids were forced to acquire more elaborate knowledge about predators' habits (Stanley 1996). These developments were associated with changes in the size and social organization of the earliest hominid groups.

The Australopithecines probably showed a mode of social organization and behavior like that of apes, being but one phase removed from the LCA, whose behavior presumably, but arguably, resembled that of chimpanzees. Fossils of the earliest hominids show that their brains were not much larger than those of chimpanzees. Moreover, little is known about the nature of their social behavior, although we can assume that they showed more complex social intelligence, a greater capacity for self-awareness, self-conceptualization, and evaluating the intentions and mental states of conspecifics. According to Donald (1991), the earliest hominids had an episodic culture: their system of cognitive representation did not include the power of imagination—the power to range freely in time and space. Their sense of awareness, social relations, and planning and strategic behavior with respect to the environment was locked into the here and now. Australopithecines may well have shown behavioral pathologies like those of chimpanzees.

EARLY REPRESENTATIVES OF THE GENUS HOMO: HOMO ERECTUS

A habitat of more arid grassland and woodlands, with seasonal climatic changes and variation in flora and fauna, encouraged greater adaptiveness among the

Australopithecines. Early representatives of Homo had a less specialized dental structure, had larger brains, and walked more like anatomically modern humans. They relied more on tools and began to eat meat. Seasonal stress, restricted plant resources, and less plentiful game meant that Homo erectus was forced to search for food and hunt over wider areas, thus necessitating greater cooperation among males. Males in alliances could better defend females against the threat of roving males and defend the group against hostile groups. Stronger male kin associations within and between groups may have enabled the formation of social lineages, or relationships among kin of the same sex living in a different group.

As brains increased in size, they required a longer period of maturation after birth. As a result, infants grew more slowly, were more dependent, and took longer to mature. Infants were thus more vulnerable to predators, hostile groups, and infanticide. The result was stronger male-female bonds for protecting the young. Foley and Lee observe: "The result of a specific environmental stress (seasonality), a shift in resource base, and changing life history parameters associated with the costs of producing offspring for H. Erectus would be the simultaneous buildup of male kin alliances (competitive and cooperative inter-group effects) and specific male-female links (polygamous rather than monogamous) resulting in a higher level of complexity of both inter- and intra- (specifically male) sex relationships" (1989, 905).

Around 2 million years ago, early hominids—members of the genus Homo, in particular Homo erectus—showed further advances in anatomy favoring even more efficient ways to survive and reproduce. According to Donald (1991), they had a mimetic culture that included communication through gestures. Bickerton (1990, 1995) suggests that they had developed a protolanguage and quite possibly a capacity for game playing. Mimetic skills, the ability to initiate conscious, intentional, but not fully linguistic forms of communication, may have helped the early Homos to share information about relevant concerns. In contrast to the Australopithecines, who are assumed to have had an episodic culture with a focus purely on the here and now, early varieties of Homo had probably achieved a capacity for abstraction that enabled them to recall the past and envision the future (Donald 1991). A sense of group identity implies that members shared a past, an awareness of a common purpose, common vulnerability, and a rudimentary ability to appraise, conceptualize, and communicate about the physical environment and, what is most important, about the group. Along with the power of abstraction, they could have developed a more elaborated, but still elementary, system of categories about mental states and emotions, and psychological states could have been more clearly linked to ongoing social relations within the group. Finally, since sickness and healing routines must have been more elaborated, as compared to the Australopithecines, involving care and protection for the diseased and injured, they could possibly have evolved rudimentary ideas about psychological suffering and social tolerance for aberrant social behaviors when linked to social stressors, social losses, and disease or injury.

ARCHAIC AND EARLY ANATOMICALLY MODERN HUMANS

Conjugal families organized patrilineally still had to take a critical step before the human form of social organization could develop. This involved maintaining relationships among dispersed offspring, which allowed the formation of alliances between individuals related by "marriage" and living in the same or different groups. Intergroup affinity, then, emerged as a revolutionary pattern of social organization.

The trend toward stronger male kinship relations led to increased recognition among individuals based on kinship. As individuals began to live longer, enhanced male-kin relationships encouraged the formation of lineages of adjacent male generations (patrilineal) which, in the long run, led to lineage recognition and social relationships organized across space and time. Associated with these trends came the formation of stronger bonds between a particular male and a number of specific females. Foley and Lee write: "On these grounds it may be argued that polygynous male family groups occurring within larger male kin lineages characterized the social organization of the ancestors and earliest representatives of modern humans" (1989, 905). This type of social organization, when mapped onto a distribution of social states based on sexual and kinship status, suggests the pathways that may have led to the different forms of social arrangements found in human societies cross-culturally. With respect to such patterns of kinship and cooperation, Rodseth et al. point to four important trends: "(1) Males maintain consanguineal kin ties, even with matrilocal residence. (2) Females usually maintain consanguineal kin ties, even with patrilocal residence. (3) Males cooperate in conflicts, both social and physical, against other males. (4) Females also cooperate but rarely do so in physical conflicts of females against females" (1991, 230).

A behavioral trend in the line of evolution leading from the early hominids to fully anatomically modern humans was the dispersal of females, the development of coalitions of male kin protecting territories, and increased female bonding. This trend was associated with changes in mating structure toward stronger female-male pairs and more stable relations, a greater investment in child rearing by both males and females, and larger groups. Groups became more structured along kinship lines and grew more hostile toward other groups. Eventually, this culminated in family units banded together in patriarchical groups with increased substructuration along sexual and kinship lines.

The behavior and cognition of Archaic humans, particularly the Neanderthals, is a controversial topic (Mellars 1996a, 1996b, 1996c; Stringer and Gamble 1993). Scholars disagree as to whether this species of Homo showed full linguistic and cultural competence. Their remains offer very little evidence of creative artistic achievements. Their stone tools and artifacts were also much less varied and refined compared to those of fully anatomical moderns, and they apparently did not carve and work on bone and ivory. Yet they were equally adept at hunting and finding raw materials for tools (Kuhn 1995; Stiner 1994). Neanderthals seemed to be as proficient or intelligent as the moderns in planning and social strategy. Neanderthal societies spanned at least 225,000 years and were able to successfully colonize regions with high latitudes and cold climates during the glacial periods, areas hitherto not settled by hominids. They may have mastered the habitat and adapted successfully as a result of improvements in social organization, morphological and physiological adaptations, and better planning for subsistence activities (Gamble 1993, 1996). On the other hand, the level of adaptation of the Neanderthals could have been due largely to cognitive, intellectual achievements that included a fully evolved capacity for language. Hayden (1993) reviews evidence that to him suggests relatively advanced linguistic and cultural competence among Neanderthals, while Davidson (1991) and R. G. Klein (1995) emphasize that these capacities and traits are features confined to the moderns they assume replaced the Neanderthals.

If Archaic humans or the Neanderthals had language, then the puzzle is why they seem almost totally devoid of culture, symbolic expression exemplified in cave painting, figurines, and carved artifacts and tools, such as the moderns displayed beginning around 40,000 years ago. The capacity for language would appear to go hand in hand with that of culture; in fact, some anthropologists have much difficulty distinguishing between these two general and basic properties of human mentation and psychological behavior. Nevertheless, it is possible that even if they had a fully evolved language, Neanderthals were simply unable to develop a mature capacity for culture that included narrative ability, myths, and symbolic expression. This might have been owing to other cognitive limitations or the pressures and demands of their extremely harsh living conditions.

THE HUMAN COMMUNITY

A community is defined as the largest group of individuals who maintain long-term relationships. It is an important and peculiarly human group, even if members often remain alone or connect with just a few members, and the whole community rarely assembles face to face. Typically, members of a human community come together in temporary combinations. For this reason, a community may be defined as "a closed or semiclosed social network, larger than a conjugal family, with fission-fusion subgroups" (Rodseth et al. 1991).

Since most humans are part of a family, a community, and the various subgroups within it that repeatedly come together and break apart, shifting loyalties along with stable alliances characterize the human condition. Hence the human form of social organization entails social and interpersonal tensions between individuals at different levels of social organization—specifically, the family and the community—and between these two more or less stable units and the many changing subgroups of everyday life.

Human beings maintain these patterns of social relations as a result of sexual breeding, consanguinity, kinship, and the formation and breaking apart of myriad subgroups, despite the absence of continued or frequent face-to-face interactions. Such a fluid arrangement of social relations undoubtedly relieves the tensions intrinsic to human social organization by allowing some perturbations to dissipate, yet it can also accentuate such tensions by not providing opportunities for the resolution of entrenched conflicts. Shared rituals serve to unite individuals of a community who generally maintain relations in absentia. An order based on culture and shared patterns of meaning reflected in symbols reminds individuals in a community of the underlying conventions and values that guide social activities and behavior in general. Culture may also be viewed as a means of reducing strain and tensions in human groups through shared patterning of values and meanings that bring order to social practices.

Relating Social Ecology to Psychopathology in Hominids

To describe the social ecology of hominids is to underscore the socially rooted character of their behavior, including psychopathological behavior. An individual's social circumstances provide a basis for understanding normal behavior, and such understanding is a necessary baseline against which to describe irregularities and possible deviations. Normal behavior not only involves social relations but also reflects how individuals solve adaptive problems posed by the ecology that affect the rhythms and patterns of everyday life and that in the long run determine reproduction and survival.

The behavior problems of nonhuman primates can be used as a point of reference against which to formulate possible varieties of psychopathology in early hominids. These could have included fears and hardships; more cognized forms of emotional distress, such as a sense of hopelessness, helplessness, victimization, or demoralization; associated breakdowns in social behavior; and social interactions and conflicts with group mates, all arising from ecological hardships and stressors. Constraints on foraging or threats from predators could cause a mother to neglect her infant, for example, or simply increase the level of irritability and harassment within the group, affecting the adaptability, well-being, and fitness of its members, contributing to psychopathology.

Stress can also arise from circumstances directly tied to interpersonal relationships and intragroup dynamics, such as mutual antagonisms, conflicts over social status and rank, and competition over food. Mating and reproduction, of course, are aspects of social life replete with tensions, contradictions, and imperatives that can contribute to satisfaction as well as misery. Symons (1979), Daly and Wilson (1988), Buss (1995a, 1995b, 1999), and Geary (1998), among many others, document the impulsive power and social psychological importance of sexuality, mating, and reproduction. It was during human evolution that these functions and their repercussions within the group began to form basic patterns of human society.

Characteristics of a group's social ecology play an important role in how emotional perturbations and psychopathologies are played out. De Waal (1996) considers the effects on group stability and continuity, and also how group mates might deal with extremes of temperament in the expression of aggression and violence. Social avoidance, lack of grooming, failure to respond to a need for comfort, refusal of food upon begging, rejection of support in the event of harassment, and punishing relatives or offspring are examples of negative sanctions that might be imposed on a group mate exhibiting psychopathological behavior. Or, by contrast, positive social responses can also be envisioned. Concentrating on positive and negative cognitions and behaviors might be a reasonable method of classifying and understanding social responses to psychopathology. (See Chance 1988a, 1988b; Öhman 1986, 1987.) Readers are referred to a discussion of the importance of the agonic and hedonic mode by Chance (1988b) and the complementary discussion on the classification of the emotions as involving positive/approach compared to negative/avoidance tendencies by Öhman (1986, 1987).

Clues reflected in the social responses of onlookers but nonparticipants in the incidents of infanticide and cannibalism among chimpanzees described earlier suggest strongly that groups of LCAs, and certainly later hominids, would have responded to infringements of social psychological conventions classifiable as psychopathology, though perhaps not as extreme as those manufactured in the laboratory, in symbolic terms. In other words, group mates might respond to psychopathology through culturally meaningful forms of social behavior framed in a mimetic or protolinguistic form (Donald 1991). The topic of how groups of LCAs, hominids, and varieties of Homo might respond to aberrant social behavior analogous or homologous to human forms of psychopathology touches on the question of a possible adaptation for sickness and healing (Fábrega 1974, 1975, 1997).

Causes of Psychopathology in Early Hominids

Observations of monkeys and chimpanzees help us visualize behavioral disturbances as they might have occurred during earlier phases of human evolution. Ethologists emphasize that environmental hardships and stress can elicit alarm calls and other forms of emotional communication that can arouse fear, anxiety, and terror in group members, depending on the personality or reliability of the sender's actions (Cheney and Seyfarth 1990). They often retreat, seek to avoid threat, or seek a position of safety. With respect to stressful events within the group or conflicts over rank, changes in an alliance network, competition over mates, undue brutality toward the weak—especially infants and juveniles—nonhuman primates seem to show politically motivated behavior. Anthropologists such as Power (1991) and paleoanthropologists such as Wrangham and Peterson (1996) draw attention to outright brutality and violence toward females within a group, as well as toward males or females of neighboring, competing groups. These social psychological routines, which Power equates with psychopathology, are explained as stemming from environmental changes introduced by the ethologists or as aberrations against innate propensities by demonic males (Power 1991; Wrangham and Peterson 1996). Hardships and social difficulties often escalate into loss of control, physical attack, and sexual assault. This causes terror, probably anger, and a sense of powerlessness in group mates, which in turn leads to avoidance, ostracism, banishment, punishment, or attempts to appease the culpable group mate, depending on the social rank and support network of observers and participants. Less striking, more subtle forms of behavior reaction, involving fear, timidity, and emotional dyscontrol, for example, can be surmised (see earlier review of studies by Schneider). In general, responses by group members would have depended on the level and type of psychopathology and on kinship, mating preferences, age, and membership in established alliance networks. One can assume that such scenarios were common during early phases of human evolution.

To be sure, it is problematic to claim that the behavior just described is evidence of psychopathology. We must avoid the pernicious trap of undue medicalization of social behavior and of the psychiatric, properly condemned by many social critics (see Conrad and Schneider 1980; Fox 1988a, 1988b; Kittrie 1971; Szasz 1961). The danger of anthropomorphism is a related bias when crossing phylogenetic boundaries, a form of speciesism in reverse. There are great differences between the clinical and the evolutionary points of view; yet at some point during human evolution, varieties of psychopathology must have become social and cultural realities as the capacity for social cognition evolved.

The kinds of behaviors and responses to stress that primatologists and comparative psychologists see as aberrant or abnormal must have been not only present but also common among the earliest hominids, the Australopithecines, and among varieties of Homo. Some of the incidents in question constitute examples of psychological trauma and would likely have produced behavioral pathologies associated with states of dissociation. Many groups probably broke apart or fissioned as a consequence of such traumas and emergent psychopathologies.

However, one can presume that under normal conditions later hominid groups showed a sense of group identity and coped with such problems in a more socially deliberative way. Unless a group were strained by extreme hardships that literally threatened its integrity, crises caused by psychopathology would likely have been met by some form of intragroup accommodation involving symbolic mediation and interpretation, such as a change in group dynamics or group relations, attempts to resolve the conflict, or efforts to support, mediate, and heal.

Acute behavioral disturbances, or behavior classifiable as psychopathology, might not always have evoked responses along the lines stipulated by exponents of strict evolutionary theory. Some scholars argue that disturbances would have been resolved by the strict economy of natural selection and that the introduction of concepts about symbolization in relation to sickness, healing, and psychopathology reflect ethnocentric extrapolations from social and cultural theories. They would equate recognition and handling of psychopathology with fully evolved language and culture and confine its prevalence to the late Upper Paleolithic (Davidson 1991; R. G. Klein 1995; but see Clark 1997 and Hayden 1993). On the contrary, I suggest that during later phases of human *biological* evolution acute behavioral disturbances were not only likely to evoke responses mediated by symbols, but also in some instances to be seen as forms of sickness. From the standpoint of group dynamics and functioning, the meaning of a behavioral pathology determines how it is handled as a social reality. As human culture and cognition evolved, early hominids learned to interpret social happenings in terms of group dynamics, group history, and accumulated knowledge. This brings up the definition of culture.

A Social Ecological Approach to the Definition of Culture

This chapter began by emphasizing the importance of viewing and interpreting behavior in its natural setting, in concrete, operational terms, and as constrained by external factors. While forms of emotional communication and early forms of language, cognition, and social traditions of behavior must also have influenced the social adaptation of early hominid groups, many social activities and routines were determined by sheer environmental contingencies. Climate, physical terrain, the availability of food, foraging routines in a given habitat, forms of social organization, pressure from social carnivores and competing groups, and demographic factors such as group size, gender and age structure, birth and mortality rates, were all important influences on behavior, and therefore on psychopathology. But since we are attempting to understand the dynamics of highly social groups of individuals with emergent human capacities for communication and cognition, we need to consider the development of culture among early hominids.

Culture is a concept usually applied exclusively to modern human groups and understood as a general capacity for symbolization expressed in language. In this definition, culture encompasses traditions, beliefs, myths, rituals, narratives about the group and the behavioral environment, and rules and standards of social behavior. This may be referred to as symbolic culture (Chase 1999). However, those concerned purely with behavioral ecology rely on a broader definition of culture. In attempting to reconstruct the behavior of early hominid groups by extrapolating from the behavior of the higher apes, who constitute the best model of the last common ancestor between the hominid and pongid line, one must take a more generalized and abstract approach. Since psychopathology is ordinarily conceived as occurring in a meaningful social or cultural context, the attempt to understand what form psychopathology may have taken during the prehistoric eras when human beings were still evolving requires a more flexible concept of culture.

Whereas behavioral ecologists understand psychopathology purely as psychological and social behavior that compromises fitness, I propose a broader approach, that is, to examine the changing contexts in which social and psychological behavior took place during evolution and how it evolved in relation to culture. To formulate a group's culture in purely ecological terms, one should first consider basic biological characteristics of group members, such as gestation length, period of infant dependency, age at puberty, age at first reproduction, age at weaning, duration of fertility, life span, and brain size. Evaluating their power of cognition would include how they represented and processed information about the physical habitat, their social intelligence, and their symbolic capacities, most particularly, a capacity for communication and eventually language. Culture as a tradition of behavior is also affected by ecological factors, physical characteristics of the habitat such as the amount of rainfall, water sources, temperature, and the availability of food, which affect how far members of a group need to range in order to subsist. In conjunction with life history factors, environmental ones determine the size of groups, their composition (age structure and sex ratio), and social organization (Parker and Russon 1996). All together affect what is learned, the kinds of learning possible, the mode of teaching that takes place, and how information is transmitted within and across generations.

The character of the acquired information and how it is organized make up the group's local natural knowledge. Since genetic factors affecting cognition, together with the physical environment, condition what groups are able to learn, it follows that if one is able to determine how one group's routine of behavior differs from that of another group that shares roughly the same habitat, that difference can be qualified as a tradition of the population. If two groups share the same gene pool and habitat but differ in social behavior, then the differences can be said to be traditions (Kummer 1971). Groups with local traditions of behavior and about which one might infer distinctive forms of social knowledge can be said to display a culture. Boesch (1996) reviews how one might assess chimpanzee culture and describes behavioral differences in chimpanzees that he believes qualify as culture.

Parker and Russon (1996) summarize what researchers have learned about the cognition of the great apes, their behavior in natural and captive environments, the kinds of learning and teaching they are capable of, the contexts in which they teach and learn, and the characteristics of the knowledge they acquire and how they use it. (See also Russon, Bard, and Taylor 1996.) Parker and Russon then present a definition of culture useful for making sense of psychopathology from an evolutionary point of view: "Cultures are representations of knowledge socially transmitted within and between generations in groups and populations within a species that may aid them in adapting to local condition (ecological, demographic, or social). These representations of knowledge depend upon various cognitive processes of individual learning and invention as well as social processes that transmit information in the population over time" (1996, 432–433).

This definition serves as a background against which to develop an evolutionary-ecological model for understanding differences between species in behavior and levels of cultural adaptation. Parker and Russon divide cultural evolution into four phases: (1) Preculture, as found in monkeys and other social mammals, comprising limited social transmission of representation of local knowledge based on simple mentation and sensorimotor abilities; (2) Protoculture, typical of chimpanzees, which involves true social transmission of social knowledge based on true imitation, limited demonstration teaching techniques involving late sensorimotor, early symbolic-level cognition, and theory of mind capacities similar to those of human children involving simple pretend play; (3) Ur-culture, typical of Middle Pleistocene hominids, which involves more complex forms of representation of local, socially based knowledge and more advanced symbolic abilities—that is, late preoperational and early concrete operations, as in the Piagetian scheme; and (4) Eu-culture, as manifested in modern humans.

Parker and Russon conceive of culture as ecological; it "assumes the existence of a stable, inter-generational community of individuals of the same species whose members interact persistently and regularly in a variety of natural or semi-natural settings . . . the same conditions that are prerequisites to forms of reciprocal altruism" (1996, 439). This definition and the model based on it take into account the diffusion of information by adult or semiadult individuals moving between groups and "implies that the local adaptations have arisen through social transmission of novel behaviors invented by members of the group or through importation of novel behavior into a group by immigrants from other local groups" (1996, 439). Parker and Russon's model attempts to integrate life history and cognitive abilities, foraging and feeding strategies typical of the species, and some of their typical social environments.

With respect to the structure and content of psychopathology, when one adopts a social and ecological slant on culture, one excludes or minimizes the influence of the content of mental phenomena that affect the behavior of afflicted persons and that shape how others interpret and respond to them. In other words, the meanings of such things as social contacts, obligations, alliances, roles, orientations, dispositions, and identities, which during later phases of evolution place psychopathology within a social and cultural frame, in the traditional, symbolic sense of culture, are left out of focus when one assesses culture within a social and behavioral environment. One ignores the meanings and interpretations of what one must see as altered routines, disturbed patterns, divisive exchanges, and outright breakdowns in behavior. Nothing is important beyond the structure—or lack of structure—of an incident of psychopathology, how it contributes to—or works against—survival and mating, its costs or even potential benefits to the individual or kin, and how it is resolved in relation to the group's survival. Psychopathology in the present context, then, would be more a locus or node of irregularity in the group's social fabric occasioned by an individual's behavioral anomalies and less a center of disordered phenomenology compounded by shared symbols laden with feelings about self, the world, and the social universe, however configured and enacted in the group. The latter aspect of culture does not come to the fore until later phases of evolution.

Glimpses of Psychopathology among Early Hominids

At the early, or ape, end of the evolutionary continuum, psychopathology must have resembled idiosyncrasies if not actual abnormalities of behavior like those described in feral communities of the more evolved nonhuman primates (de Waal 1989, 1996; Goodall 1971, 1977, 1986, 1990; Kummer 1995; Troisi et al. 1982, 1983, 1984), and like abnormalities manufactured in laboratory settings. The first set of studies provides ample evidence of personality and behavioral eccentricities, anomalies, and inconsistencies, as well as some frank episodes of crisis and breakdowns in social behavior, whereas the manufactured behavior reveals the vulnerability of the hominid line to disturbances as a consequence of adverse developmental experiences and physical causes such as lesions, toxic conditions, and infections of the central nervous system. Some manufactured syndromes produced profound maladaption that would likely have caused the individual to fail to thrive in a natural setting. More subtle syndromes appeared in mothers of young infants denied constant and predictable access to food. The effects disrupted maternal-infant relations and bonding and produced levels of brain chemicals that resembled the biochemical profiles of human adults with posttraumatic stress disorder (Andrews and Rosenblum 1988; Bremner et al. 1998; Coplan et al. 1992, 1995, 1996; Coplan, Rosenblum, and Gorman 1996). Conditions creating these psychopathologies were similar to what these animals, bonnet macaques, and other nonhuman primates would likely encounter in natural settings. It seems plausible, therefore, that the manifestations and consequences of these behavioral disturbances suggest what early forms of psychopathology among higher primates would look like in the wild.

With a little imagination, these and other animal models of psychopathology can be projected into the earliest phase of the human evolutionary past. Psychopathologies among the very earliest hominids could well have been manifested as breakdowns of childhood or adolescent behavior stemming from varying degrees of separation from a mother and/or peers (Clarke and Schneider 1993; Clarke et al. 1994; Kaplan 1986); fearfulness, timidity, and lack of emotional control as a consequence of abnormal developmental experiences and intercurrent stress (Kraemer 1992); behavioral changes resembling human psychiatric disorders such as psychosis, depression, and anxiety produced by chemical lesions (Ellison and Eison 1983; Harris 1989; Healy 1987; Iversen 1987; McKinney and Bunney 1969; Öhman 1986; Oliverio, Cabib, and Puglisi-Allegra 1992; Weiss 1980); breakdowns in the behavior of juveniles and adults as a consequence of experimentally produced stress; behavioral and endocrine changes associated with status conflicts, repeated social harassment, or sustained opposition and antagonism (Sapolsky 1987, 1990a, 1990b, 1998; Sapolsky and Ray 1989); and breakdowns produced by disease or physical lesions to the brain (Franzen and Myers 1973; Fuster 1980; Myers 1972; Squire and Zola-Morgan 1991).

Individuals suffering from severe forms of psychopathology would probably not have survived very long in natural communities, even though some group mates may have attempted to alleviate the suffering or to heal the afflicted member (Fábrega 1974, 1975, 1997). On the other hand, those suffering from less extreme varieties of psychopathology, analogous to the above-mentioned disturbances but caused by less severe forms of social and environmental stress, might have been able to persist. The manifestations would have resembled those described by researchers, only one needs to picture them as played out in forms of abnormal behavior resulting from conflicts and rivalries over rank and status and/or involving deviations in temperament or personality. Possibly some incidents of psychopathology would result from overwhelming, sustained stress, sometimes in diseased and physically compromised individuals. There is clear evidence that archaic humans survived and probably remained integrated in groups despite what appear to have constituted disabling physical lesions, and this suggests that, depending on types of symptoms, some individuals showing psychopathology also have survived (Trinkaus 1995). This scenario may be applicable to even earlier phases of human evolution.

One tactic in primate studies involving behavior and cognition is to compare members of a species, say chimpanzees, with human infants and children (Heyes and Galef 1996; Parker, Mitchell, and Boccia 1994). The emergence of language and development of a theory of mind during infancy is a classic example. Such a comparative approach suggests that one might be helped to visualize the character of psychopathology among the earliest hominids by drawing attention to how limitations in intellectual and cognitive development affect the expression of psychiatric disorders among selected contemporary populations, such as psychopathological behavior among mentally retarded children or adults, especially those compromised by severe traumatic stress and developmental disabilities.

Mentally retarded individuals are generally held to be susceptible to all varieties of emotional, behavioral, and personality disturbances. Yet identification and diagnosis of such disturbances in mentally retarded adults, on whom more emphasis will be placed for heuristic purposes, is difficult. Standards for diagnoses are based on average adults, and, one may add, those of distinctive ages, cultures, and ethnicities, whereas intellectually compromised individuals express and show psychiatric disorders in different ways. Sovner (1986) identifies the following four nonspecific factors associated with mental retardation that influence the character and manifestations of psychopathology and diagnosis: (1) intellectual distortion or concrete thinking and impaired communication skills, such as the inability to label one's own experiences; (2) psychosocial masking or impoverished social skills and life experiences, such as unsophisticated presentation and lack of poise, easily misinterpreted as nervousness; (3) cognitive disintegration or stress-induced disruption of information processing, such as bizarre presentation and psychotic-like states produced by stress and thus misdiagnosed; and (4) base-line exaggeration or increase in severity of previously maladaptive behaviors, which makes it difficult to diagnose established target symptoms.

In those whose capacity for abstraction, introspection, and sheer communication is limited, allegedly well-established clinical conditions are more variable and less clear-cut. It is often impossible, for example, to determine the clarity of thought of an intellectually compromised person, or to know whether false beliefs or altered perceptions are part of the clinical picture. A mentally retarded person's mood and emotional state can be particularly difficult to ascertain. Anxiety, fear, and mental perplexity can easily disorganize and compromise cognition and the organization of behavior. Putative mental or brain-related intellectual deficits can cause any number of altered behaviors. For example, anxiety and fear often dominate the clinical picture, along with restlessness, failure to control emotions and impulses, and hyperactivity, which makes it difficult to identify special criteria involving mental content needed for valid diagnosis. Criteria of diagnosis have to be modified to accommodate for the ambiguity, complexity, and variegated nature of psychopathology among the mentally retarded (B. H. King et al. 1994; Sovner 1986). Moreover, even expert diagnosticians may be biased as a result of associating psychopathology with mental retardation (Reiss, Levitan, and Szyszko 1982).

Clearly, intellectually handicapped adults or children cannot be equated with hominids and earlier varieties of the species Homo. A compromised human brain is anatomically different from that of primates (see Deacon 1997), and the linguistic, cognitive abilities of intellectually handicapped humans are qualitatively different from those of the higher primates. Clinicians working with intellectually handicapped adults and children, as well as behavioral pediatricians, can sensitively point to myriad subtle, compelling, and engaging features of psychopathology among their patients that reflect fundamental human capacities and aspirations. Thus, to uncritically equate such human populations with primitive man is grossly improper. However, there is a tradition of exposition and even empirical research that compares the cognitive, communicational abilities of nonhuman primates with infants and juvenile humans which provide models of intellectually handicapped adult human beings (Cheney and Seyfarth 1990; Dennett 1987, 1995; Tomasello and Call 1997). We could perhaps learn something about the origins of psychopathology, then, by examining emotional and behavioral disturbances in very young children and individuals with different degrees of cognitive impoverishment or intellectual handicap, just as behavior suggesting psychopathology in higher apes such as chimpanzees has been used to draw inferences about possible varieties of psychopathology among the earliest hominids.

Given these limitations, it appears that forms of psychopathology among individuals with limited cognitive capacities (or limited intellectual function, emotional expressiveness, and overall social psychological functioning) manifest globally complex and mixed behavioral, emotional, and cognitive changes rather than discrete, clearly manifested symptoms that can be linked to the standard mental faculties Western-trained psychiatrists and psychologists find to be disordered. Rather than exhibiting discrete, singular mind/brain dysfunctions that fit classic diagnostic criteria, retarded people exhibit rather global alterations of social and psychological behavior, such as breakdowns in cognition, emotional disregulation, strong emotionality, and a failure to perform ordinary, routine tasks and activities.

Strum and Latour (1987) contend that social action in early hominid communities was complex in nature (see also Gamble 1996). By this they mean that the reconstruction of social order as a result of stressors like those causing and posed by psychopathology was likely to have been coarse and truncated, undertaken "only with limited resources, their bodies, their social skills, and whatever social strategies they . . . [could] construct." Such social skills "are limited to negotiating one factor at a time" (Strum and Latour 1987, 790). In contrast, at the Homo sapiens end of the evolutionary continuum, social life is extremely complicated. "Something is 'complicated' when it is made of a succession of simple operations. . . . Such complicated social life draws on the use of symbol, place and object to achieve this simplification" (791).

Extrapolating from this formulation, one might speculate that among members of evolved Homo complicated psychopathologies would be configured and enacted in a social and cultural ensemble that was differentiated, meaningful, and composed of several simple operations or units of behavior entailing varying amounts of display, gesture and gesticulation, grooming or avoidance, aggression or timidity, anger or reconciliation, language, performance, sign, and symbol. On the other hand, in the event of a dramatic behavioral perturbation causing or stemming from psychopathology, hominids and early varieties of Homo are likely to have displayed complex psychopathologies. Here, one would presumably see global, condensed compounds of behavior that reflected general and compelling features of a social situation and which perhaps fused a large number of personal and group concerns. Members of the group could cope with such behavior in a decisive manner and by means of a narrow range of options that involved many group mates to quickly reestablish stability.

At least four additional factors may have influenced how psychopathologies were enacted during early phases of human evolution. First, the exigencies of sheer survival limited what could persist and thrive in a community. Because social cognition and social organization among the earliest hominids was still unelaborated, comparable to that of the last common ancestor (inferred from observations of contemporary chimpanzees), it is doubtful that deviant or asocial infants would have survived, nor would adolescents or adults who displayed behavioral breakdowns. Second, because efforts at healing could not have been very complex or elaborate—again, judging from chimpanzee behavior (Fábrega 1974, 1975, 1997)—only forms of psychopathology that communicated clear forms of suffering, debility, enfeeblement, need, or somatic manifestations would have elicited affiliative, tolerant, healing responses. Third, while many psychopathologies interrupted the flow of behavior in a group and compelled attention, many did not. Some behaviors caused by disruptions in development and the social environment, such as communicative deviance, social pathology, irregularity of temperament, or personality disorders, while played out in complex networks of social activities, might not have necessarily grossly interfered with ongoing social action. The latter varieties of psychopathology would have been enmeshed imperceptibly in everyday community activities. While also socially complex, these forms of psychopathology should be regarded as socially invisible in the group, camouflaged by behavior relating to foraging, mating, parenting, and group activities directed at avoiding predators and seeking safe living places. Finally, the effects of psychopathology on individual fitness and increasingly, as hominids advanced toward protoculture, on the safety and integrity of the group would have determined how psychopathology was manifested and endured. In other words, the cost of psychopathology, given a group's prevailing social ecology, would have influenced how it was configured, enacted, and handled by group mates. My analysis here can be criticized as medicalization and anthropomorphism, to be sure, but its logic merely derives from scientific generalizations in ethology, primate cognition, paleoanthropology, and archeology; and it is not inconsistent with general tenets of evolutionary psychologists who promote the idea of psychological adaptations and algorithms.

Intrasexual Conflicts and the Conditioning of Psychopathology

Competition within the sexes and mate preference are the two main parameters that describe and condition sex differences in mating and reproductive behavior in primate groups. The same parameters can be assumed to have applied to hominid groups during evolution. In primates, differences in body and canine size are generally interpreted as indicating how sexual selection was expressed in male-male competition for social dominance and for mates, whereas female selection for winners of these rivalries played a dominant role in the way reproductive strategies are realized. Similar patterns may have characterized early hominid groups, causing one-on-one male competition and aggression. With the emergence of varieties of Homo, especially Homo erectus, sexual dimorphism lessened somewhat and came to approximate that of humans. Nevertheless, sexual dimorphism played an important role in conditioning how hominid males and females related to one another. Male behaviors—such as desire for mating variety, desire to control female sexuality, desire for certainty of paternity, alliances with other males, paternal provisioning—and female behaviors—such as competition for the dominant male, desire for protection and resources from males, desire for social support from kin and friends—figured centrally in the patterning of male aggression against females, in sexual coercion, and in patriarchy more generally (Smuts 1992, 1995; see also Geary 1998).

Styles of mating and parenting behavior depend heavily on the context of the group. At the same time, in pursuing their respective reproductive strategies, males and females created and perpetuated relations of conflict and stress both within and between the sexes. During phases of human evolution, such conflict must have affected sex differences in mating and parenting behavior specifically, and social and psychological behaviors within the sexes and between males and females generally. Consequently, how the sexes oriented to each other and to potential and actual partners can also be assumed to have conditioned manifestations of psychopathology within hominid groups.

During early phases of human evolution, there must have been competition between individuals, and later, perhaps, among varieties of Homo characterized by multimale and multifemale groups, there may have been coalition-based male competition for female mates. Males competed for control of resources with which to secure mates, as well as for more aggressive control of female reproduction, and linked to this competition were females' preferences for mates. In other words, females preferred dominant males as potential mates, since they were more likely to provide food for them and their offspring and to protect them from violence by other males. Sex differences in behavior may have devolved from more conflictual factors linked to sexual selection per se, such as male-male competition and female mate preferences, as compared to less conflictual ones, such as a sexual division of labor based purely on subsistence activities (see Geary 1998).

It follows that sexual jealousy and rivalry likely caused tension and stress, social conflict, hostility, and outright emotional distress in groups of hominids. Females and their offspring undoubtedly experienced psychological trauma as a more or less regular accompaniment of relations between the sexes, and such trauma was likely an important cause of psychopathology in those who were more vulnerable. Sexual rivalry would not only be a potential cause of psychopathology but undoubtedly shaped the content of many episodes of psychopathology, regardless of cause. Males' suspicion of cuckoldry or infidelity, and guarding and monitoring their mates were common sources of worry, while females must have experienced similar anxieties, as well as fear of sexual coercion and control. Such tensions undoubtedly culminated in male-male aggression, as well as male-female aggression and sexual assault. Added to this cauldron of social and psychological tension was the constant threat to the dominant male's authority, or assaults by male coalitions of kin or other closely related individuals in a conflict over power, which aroused fears of aggression and violence directed at infants, if not outright infanticide. Such an atmosphere of threat and potential violence must have contributed directly to a vulnerability to psychopathology on the part of both males and females, in particular during infancy and childhood, with potentially harmful effects on later functioning.

During later phases of human evolution, and possibly even earlier in favorable social and behavioral environments, more stable friendships and affiliations between males and females probably emerged. An exchange could have developed whereby the male provided food and protection, while the female reciprocated with sexual favors and/or exclusivity (Smuts 1985, 1992, 1995). At some later phase, a move toward more stable, long-term male-female relations, and in some instances monogamy, may have supervened. Such a development is assumed to have been accompanied by other changes, such as concealed ovulation, continual sexual receptivity by the female, and a greater paternal investment in rearing offspring (Geary 1998). The establishment of longer-lasting, more stable male-female relationships may have moderated and lessened the group tensions associated with mating and thereby lessened the prevalence and especially the intensity of social crises, conflicts, and violence surrounding sexual behavior. This is, of course, highly contestable: Many cross-cultural studies of patriarchy, male aggression toward females, and sexual coercion suggest that themes prevalent during human evolution continue to this day, often supported by male-governed social institutions (Smuts 1992, 1995).

One can assume that psychopathology associated with acute disturbances involving male-male and male-female violence instigated by conflicts over power, resources, and mates, and reducible to competing sexual strategies, became less frequent and intense with the emergence of more stable affiliations between males and females (Geary 1998; Smuts 1992). This assumes similar variations in environmental stability, pressure from predators, availability of resources, and size of group. As crises caused by sexual competition diminished, there would also be less psychological and interpersonal distress and agony. However, because of the biologically based reproductive strategy of males to seek multiple mates, the potential for polygyny that seems to prevail among nonhuman primates and humans generally, which is likely to have existed among hominids and earlier varieties of Homo, and the male's difficulties in guarding his mate, owing to concealed ovulation and extended sexual receptivity in females, sexual relations and strategies must have been continuing sources of worry and agony for males, and fear and even terror for females.

Therefore, even as male-female affiliations became more stable and longer lasting, differences in sexual behaviors and tensions continued to shape the concerns of both males and females. This added to the range of nonreproductive stressors inherent in group life that must have produced psychopathology in vulnerable individuals. In brief, social and psychological behavior, and particularly instances of possible psychopathology, continued to be modulated by concerns linked to sexual selection and competition. However, as males began to share in the work of infant and child care, and thus could be more certain of their paternity, other conflicts rose into prominence. For example, differences between males and females regarding choice of mates, adequacy of resources, and strategies of parental investment in the care of dependent offspring undoubtedly contributed to continuing tensions and conflicts (Geary 1998).

Social and psychological behaviors in males and females arising from differences in reproductive strategies should be viewed as a principal (if not dominant) leitmotif in hominid group activities and in the everyday concerns of individuals. Consequently, they must have also played a prominent role in causing and shaping psychopathological behavior during human evolution. Research conducted by anthropologists and psychologists influenced by evolutionary theory gives much attention to the important influence of strategies developed during evolution on contemporary human personality development and family dynamics. As an example, a range of adult behaviors have been explained as an outcome of ecological conditions prevailing during a child's early development; specifically, the consequences of a father's absence compared to a father's presence (Draper and Harpending 1982, 1987, 1988). Experiences during the first five years of life, a sensitive period of learning and development, are thought to affect differences in cognition, emotion, attitudes, and behavior tendencies of males compared to females in mating and parenting specifically and social coping in general. In essence, a child's perception of the ecological environment is said to contain clues that entrain developmental programs in a distinctive route that proved to be adaptive during phases of human evolution. It is not difficult to see the roots of particular forms of contemporary human psychopathology, such as antisocial behavior and sociopathy, as related to thwarted developmental programs in a child's social and cultural environment (Harpending and Draper 1988). Consequently, analogous forms of psychopathology can be assumed to have arisen during human evolution, at least during later phases.

Developmental psychologists have also linked problems of mating and parenting with contemporary family conflicts and violence, including child maltreatment and abuse, in both industrial and nonindustrial societies. While evolutionary factors may be ultimate causes of these forms of psychopathology, factors linked to contemporary ecological stress caused by economic deprivation, overcrowding, unemployment, and ineffectual parenting and poor interpersonal relations are seen as proximal causes (Burgess and Draper 1989). Family violence and child maltreatment are forms of maladaptive social behavior that are strongly correlated with individual psychopathology. Thus, given the universal possibility of ecological stress in all forms of hominid sociality, similar forms of psychopathology can be assumed to have prevailed during human evolution.

Male Sexual Coercion: A Test Case of Psychopathology during Human Evolution

As noted earlier, evolutionary biologists and psychologists have a very different point of view from that of clinicians regarding social and psychological behavior. The issue of male sexual coercion is an example. Male control of female sexuality and frank sexual coercion of females leading to physical aggression and violence—sometimes including the killing of infants or child maltreatment—are common among nonhuman primates and could well have prevailed during human evolution (Geary 1998; Hrdy 1979, 1981; Smuts 1992, 1995). A host of issues of evolutionary import are linked to such behavior, such as differences between the sexes in mating choice, mating strategies, and parental investment in offspring, as well as male fears of female promiscuity, cuckoldry, and concerns about paternity. Such anxieties are held to explain hostility and violence toward females, including rape, homicide, infanticide, and child abuse. In contemporary human populations, such behavior is widely recognized as evidence of severe social malaise, as signs of a society's moral and ethical breakdown. For clinicians, it can signal the presence of distinctive types of male and even female psychopathology (for example, Daly and Wilson 1988).

Sexual coercion and associated behavior are found in the living ancestors of Homo sapiens, or nonhuman primates, and have been observed in a number of societies with different types of social organization, size, complexity, and culture (Smuts 1992, 1995). Because such behavior is widespread, recurrent, and embedded in a variety of social institutions that seem to favor males over females, it presumably reflects conflicts of interest relevant to fitness that have existed throughout human evolution. From one point of view, then, the behavior makes sense in light of evolutionary biological theory, while from another it can also be understood as contributing to, if not evidence of, psychopathology. Let us examine the differences between the approach to such behavior taken by an evolutionary biologist or psychologist compared to that of a clinician. Relevant questions are the following: At what point during human evolution can one safely and unproblematically add to an evolutionary interpretation a clinical one that would interpret such behavior as psychopathological? If a pattern of maladaptive social and psychological behavior of an individual hominid appears to drive forms of sexual coercion and/or violence, can the behavior and the individual be qualified as psychopathological, in spite of the seeming rationality and naturalness of the behavior from the standpoint of evolutionary theory? At what point during hominid evolution, and for what reasons, can repeated acts of sexual coercion be explained clinically as well as in evolutionary terms? When might an external observer of a hominid group appropriately conclude, on the basis of objective, evolutionary theories of behavior, that varieties of sexual coercion are normal and natural, not abnormal or psychopathological? Does the hominid group's level of cognition and culture play a role in how the behavior should be characterized? If so, what factors should be taken into account?

According to the theory of human evolution, individuals are products of a developmental process that has endowed them with psychological mechanisms and programs of behavior designed to help them survive and reproduce. These algorithms are partly innate and partly learned from the social group, until cultural rules of conduct are internalized. Early in the evolutionary process, the algorithms can be assumed to have been physiologically and behaviorally compelling and activated by fundamental biological motivations. Their outright expression could be balanced or held in check by imperatives dictated by qualitatively different algorithms competing for expression. These balancing imperatives of the actor were similar to algorithms underlying the behavior of group mates or conspecifics that were associated with individual and group responses to prevailing environmental circumstances and local contingencies. Changes in group composition or instabilities in the social ecology could disrupt the balance and integration among competing programs of behavior, causing vulnerable individuals to contravene established conventions. In short, behaviors having compelling biological functions and goals fundamental to survival and reproduction were in competition with and had the potential to thwart, overwhelm, and injure group mates and conspecifics.

Whereas sexual coercion must have been commonplace, or frequently attempted, during human evolution, it might have been held in abeyance, modulated, and regulated by other motivations and goals of group mates, depending on circumstances. While the logic of evolution in a highly social species demands the constraint of the unbridled expression of behaviors such as those considered here, they invariably burst forth into the group's social spaces under pressured, challenging ecological circumstances. Each individual expresses a mixture of such imperatives and unique endowments that modify their expression and intensity. These would include, for example, differences in cunning, physical attractiveness, temperament, personality, strength, and intelligence. The social and psychological behavior of any individual is affected by all these factors, and such behavior could include forms of sexual violence and coercion, as well as psychopathology. Such behaviors, in effect, express basic and fundamental routines and programs related to reproduction and survival.

In one light, the individual whose behavior includes sexual coercion and violence and/or infanticide can be regarded as expressing natural functions and seeking goals consistent with the pursuit of survival. (My example presupposes a male individual, but a female could show similarly aggressive behavior or mate preferences that stoke male jealousy.) A male's failure to assess accurately the potential responses of males placed lower in the hierarchy, or the character of the bond between the female target and her kin, friend, or would-be partner, would thwart his reproductive needs. It is enough to explain his behavior as motivated purely by evolutionary considerations. However, a series of futile attempts at hostile matings or infanticide, especially if they lead to personal injury or social isolation and rejection by the group, could suggest that he lacks resourcefulness or adaptiveness. Similarly, if his repeated actions correlate with an elevated level of testosterone or a lower level of serotonin, he might even be seen as deviant or maladaptive, if not incompetent. Deficiencies of social intelligence, problems involving the regulation of attention and emotion, or a bad temperament could also explain the actor's repeated and futile efforts to obtain a mate and reproduce (Rutter and Rutter 1993). In short, while a pattern of behavior might conform to evolutionary imperatives, it might also, given extenuating circumstances, fit the clinician's diagnosis of psychopathology.

The restraints that regulate more or less innate reproductive strategies of behavior are an important consideration in determining whether sexual coercion is normal or pathological. On the one hand, at the ape end of the human evolutionary continuum, pure success in outcome, regardless of the situation, or an effective balancing and modulation of biological functions and goals in light of contingencies tied to social ecology would appear to explain incidents of sexual coercion and violence. On the other hand, these outcomes would appear to logically preclude a qualification of psychopathology. A clinician who observed syndromes of behavior involving sexual coercion would probably have little difficulty in raising the question of psychopathology, a reasonable diagnosis if any of the extenuating conditions listed earlier prevailed. The behavior, if repeatedly thwarted, would be seen simply as causing distress and social-psychological impairment, as well as being antisocial and indeed socially destructive. It should be emphasized that the same logic would apply to any number of nonreproductive behaviors tied to evolutionary imperatives, such as those of foraging and hunting, avoiding predators, and relating and competing with others in general.

One could argue that behavior cannot be called pathological before a group's cognition and culture have developed concepts of appropriateness, rights, responsibilities, social obligations, and rules, or before the emergence of notions of human malevolence, suffering, and disease and injury. That is, not until values, morality, and distinctive patterns of mental attribution and social accountability have emerged should the possibility of psychopathology be allowed to nullify a satisfactory explanation of behaviors such as sexual coercion and violence, if these are based on the logic of evolution. Much of the behavior of nonhuman primates, and, by extension, of hominids, that might appear psychopathological in human groups are the result of more or less innate evolutionary routines and programs, that is, behavior designed by natural selection and governed largely by genetic considerations. Human culture considered in its symbolic sense—as a system of social symbols and their meanings—can be viewed as partly designed to modify the unbridled pursuit of biological functions and goals. Thus human moral considerations, realized in the form of systems and values involving approbation and disapprobation, are the basis for characterizing behaviors such as sexual coercion, violence, and infanticide (no less than stealing or homicide) as psychopathological. The science of psychopathology and institutions designed to prevent, undo, and meliorate antisocial, aberrant human behavior can be viewed as extensions of a society's systems of morality that have evolved culturally to explain and control routines of behavior rooted in evolutionary algorithms.

The clinical versus evolutionary quandary over the meaning of behavior resembles a more general dilemma inherent in the study of psychopathology: the problem of cultural relativism. To say that behavior cannot be called pathological until a system of shared rules and conventions has evolved in a group of hominids suggests that one must take into account the social meanings of such rules and conventions before applying a diagnosis of psychopathology in the first place. If a society's institutions condone forms of sexual domination, control, and even coercion and violence, under appropriate circumstances, then such behavior could be said to be normative and prescribed by prevailing standards and values. To call behavior psychopathological, the cultural relativist would suggest, is to condemn and disapprove, and even to sanction therapeutic intervention, yet the members of the society in question, using extant community standards, may not judge the behavior as in any way pathological but rather as conforming to local conventions.

Arguments of this type have moved theorists in psychology and especially psychiatry to further medicalize psychopathology. This has involved identifying biological markers or indicators of psychopathology and making claims about universal systems of psychiatric diagnosis and classification applicable across societies, regardless of cultural conventions. If criteria for psychiatric disorders could be rendered in a culture-free language to describe deviations in physiological, neurobiological, neuropsychological, and neurovegetative behaviors, and if—a very big if—clinicians could apply such criteria to syndromes of behavior and distress unproblematically, regardless of local cultural conventions, they would be spared the dilemmas implicit in cultural relativism. However, such an escape from local contexts and meanings is difficult, and a society's systems of symbols and their meanings must always be taken into consideration before making a claim of psychopathology, as I have argued elsewhere (Fábrega 1994a, 1994b, 1995). Thus, a continuing controversy in psychiatric theory has its counterpart in the clinical versus evolutionary quandary over how the concept of psychopathology is to be used.

Comment

When conceptualizing psychopathology purely in relation to ecology, as a concrete, mechanically situated incident in a group of hominids, keep in mind that one describes the condition of an individual whose evolutionary grade is variable. When such a condition is realized, or can be labeled or diagnosed by a suitable observer, it has at least five measurable properties: (1) potential causes, including genetic, innate, developmental, ecological, and social factors (the immediate precipitant of an instance of psychopathology can also be thought of as involving social stress); (2) manifestations or a descriptive profile, including such things as cognition, emotion, social behavior, and physiology, that is, bodily manifestations; (3) a temporal record or profile that indicates whether onset, or recognition, was abrupt or gradual, whether it was resolved quickly or slowly, and overall duration; (4) the individual's relationship to the group and group relations linking individuals promoting, or discouraging, harmony, acceptance, and approval; and (5) actual or potential effects on fitness, including mate selection, fecundity, reproduction, parenting, mutual behavior, foraging and subsistence, food procurement and processing, avoiding predators, and aggression.

A basic yardstick for calibrating psychopathology is to evaluate an individual's effectiveness in achieving biological goals. Clearly, a hominid's success in the social arena, as well as in foraging and hunting activities, would have been a consequence of many personal attributes other than the relative absence of psychopathology. And, of course, merely equating the degree of biological resourcefulness with behavioral normality and absence of pathology is to make an anthropomorphic judgment that has been philosophically challenged (see Kennedy 1992). These caveats aside, it is still constructive to push a strictly social and behavioral slant on psychopathology in the attempt to seek possible clarification of evolutionary origins.

A hominid's general state of health, physical size and strength, stamina, quickness and agility, sensory acuity and responsiveness, sheer neuromuscular coordination, and the intrinsic organization of the central nervous system are undoubtedly correlated with biological resourcefulness—the ability to cope with stress and to promote welfare and fitness (see Broom and Johnson 1993; Dawkins 1980, 1990; McFarland 1989, 1993). Individual variations on these measures undoubtedly enter into the estimation of an individual's temperament, social intelligence, and even personality, a term that many primatologists and comparative psychologists are wont to use to explain an individual's degree of general resourcefulness. Given the many influences on behavior, it would be nonsense to attempt to set apart subtle forms of psychopathology from an individual's relative lack of ingenuity or social intelligence in pursuing subsistence needs and social goals. At best, it would require keen observational skills and scientific knowledge about a hominid or nonhuman primate group to be able to invoke psychopathology as a factor in explaining an individual's well-marked deficiencies and failures in the group.

If one were to offer a hypothesis of a putative psychopathology during the phase of evolutionary development associated with very early hominids close to the pongid line, one would have to say that its manifestations would consist of abnormal motor and social routines, including forms of emotional behavior that occur in relation to internal, brain-related or physiological, changes and are reflexively but inappropriately elicited as a result of failures to properly interpret the stimuli arising in the physical habitat or from others' actions. The symptoms would tend to undermine normal social routines and impair the individual's ability to efficiently meet subsistence and mating needs. Impaired well-being, particularly the inability to perform even automatically desired, required, designed, innately programmed adaptive routines—feeding, seeking shelter, avoiding dangers—would have to be qualified as suffering, as Dawkins (1980, 1990) argues. Emergent in these early hominids were symptoms or manifestations of psychopathology carrying a symbolic or communicative significance, either to self or others. In short, emergent and perhaps implicit in a social and behavioral ecology of psychopathology among hominids were conditions of sickness, stress, social conflict, and even suffering considered as self-identified representations that could have been related to possible causal factors, thought about, and expressed to group mates. In certain respects, then, varieties of psychopathologies were not mere mechanical, mindless behavioral actions or routines, poorly articulated and poorly integrated to notions of self-identity and social cognition more generally. Finally, whereas the evolution of culture and cognition certainly introduces complexity into the understanding of behavior, as illustrated earlier, it cannot eliminate the dilemmas inherent in the tension between clinical versus evolutionary accounts of behavior. Such quandaries about normality and abnormality continue to plague modern, biomedical psychiatry.

- Social ecology describes the relationship between the physical, biotic environment and size, composition, and structure of a nonhuman primate or hominid group; behavioral ecology adds to this characteristics and patterns of social and psychological behavior.
- Paleoanthropologists devise models of the social and behavioral ecology of hominid groups that provide an evolutionary baseline or stage for normal behavior onto which one may project hypothetical pictures of psychopathology.
- Hypotheses about social behavior including psychopathology during human evolution should ideally integrate aspects of physical environment, group demography, social relations, and evolving competencies in the areas of language, cognition, and culture.
- Given the sparse archeological remains, scenarios of early hominids are abstract, static, formal, and relatively devoid of details about social relations and culture, but those about Archaic humans are richer, although their interpretation is highly contested.
- Hypotheses about the activities and happenings in hominid groups during phases of human evolution allow one to construct hypothetical scenarios of what their psychopathologies would have been like.

- Early varieties of psychopathology on the evolutionary terrain were probably highly emotional, consisting of global, acute, and relatively condensed behaviors, and responses to them based on symptom intensity, group dynamics, and existing ecological pressures.
- Starting with later varieties of Homo and Archaic humans, it is likely that psychopathologies were more finely graded from the standpoint of language, cognition, emotion, and culturally interpreted social and ecological happenings.

Chapter 10

The Content of Psychopathology during Evolution

Since psychopathology shows up as social and psychological behavior, we must consider characteristics that underlie and shape such behavior. As a function of changes in cognition and culture, behavior of hominids more and more came to resemble that of humans. This chapter will discuss some of the changes in culture and cognition presumed to have taken place during different phases of human evolution, focusing on corresponding changes in how incidents of psychopathology were configured and enacted as a consequence.

Modeling Cognitive Changes in Animal Behavior during Evolution

Drawing on knowledge from general biology and comparative psychology, Graham Richards (1987, 1989a, 1989b) proposes that behavioral guidance based on information copied from environmental objects and events is the root process explaining the change from animal to human behavior, including phases and even stages of hominid evolution. Richards's model is termed *physiomorphic learning* and is conceived of as a generalized form of imitation. He begins by characterizing the lives of animals, like the primates', as "consisting of a sequence of . . . life routines: obtaining food, grooming, finding or creating shelter, avoiding predators, mating, etc. These life routines themselves comprise sequences or 'programmes' of behaviors on the completion of each of which the animal is in a particular state from which it can proceed to the next" (1989a, 245). He views these sequences as "highly genetically programmed" but capable of being interrupted by environmental events that can shift the animal into alternative life routines, the original routine being held in check for later completion.

Richards's model is highly consistent with postulates of decision theory as applied in ethology (McFarland 1977). In this perspective, an animal's decision-making apparatus, designed by natural selection, enables it to operate on the basis of an objective function that describes the animal's own internal states and which it controls by means of built-in design principles to regulate behavior. It is also a cost function that describes the environment in terms of the availability of what it needs, or impediments to fulfilling these needs, so as to meet its goals. "Thus we can expect the laws governing learning to be designed in such a way that they tend to shape the objective function of the individual, and make it more like the cost function of the current environment" (McFarland 1977, 20). McFarland provides a synchronic, static, and formal account of how animals behave in natural habitats, whereas Richards's formulation is diachronic, more dynamic, and more descriptive. Richards's model, moreover, gives an account of a distinctive class of animal behaviors, how the behavior of primates and hominids might have changed during evolution.

According to Richards, in more evolved animals like mammals and especially primates, life routines come to be implemented so as to allow flexibility, resulting in behavioral variations that enable the animal, if a basic program is blocked, to copy or draw from the environment information needed to reach a particular desired completion state. The copying of environmental information can involve the behavior of conspecifics, the properties of objects—for example, the hardness of stones serving as a clue for how to shatter nuts or bones—or any environmental event or fact that produces what the animal desires or finds useful. Richards represents an imitative behavior sequence by schemata of the form X causes Y: "Basically, in imitation, what is learned is that a certain behavior X has an outcome Y, this sequence having been identified in the environment, X being within the organism's capacity to copy (or co-opt, e.g. in the case of using a stone tool the organism is co-opting the properties of broken stone into its own repertoire) and Y an outcome which furthers a life-routine" (1989a, 108). Thus, a particular hominid's behavioral repertoire consists of a proliferation of such schemata. This description probably applies to the last common ancestor and certainly covers the earliest stages of hominid behavior.

Richards contends that progressive imitation and incorporation of environmental information, also termed imitative modeling, eventually leads to a change in how the organism stores information. "Instead of tagging imitatively derived schemata to life routines, it is simpler to reverse the process and tag life routines to schemata" (1989b, 246). The result is that "instead of the environmental phenomena's being given meanings in terms of the organism's motivational states, the organism's motivational states are given meanings in terms of environmental phenomena. 'Shelter means leaves' reverses to 'leaves mean shelter,' 'water carrying means gourds' to 'gourds mean water carrying,' and so on. . . .

[This] represents a radical reversal of the way in which all animals have hitherto related to their worlds" (1989b, 246).

This reversal ushers in what Richards defines as the third stage of a five-stage process of human evolution: "A hominid living in this mode will be experiencing the environment as a (still by our standards restricted) catalogue of instructions—big leaves are read as 'gather for shelter if you haven't already.' There is no need to wait for the 'leaf-gathering' moment in the routine or to represent the leaves in detail internally. The satisfactory enactment of life routines is ensured by embedding them, so to speak, in the environment itself so that they all remain continually accessible instead of chaining consecutively; . . . the external world (including, of course, other hominids) becomes a 'library' of behavioral modes and existential properties" (1989b, 246).

Richards elaborates these stages with reference to morphological changes in the hominid line—in particular, bipedality, hairlessness, and manual dexterity, all of which facilitate the creation of a midday foraging niche that enables effective competition against carnivore predators. The fourth stage involves coping with the behavioral diversity and complexity associated with the interruptions and detours involved in enacting life routines, which become longer, more convoluted, and intertwined. The developments during Richards's stage 4 involve a progressive expansion in an organism's sense of inner control and agency in the pursuit of life routines, an awareness of its needs, pursuits, and goals:

> As the ratio of genetically-wired-in behavior to imitatively-based behavior gets smaller, as the actual enactment of life-routines becomes more flexible and complex, the role of the original, fully-elaborated, "wired-in" life-routine becomes less and less important. What was originally the central point of reference for all behavior [that is, a particular life routine], from which X-Ys were but short detours, becomes increasingly redundant. What remains significant is the motivational basis for the life-routines. (1989a, 110)

> In this fourth stage the situation can again be dramatically simplified by abandoning direct life-routine tagging altogether. We now have a mode of learning, of behavioral amplification, based on the simple rule "For any outcome identify an operation that yields it and copy." Life-routine tagging *as an essential aspect of acquiring new behavior* becomes *obsolete* once this radical generalized solution has been recognized. . . . All that may now remain of a previous life routine is its end state. The programmed sequences of component behaviors become redundant. The individual is required only to know what he or she wants; getting it is simply a matter of knowing what, in the present circumstances, can bring it about. (1989b, 247; emphasis in original)

> Under these circumstances the exploration of the X-Y possibilities in the environment acquires an imperative of its own. They are no longer

interesting only insofar as they can be immediately related to a particular life-routine but because *any* X-Y might prove valuable for some purpose or other at some time in the future. The advantages to the individual, and *inter alia* to the group, of maximizing their repertoire of such "knowledge" would soon have become apparent. (1989a, 110)

Stage 4, then, involves the advent of a *general problem-solving strategy independent of any life routine* that can be equated with the behavior of self-aware and group-aware organisms acquiring, storing, and using knowledge derived from the environment (including conspecifics) to maintain themselves and their kin, as compared to merely carrying out life routines.

The fifth stage involves a further refinement by means of lexical encoding and the development of syntax to cope with "the more rapidly expanding range of environmental phenomena being assimilated into the behavioral repertoire and the behaviors themselves. . . . The important outcome of Stage 5, in addition to syntax, is that knowledge storage again shifts location, from the natural environment to the cultural one" (1989b, 248). Here, then, Richards is visualizing modern humans who have more or less acquired a full linguistic capacity and culture. Rudimentary forms of language and culture are undoubtedly found in stage 4 and perhaps even stage 3, as the organism is progressively freed from the need to innately tag life routines and learns, internalizes, and stores information from the environment as emergent knowledge structures or pools of information. However, internalized pools of information drawn from the environment, now linguistically and symbolically coded and integrated, and which entail self-centeredness and group-centeredness and reference, become more and more determining influences in behavior during stage 5.

Richards's model, then, allows one to visualize progressively more elaborated forms of hominid behavior. Starting with mammals and moving on to nonhuman primates and eventually to the earliest hominids, he illustrates how the behavior of Homo erectus, Archaic humans, and eventually modern humans progressively changed. Among the very earliest hominids, life routine behavioral chaining is largely under innate, genetic control. Then during the Homo erectus stage, evolution brings a lessening of these controls and more environmental information is incorporated into behavior. Eventually, the individual is released from rigid, innate controls and through progressive copying and imitation, and a shift in the location of clues on which to carry out life routines, learns to use information stored in the environment as a basis for carrying them out. Eventually, individuals learn to turn inward and, instead of innately pursuing behavioral programs associated with life routines, rely on their self-perceived needs, or motivational states, as a basis for behavior. Linguistic competence—lexical encoding and syntax—characterizes the final stage of evolution, which entails an elaborated form of self-awareness and self-conceptualization, although

Richards acknowledges that emergent varieties of psychological awareness and efficient communication can be associated with earlier, oral modes of communication. An important development in this process is culture.

Richards (1989b) holds that culture emerged as a consequence of language. Both are construed as devices for amalgamating and domesticating the diverse behaviors made possible by evolution. In essence, culture is a template of symbols and their meanings that enabled humans to integrate and focus individual behavior on the group. Richards contends that the enormous amount and variety of physiomorphic copying and imitation made possible by the protean nature of human morphology and motor behavior (the means by which environmental information is modeled, copied, and imitated) constituted a threat to social order. In other words, individual behavioral variations and the acquisition of knowledge made possible by physiomorphic learning posed a major challenge to the organization, integrity, and stability of hominid groups. Groups were threatened with fragmentation as a result of behavioral diversity. Culture was the symbolic glue that kept humans centered on matters of importance to the group, which kept the group from fragmenting.

We can use Richards's formulation as a general framework for obtaining a panoramic view of the revolutionary changes in hominid behavior. His model is nothing if not broad and comprehensive, for it incorporates notions of life routines; selfhood, a process that also involves the development of psychological language); behavioral variation; and the emergence of language, culture, religion, cosmology, and mythology in establishing the human mode of adaptation (Richards 1989b).

Physiomorphic Learning as a Model of the Structure of Psychopathology during Evolution

Richards's frame of reference, which he terms *physiomorphic learning*, can be used to obtain a general picture of psychopathology during human evolution. Psychopathology involves a behavioral ensemble brought about as a result of an abnormality or disturbance. This cause would include, for example, a lesion, a disease, injury, or pathology, or a metabolic/physiological perturbation produced by social or environmental stress that affects developmental processes or the executive system of a vulnerable organism.

Psychopathology during Richards's first two stages of innate life routine chaining would consist of breakdowns in the capacity to carry out basic life routines such as finding shelter, feeding, or mating. The breakdown results in a failure to effectively use information from the environment so as to insert it into a life routine and thus reach goals dictated by innate programs. As an example, a feeding routine may be triggered by a physiological or chemical signal or an external stimulus. Appropriate sensory and motor actions are directed at looking for plants or small mammals to eat. However, the routine may fail because the individual does not draw on the target information presented by plants or mammals. In other words, a food resource coded in information terms will not be properly inserted into the sequence of life routine chains. Likewise, a mating routine may be triggered by a hormonal signal but not properly executed; the appropriate partner is not approached or is approached through the wrong routine. In Richards's scheme, then, a breakdown of behavior, the manifestation of a putative psychopathology, consists of faulty chaining of life routine units. Parts of unconsummated life routines may persist, but overall coordination and execution would break down. The ability to shift across routines or keep an interrupted routine in abeyance might also be impaired.

At stage 3 of Richards's model, wherein the environment is now the stimulus for organizing and carrying out life routines, the structure of these routines may not necessarily show a breakdown in organization or integration in response to environmental information. Instead, individuals may use it for inappropriate ends; for example, they may collect environmental objects without tagging effective routines to them. In this instance, the environment might trigger the wrong life routine, which might be produced without the fragmentation of the routine, as was the case earlier. On the other hand, since neural structures are built upon phylogenetically older structures, there may well be remnants of earlier dislocations in the chains of a routine that in turn would be improperly linked to environmental inputs. This would mean that information relevant to one routine might be used to carry out another, so that the latter would be wrongly constructed or fragmented. Moreover, psychopathology could affect more than one life routine; more than likely, all will be compromised, but perhaps in different ways. For example, a disturbance consisting of the ineffective use of environmental information, which includes conspecifics, would prevent an afflicted hominid from making effective use of the ranking system and from responding to communications about group coalitions for defense against enemies.

Psychopathology during this stage, then, might involve slowness, puzzlement, disorganization, and perhaps even an awareness of failure to cope and thrive, the results of an abnormality that disrupts the individual's capacity to use environmental information to pursue life routines. Since normal group mates share a capacity for self-awareness and social attribution, they may also be puzzled by the psychopathological behavior, in which case we should take into consideration their responses and labeling of the psychopathology. Perhaps recognition of failures, deficiencies, and impairments may cause group mates to accommodate to an individual's needs, expressed as food sharing, grooming, supporting, and helping. Such accommodations have been described for chimpanzees (Goodall 1986) and even for rhesus monkeys (de Waal et al. 1996).

During Richards's stage 4, individuals have solved some of the problems of behavioral diversity. As a result, long and convoluted life routines are set in motion by environmental clues as desired end states are identified from motivational clues. Thus individuals can carry out their life routines more effectively in relation to conscious decisions about goals. They know and are able to pursue desired ends by balancing information about (1) personal needs, (2) group needs, (3) clues from the environment, including (4) specific information about planning for future and current life routines, now properly termed needs and desires, and (5) pools of information accumulated from prior transactions. In the event of a psychopathological incident, a particular outcome may be selected—now a plan, purpose, or goal—but the operation to carry it out would be faulty; or operations may be implemented, but seemingly geared to the wrong outcomes; or the individual may show deficiencies in identifying, selecting, or carrying out purposes altogether.

Since desired goals of selected life routines can now be thought of as social and group-defined goals and needs, psychopathologies are now manifested in behavior that can contravene the needs of group mates and of the group itself. Others may see behavior as disjointed or dilapidated and label it as deviant, abnormal, unsound, or unproductive. A well-developed capacity for self-awareness and self-conceptualization implies that individuals, depending on the type and severity of the disturbance, may show some understanding about the bizarre manifestations that accompany a psychopathology. Alternatively, if individuals have become highly socialized and repeatedly enter into core configurations that integrate individual and group needs (Caporael n.d.), the affected individual may not have personal insight into patently deviant, abnormal behavior.

Individuals located at stage 5 of Richards's scheme have resolved the problem of information storage and processing involved in balancing the self's and the group's needs and goals by the development of language. They can use the language and culture of the group not only to pursue desired end states but also to remain integrated within the group. In other words, the individual is anchored by an understanding of the self, of the group, and of its system of knowledge by the symbolic glue provided by language and culture. In this stage, then, in addition to breakdowns in the performance of overt actions, as at earlier stages, psychopathology will now be manifested in disturbances in how language is implemented and applied to behavior—particularly, how the glue of culture is used to define the self, others, the behavioral environment, and the group. Psychopathology can now produce not only faulty motor behavior chaining, disjunctions between chains and end states, or social behavior disturbances and illogicalities, but also confusion about beliefs, rituals, myths, and existential rationales. Psychopathology now enters the individual domain, a person's social, cultural, and historical placement.

This account leaves out how physiological manifestations of psychopathology figure in the equation. Fundamentally, one has to assume that behavioral ensembles are the products of hierarchical programs that regulate the individual's internal environment as well as overt actions and psychological experience, so that psychopathologies also incorporate bodily changes, given psychosomatic and somatopsychic unity (Damasio 1994, 1999).

The Evolution of the Modern Mind and Psychopathology

In discussing the origins of the modern mind and culture, the psychologist Merlin Donald (1991, 1993a, 1993b, 1995) formulates modes of cognition relevant to the evolution of modern thinking, memory, and culture. He posits an episodic form of cognition in apes, an early, prehuman form of mimetic culture and cognition that he associates with Homo erectus, a mythic-narrative-linguistic mode in Homo sapiens sapiens, and a theoretic technologically influenced mode in contemporary populations. This section notes the potential relevance of the mimetic phase of cognition. Donald's formulation relies on different concepts and emphases from Richards's, discussed above. While one can see a correspondence between the ideas of Donald and Richards, I do not attempt to do so here.

According to Donald, Homo erectus is assumed to have acquired the capacity for voluntary retrieval of information, cognitive rehearsal, and symbolic thought in terms of mimesis. This is a distinguishing landmark, he believes, since it contrasts sharply with the episodic culture or mode of cognition found in the great apes. Mimetic cognition involved voluntary retrieval of memories and knowledge, and a capacity for self-conception and social attribution. Whether this cognitive transformation was entirely prelinguistic or necessarily involved a form of verbal language is a controversial question (see Bickerton 1990; Gallup 1982; Povinelli 1993; Povinelli and Cant 1995). Donald suggests that the emergence of language required a prior development that involved a system for motor production: "Before they could invent a lexicon, hominids first had to acquire a capacity for the voluntary retrieval of stored motor memories, and this retrieval had to become independent of environmental cueing. Second, they had to acquire a capacity for actively modeling and modifying *their own movement*. Without these two features, the motor production system could not break the stranglehold of the environment; . . . the system must first gain a degree of control over its own outputs before it can create a lexicon or construct a grammatical framework governing the use of such a lexicon; . . . a revolution in nonverbal motor skill would have had immediate and very major consequences in the realms of representation and social expression . . . [and would] account for many of the nonverbal skills of modern humans" (1993b, 740). Such a cognitive transformation is assumed to have given hominids the ability to rehearse and refine motor actions voluntarily, guided by the body's perceptual schema in its environment, thus yielding an implementable self-image. Mimetic skill or a capacity for mimetic cognition that could use any part of sensory-motor and even visceral systems to construct representations means that what were previously only stereotyped, environmentally reactive emotional and social communications were now rehearsable and could render intentional mechanisms more visibly compelling in a social and psychological sense (Donald 1993b).

This describes a psychology and culture of early hominids characterized by action-centered cognition and gesturally realized expressive abilities strongly preadapted for language. Improved ability to retrieve and represent information about the self and previous social experiences would presumably lead to a refined capacity to plan, formulate intentions, attribute individual identity to group members and, more generally, to experience others as psychologically differentiated and personally vital. Given existing skills in social and emotional communication, such individuals would be capable of undertaking more elaborate social construction, social planning, social rehearsal, and a general review of ongoing social activity. They could remember past events and have a picture of group composition in which the self was differentiated from others. More refined sharing of individual, interpersonal, and group experiences, even if only formulated holistically in terms of actions and discrete physical accomplishments, would allow for greater emotional and psychological intimacy. Individuals with voluntarily retrievable memories would be able to construct psychological representations of personal and group experience, and thus could better formulate and express emotions. More intense bonding would render them more susceptible to the emotional and existential implications of disease and death.

In Donald's formulation, self-conception is anchored in the capacity for motor programming of action, in particular the voluntary retrieval of prior events involving the self. Interestingly, Povinelli and Cant (1995) offer a provocative hypothesis about the relationship between motor programming and the development of a sense of identity. Their presentation emphasizes a specific ability keyed to a special set of actions. They "argue for a distinction between locomotor systems governed exclusively by highly stereotyped action schemata, and those that exhibit a tremendous degree of flexibility, both during their use and between each successive use" (404–405). In this interpretation, environmental hazards associated with arboreal clambering was the environmental problem solved by developing new locomotion strategies that brought a sense of personal agency, which, in turn, presupposed self-identity. First of all, "an awareness of personal agency presupposes an identity onto which that agency can be mapped; . . . if one were to evolve a sense of personal agency, one would need to first evolve an explicit set of core self-attributes. . . . It also seems clear that the very first aspect of the self to be conceptualized would be the definition of the self as agent" (410–411).

Nelson (1996) uses the idea of mimesis to formulate how human children develop language and cognition by chunking sensory information into events. In other words, during the ontogenetic development of language and cognition, the mimetic form of cognition—which phylogenetically enabled the creation and representation of knowledge and motor action sequences through bodily expressions in earlier varieties of Homo—is manifested in how, as significant events are stored in memory through language, a child comes to think, remember, acquire knowledge, and execute meaningful motor actions, in particular those involving speech. Her account is highly consistent with that of Tomasello and Call (1997) regarding the intersubjective, intentional basis of cultural learning discussed below. To Nelson, the representation of environmental events is mirrored in the child's use of language and symbolic thought, and mimesis is the substratum for the human capacity in children.

CONCEPTUALIZING PSYCHOPATHOLOGY IN MIMETIC TERMS

Comparative psychologists and ethologists who study chimpanzees in natural habitats report observations about group activities and personal expressions and behaviors that appear analogous to many human reactions (B. J. King 1991; J. E. King, 1999). Compared to chimpanzees, early hominids would have demonstrated an additional capacity for voluntary retrieval of representations pertaining to a shared past, the self, and a global and sketchy sense of others' personalities. Consequently, all of the features of psychological and social experience attributed to chimpanzees in these more evolved early hominids would have been not only more differentiated and refined but also more persistent and more influential on morale, motivation, demeanor, and social disposition.

Donald's model of cognition and culture in early varieties of Homo can be used to offer a more elaborate interpretation of earlier varieties of psychopathology. Even if it is unlikely, the forms of psychopathology described earlier for the higher primates and early hominids theoretically could have occurred under natural, feral conditions. Mimetic skill implies a capacity to visualize the self and the group in terms of past activities and future possibilities. Relationships, significant others, and ecological events could have been represented in a psychologically and emotionally more elaborated way. Hence, individuals would have been more vulnerable to the consequences of failure, environmental calamities, and hardships caused by rivalries, antagonisms, rejections, and unstable relationships. The rudiments of such human emotions as disappointment, worry, distrust, deference, respect, love, compassion, anger, grief, and, of course, envy would have been possible.

A group whose members are capable of mimetic representation might have shared elements of culture, which would extend the picture presented in chapter 9, on social ecology. Mimetic cognition implies a capacity to share knowledge about the physical and social environment, about oneself and group mates, and about the past. Something like a group tradition and behavioral scenarios that incorporated past and future would emerge. Order, organization, shared knowledge, shared perspectives about actions completed and contemplated, shared activities for subsistence and leisure, and a more or less articulated cognitive representation of the world and the group, all point to an emergent symbolic mode of life that contrasts sharply with that described for monkeys and chimpanzees.

It is important to emphasize that Donald sees mimetic culture as largely nonverbal and explicitly nonlexical and nonlinguistic. In other words, while individuals who use mimetic culture may be able to represent and communicate about phenomena through motor behavior and orally produced sounds accompanied by whole body movement, they cannot generate true lexical units and coordinate them syntactically. Mimetic culture bearers may possibly have shown a capacity for protolanguage like that proposed by Bickerton (1990, 1995), but Donald seems to exclude this possibility. If they did, this would have further elaborated and complicated their cognitive states and opened possibilities for more diverse kinds of psychopathology.

To illustrate what kind of behavior ensemble could be constructed by a mimetic form of representation, cognition, and culture, Donald analyzes begging. His observations suggest the kinds of responses hominids might have shown to diseased or injured group mates, or those showing signs of psychopathology. He writes:

> [Begging] encompasses many potential instances in which the subcomponents of the event will vary. Subsets of [begging] might include: (1) different agents; (2) different actions: begging on the knees, with the eyes, crying and whining, and so on; (3) different contexts: begging in public, in groups; (4) different attitudes: defiantly, desperately, with threats. Begging serves to elicit a constant response pattern (giving or yielding) in the perceiver. But how can such varied patterns of behavior elicit common response tendencies in the first place? . . . The essential elements in recognizing [begging] include not only aspects of posture, expression, and sound but also perceived dominance relationships and most importantly, the perceived *intention* of the agent. Begging is a very powerful stimulus: it is found only in higher mammals and is easily understood despite its abstract nature. This testifies to the complexity of prehuman mammalian event perception, even before the evolution of mimetic skill. (1991, 191–192)

Donald describes a mimetic form of cognitive representation and communication that is a well-established pattern of behavior found in primates. Begging incorporates body posture, motor movements, facial expressions, gestures of hands and extremities, and most likely nonverbal sounds that communicate a sense of privation and need so as to elicit caring and consideration. For Donald, mimetic cognition is a viscerosomatic form of culture and communication that is not erased by language but survives at a lower level: "The mimetic layer of representation survives under the surface, in forms that remain universal, not necessarily because they are genetically programmed but because mimesis forms the core of an ancient root-culture that is distinctively human. No matter how evolved our oral-linguistic culture, and no matter how sophisticated the rich varieties of symbolic material surrounding us, mimetic scenarios still form the expressive heart of human social interchange" (189). Begging, then, is a symbolic and expressive behavior routine. It presupposes reciprocating interpretive cognitive agents: a communicator or sender and one who responds. Such a routine, symbolically reconfigured, can communicate need, distress, suffering, handicap, and mental anguish. Earlier varieties of Homo may have responded in this way to psychopathology in group mates, as well as to disease and injury.

Extrapolating from Donald, organisms capable of mimetic cognition and representation would manifest psychopathology as deficient understanding of the group's shared behavioral programs, that is, the group's model of reality or picture of the world. The symptoms of such pathologies would be failures to appreciate, represent, and communicate to the self or others adaptive interpretations of events in the ecology and group. Deviant behavior and unusual communicative displays or responses would be visible manifestations. Organisms in groups characterized by this form of social psychology are by definition responsive to viscerosomatic and gestural codes that are meaningful, that is, symbolic, and that mirror important happenings in the self, the group, the body, and the outside world. Hence, such organisms would be able to form elementary conceptions of need, disability, and even abnormality.

Generalizations from Studies of Primate Cognition

Tomasello and Call (1997) have offered a synthesis of primate cognition that can be used as a platform to formulate cognition and behavior of the hominids who came later, including aspects of their psychopathology. They see many primate behaviors involving the natural environment, reflecting physical cognition, as common to and growing out of the abilities of mammals in general. Many human cognitive traits regarding the natural world are said to be but extensions of mammalian/primate ones, but Tomasello and Call do not address the question of whether these have been elaborated neuropsychologically and what this might imply about the modularity of mind. At any rate, it is in the area of social cognition, involving behaviors among group members, that is, social and purpose driven, intensional objects, that, according to them, primates show their advance over mammalian cognition. As indicated earlier, Tomasello and Call judge that there are no significant differences between cognitive abilities of monkeys and chimpanzees, a position that is somewhat controversial in light of the material reviewed earlier regarding the observations of other ethologists. I frequently formulate contrasts between hominids and prehominids, using chimpanzees as my test cases, and acknowledge that there may not really exist hard evidence for this emphasis on chimpanzees. At any rate, it should be emphasized that to Tomasello and Call the observational skills of both species of primates appear to reflect a heightened understanding of interactions of others, which they refer to as tertiary or third-party relationships.

It is the contrasts that Tomasello and Call draw between primate cognition and human cognition, the "small difference that makes a big difference," that is useful to one seeking to understand possible changes in the character of psychopathology during human evolution. From soon after birth, human neonates show qualitative differences compared to nonhuman primates—for example, protoconversations and affect attunement with caregivers, skills in imitation or mimicking—suggesting an innate ability to socially and psychologically identify with conspecifics. This leads to what Tomasello and Call term a social cognitive revolution, beginning around nine to twelve months, at which point infants begin to tune into the behavior and cultural activities of adults and both elicit and allow the latter to tune into theirs, for example, gaze following and joint attention, in the process beginning to internalize aspects of mind, theirs and those of their caregivers, and aspects of the culture into which they are born, for example, imitative learning and social referencing.

From infants' social interactions and cognition in relation to objects and persons during the first year, Tomasello and Call go on to delineate other major milestones of human cognition, including (a) appreciation of social intentions/social cognitions of persons as agents and (b) learning of the symbolic implications of cultural artifacts and resources, including language, during the second year, all of which (c) enhances cognitive representation and organization, leading to (d) developments during the third and fourth year involving theory of mind reasoning and the appreciation of abstract categorization and relationships of inanimate objects, culminating in (e) the ability to conceptualize events and things, including the self from multiple perspectives, with its obvious implications for the development of a sense of self and the labeling/understanding of social emotions. Beginning with the foreshadowing of the social cognitive revolution of the first year of life, then, "Cultural learning . . . relies fundamentally both on infants' ability to distinguish in instrumental actions the underlying goal and the different means that might be used to achieve it, and on their tendency to identify with others so as to align their own goals and behaviors with theirs . . . this form of learning is foundational to the acquisition of language and other symbolic and cultural skills" (Tomasello and Call 1997, 407).

In short, Tomasello and Call imply that culture—understanding and use of material and symbolic, linguistic resources—is, comprises, and/or necessarily is fused with shared purposeful, goal-oriented communication, that is, capacity for a symbolic intersubjectivity with persons and cultural material. This is what other anthropologists and biologists refer to as the symbolic niche. Culture, in other words, is locked into an understanding and practice-oriented disposition to an intensional world. Hallmarks of what archeologists labor and argue over regarding hominid activities during the Lower, Middle, and Upper Paleolithic eras, such as the capacity for foresight, planning, constructive use of materials for fashioning tools, communication and language, the shaping of living spaces, and purposeful arrangement of corpses for burial, all are rooted and devolve from an emerging special capacity for social intimacy that Tomasello and Call emphasize. Attributions about possible cultural, linguistic, and artistic behavior based on analyses of the archeological record, a dominant topic in contemporary archeology, certainly bear a direct relationship to the implications of the social cognitive revolution at the end of the first year of a human infant's life that their interpretation of primate cognition has taken them to.

The account of Tomasello and Call allows one to posit that the distinguishing cultural properties of human cognition constitute the pillars of, at least, the psychological and social behavioral characteristics of hominid psychopathology. Specific varieties of these will incorporate and express in behavior the interactive network created between self and group mates, all of them creatures that are symbolically, meaningfully positioned in the world that forms their sense of reality. This cultural world is initially created in the interactions between infant and caregiver. Language and culture, learned through the construction of events and narratives involving self, other, and the social world, symbolically shape the content and experience of an instance of anxiety, depression, psychosis, syndrome of pain, suffering, dissociation or altered sense of consciousness, and/or experience of bodily dysfunction. A person's sense of self, embodiment, and differentiation is culturally learned beginning when as infants and, like other cultural objects of the environment, they are made the center of attention of the caregiver. This provides a nucleus for the understanding and communication of social emotions like shyness, self-consciousness, sense of self-esteem, and eventually the myriad negative social emotions that are hallmarks of established forms of psychopathology. In short, the distinguishing human capacity for cultural learning, which is manifest early in life and grows to include understandings about social intensions, emotional self-referencing and identity, social emotions, and a symbolically structured behavioral environment, forms the template for varieties of psychopathology that emerge during human evolution.

A Brief Résumé of the Evolution of Language

The role of language in thought and the evolution of language have aroused much controversy (see Bickerton 1990, 1995; Carruthers and Smith 1996; Gibson and Ingold 1993; Noble and Davidson 1996; Pinker and Bloom 1990; Tomasello 1992). Should language be viewed as having served first as a means of communication, then later used for internal thinking or speech? Or do the roots of language lie in special forms of internalized, silent modes of cognition that enhanced adaptation and later came to be expressed by means of words and utterances that were gradually structured in patterns? Another controversy involves the pace and timing of the evolution of language. Some argue for protracted gradualism (Deacon 1997; Foley 1995a; Gibson and Ingold 1993) and others for a more sudden, "all-at-once" development that links language with creativity (Chase and Dibble 1987; Noble and Davidson 1996). These views are not necessarily inconsistent. Many who argue for a late, hastened pace of evolution believe that it took place in a progressive and systematic way.

The gradualists vary as to when evolutionary change occurred and how to conceptualize the process. They hold that the many different syntactic and grammatical rules of various systems described by linguists are possibly surviving landmarks of the slow, linear refinements in the capacity to communicate linguistically with the self and others (Pinker and Bloom 1990). While not necessarily an example of gradualism, Bickerton's (1990, 1995) idea of a protolanguage that enabled the formulation and expression of a few key concepts and simple utterances is a concrete example of an early phase of linguistic competence. Some gradualists would concede that the earliest humans, such as Homo erectus, who lived approximately 1 to 2 million years ago, probably displayed rudimentary forms of linguistic communication, which, because they significantly enhanced social adaptation, were selected for and slowly refined (Deacon 1997). However, Donald (1991, 1993b) holds that a lexically driven language capacity was a later development, as do a number of other researchers, such as Lieberman (1991) and Noble and Davidson (1996).

Many researchers imply that the acquisition of language was something like crossing the Rubicon. In other words, that the evolution of human language signaled the emergence of a unitary, all-or-nothing, and special cognitive capacity. They equate this development with the transition from the Middle to Upper Paleolithic Era (about 50,000 years ago), which culminated in the dominance of anatomically modern humans with a creative and artistic culture. Paleolithic man could make artifacts and tools of bone and ivory, build complex dwellings, and execute sophisticated cave paintings, all of which are seen as revolutionary forms of symbolic expression and communication, marking a creative explosion in human prehistory (Pfeiffer 1982). Those holding the totalistic, fully emergent view would see this development as evidence of a new form of mentation spearheaded by linguistic competence. Finally, some researchers believe that the transition to modern man involved not the emergence of a new, genetically based capacity for language but the application of previous cognitive traits to communication through language and speech. Rather than viewing language as a newly emergent, brain-based, cognitive capacity rooted in mutation, for example, they see the ability to communicate via language as an example of an exaptation, the development of a new function for cognitive capacities already used in planning and social relations (Gamble 1993; Gould 1991).

Regardless of how the evolution of linguistic competence is understood, it is apparent that once this form of cognition and communication was attained, it had enormous consequences for solving problems and coping with the environmental challenges and group concerns. Even a protolanguage enabled hominids to develop better strategies for subsistence, for avoiding predators, for coordinating living arrangements, and for coping with social conflicts. What needs clarification is its overall effect on the character of psychopathology.

Psychopathology in Hominids with Protolanguage

Derek Bickerton's (1981, 1990, 1995) concept of a protolanguage can help explain not only changes in cognition during evolution but also changes in the character of psychopathology. Bickerton, who is a linguist, asserts, "There is a mode of linguistic expression that is quite separate from normal language and is shared by four classes of speakers: trained apes, children under two, adults who have been deprived of language in their early years and speakers of pidgin" (1990, 122). He postulates that protolanguage may have been an early phase of language for Homo erectus. This supposition is contestable, but the timing of something like a protolanguage is not the focus here. The concept enables us to better conceptualize cognition, culture, and communication in Archaic humans and what this implies about changes in psychological experience, especially psychopathology, during this phase of human evolution.

Bickerton outlines certain characteristics of protolanguage: (1) Utterances tend to be short, for example, "me like you," "me not like you," whereas fully developed languages may have long sentences with subordinated elements. (2) Whereas fluent speakers can utter a hundred words a minute, protolinguistic utterances are characterized by many hesitations and reformulations. (3) Protolanguage lacks the rules for connecting verbal elements such as are found in developed language. Language propositions, in Bickerton's terms, consist of a verb and several qualifications, or arguments. These expressions may refer to any participants in an action, state, or event: "While some argument roles are optional (typically, though not always, time, location, instrument, beneficiary, and so on), others may be obligatory in the case of particular verbs" (1995, 30). Some examples of obligatory argument roles include Agent ("who or what performs an action with an object"), Patient or Theme ("who or what undergoes or experiences the action"), and Goal ("who or what is the end point, or receives the result of the action"). Bickerton explains: "In language, either all the obligatory arguments of a given verb are expressed or, if one is omitted, the missing argument is recoverable by rule" (1995, 30). A speaker infers the connections between verbs and arguments, implicit or explicit, by semantic interpretation. However, protolanguage lacks such rules and mechanisms for linking verbs and arguments: "Arguments of verbs may be inserted or omitted at will. When an argument is omitted, no regular process enables us to recover it" (1995, 31). (4) Whereas in developed language roughly half of the constituent units (nouns, verbs, conjunctions, prepositions) refer to the structure of an utterance and serve to link verbs and arguments, in protolanguage almost all units "carry specific reference to categories of object or occurrence supposedly occurring in the real world. None of the units make reference to the structure of utterances, for there is no structure to which reference can be made" (1995, 31).

By means of a protolanguage, individuals could formulate and communicate about basic matters of survival and reproduction. It entails a high order of intentionality and voluntarism (rudiments of these being already present in the pongid/hominid ancestor and Homo erectus) that involves a measure of cortical control over symbolic or conceptual behavior. A protolanguage would provide a system of elementary symbols, the means for its voluntary use, and the necessary behavioral controls for using such a system of cognition. Individuals with a higher order of intentionality and voluntarism can inhibit some responses, carry out others, or conjure up a scenario (or symbolic representation) that responds to a situation. By *responses* one means physical or symbolic actions bearing a potential or real relationship to a situation or object that is conceptually apprehended. In conjuring up situations and voluntarily controlling behaviors, the individual is operating by means of introspection and intentionality.

A protolanguage provides symbols—words, images—pertaining to concepts that refer to objects and conditions in the outside world, and to states and intentions of the self and others. Functioning like a platform for physical and social action, a protolanguage implies self-awareness, self-conception, mental attribution, and a conventionalized, if restricted, knowledge about the world, the self, and the group. A protolanguage has little biological significance if it is not shared with group mates, which implies that although it has a neural infrastructure (is an adaptation or an exaptation), it also requires learned conventions shared by a social group. It furnishes a relatively simple and unstructured mental world limited to recent events; in other words, thinking and planning are restricted by the inability to freely displace referent ideas, memories, and planned actions in time and space. Lacking a developed syntax, users of a protolanguage are not able to formulate complex mental states or social scenarios, for they could not construct the necessary propositions that allow a fully linguistic representation of the world.

Individuals with a protolanguage should be capable of incorporating time—now, earlier, later—in an utterance or string of symbols. However, time could be factored into actions, intentions, or conjured-up scenarios only in a limited way. Moreover, the content of mentation and social communication is severely restricted when compared to language proper. A protolanguage, therefore, enables the symbolic encoding and mental manipulation of a world that is largely here and now, enabling individuals to conceptualize and communicate about practical everyday matters. As a medium for registering subjective states about the self in relation to the physical and social environment, a protolanguage also regulates and moderates voluntary behavior. Just as physical events pertaining to climate, food, and predators become objects for rudimentary cognition, a protolanguage opens up the domain of social and interpersonal relations.

A protolanguage would provide a basis for differentiating among a range of emotions and bodily states and to formulate impressions or basic reactions about the world, one's own actions, and the actions of others. The basic emotions of psychologists imply a body, a nonverbal basis for emotion that a protolanguage would draw on and elaborate. Indeed, a protolanguage provides a medium that expands on the system of nonverbal communication at the root of nature and culture, as discussed by Tomasello and Call (1997) and also by Segerstrale and Molnar (1997). With a protolanguage, then, individuals should be able to begin to differentiate, cognize, experience, and communicate about joy, satisfaction, fear, danger, comfort, safety, need, hunger, pain, malaise, rudimentary shame, embarrassment, guilt, agreement, distrust, opposition, and anger. These are the social emotions that infants begin to learn early in development as elements of culture, as formulated by Tomasello and Call (1997), and to the extent that ontogeny provides clues about the phylogeny of language and experience, one may assume that such emotions and associated behaviors emerged with and were made possibly by a protolanguage.

Emotions and experiences of a self-aware and culturally informed individual would not only color behavior but also motivate it. Consequently, one can surmise a range of interpersonal difficulties associated with such emotions. By means of a protolanguage it would be possible to form ideas and attitudes about group mates, allies, parents, siblings, potential mates, and offspring, as to whether they are friends or enemies, good or bad, energetic or slow, reliable or unreliable, for example. Individuals with a protolanguage, in other words, could rely on a social map that ranks kin and group mates, and an elementary dictionary for assessing their personalities.

A protolanguage, then, would also sharpen individuals' vulnerability to any number of psychological and behavioral difficulties that might be characterized as psychopathologies. For example, a projection about excessive, disruptive intragroup tension or fears about environmental conditions, such as the availability of food, the threat of predators, floods, and uncontrolled fire, become realistic possibilities and experiences. Real or imagined problems and crises, when formulated by means of protolanguage, contextualize mental experience if living conditions become especially arduous, threatening, or unpredictable. And, should the individual become ill, so that the apparatus that registers experience and modulates behavior is affected, a protolanguage enables an individual to grasp these disturbances. Mimesis, by contrast, limits individuals who are diseased, injured, or manifesting psychopathology to think and communicate about their discomfort, malady, and overall social disarticulation in bodily and emotional ways. Members of a mimetic culture can think about, model, and communicate their physical and psychological disturbances through gesture and pantomime, combined with limited vocal sounds. A protolanguage provides a cognitive capacity that enables individuals to frame their interpretations of changes in bodily and mental experiences in a more abstract and variegated idiom.

A capacity for protolanguage also allows for a richer, more elaborate way of formulating attitudes toward group mates. This would involve such experiences as liking/disliking or trusting/ distrusting, or classifying group mates in terms of dualisms such as threatening/safe, sexually attractive/unattractive, or comforter/comforted, or sensing another as potentially volatile and capable of causing physical harm. These emotional orientations become mental categories, although devoid of any extended narrative content. However, they can still provide a basis for worry and inner conflict. A protolanguage allows real or imagined attitudes or behaviors by group mates, as well as conditions in the environment, to explicitly influence an individual's well-being and comfort, predisposing the individual to psychopathology and strained group relations.

Individuals who can adapt to their environment by means of a protolanguage are by definition better equipped to survive, reproduce, and compete against others. A protolanguage evolves when individuals are preadapted for it and ready and motivated to benefit from it (Bickerton 1990). However, the capacity to formulate the world symbolically makes such individuals vulnerable to interpretations of experience that can render social behavior and mental activity disruptive and disorganizing. This is the negative side of the protolinguistic adaptation: its capacity to create fantasies and enhance worry as well as to malfunction and to lead to maladaptation, or psychopathology.

The obvious example of such a potential maladaptation is a vulnerability that can weaken the neural substructures that serve and mediate cognitive and linguistic states associated with a protolanguage. Crises such as food deprivation, malnutrition, infection, and altered metabolism for any number of reasons, such as injury or disease, can disrupt the operation of the centers that control emotions and cognition, including the machinery that produces a protolanguage. Physiologic perturbations of this type cause disruptions in the deployment of a system of symbols and means of communication on behalf of survival and reproduction.

Actual or imaginary events that portend danger or a threatening degree of strain to the individual or the group can lead to interferences in bodily and mental functions. The stress caused by awareness of hardships and dangers can derail the brain machinery that serves emotion, body physiology, cognition, routines regulating social behavior, including the mechanisms subserving protolanguage. In individuals with protolanguage, compared to those limited to nonlinguistic, mimetic forms of cognition, all of the preceding can lead to more complex and varied behavioral maladaptations and psychopathologies.

Hierarchical organization of the nervous system means that whatever perturbs the individual, wherever it is first registered or experienced, however it is propagated, and regardless of the overall disturbance that results, if that perturbation is significant and/or sustained, it will be manifested in a disturbance of behavior that can have emotional consequences. In other words, behavior is the medium through which a physical or social stressor that causes a perturbation is expressed. It is registered in the central nervous system and affects the physiological systems regulated by the brain, given psychosomatic and somatopsychic integration. Associated with this holistic response is a mental and emotional state that has subjective meaning and behavioral consequences that can bring harm to both the individual and the group. Many of these results are possible simply with social attribution, but they are heightened and broadened as a result of protolanguage.

Signs and symptoms of generalized worry are very common kinds of sicknesses to which groups are vulnerable. The individual or group's interpretation of the cause of such disturbances will be based on the prevailing ways of understanding sickness and healing. General disturbances that, by definition, are not easily localized characteristically allow for the full play of a group's cultural and social understanding of sickness and healing. On the other hand, sickness produced by localized structural lesions of the brain are much less common. They produce more discretely profiled behavioral changes—for example, weakness, seizures, paralyses, loss of the ability to verbalize, or gross incoordination. It is to be expected that disturbed behaviors produced by a neuroanatomically localized lesion will likewise be expressed in some social context and will also be interpreted symbolically, with consequent social and psychological changes; and in these instances, the behavioral disturbances are likely to elicit greater concern.

What might be absent from a condition of psychopathology in varieties of Homo who are able to regulate their lives by means of a protolanguage? By definition, a group of such individuals would lack an inventory of myths, stories, and elaborated mentalities about themselves and the world. Their inability to formulate complex propositions, let alone narratives, deprives them of a means for relating time, motives, organisms, objects, environmental conditions, actions and intentions of group mates, and group concerns in a conceptual-linguistic form that provides a rich narrative about life. For example, it is not likely that disease, injury, or psychopathology will trigger a complex awareness of one's place in the scheme of things, the meaning of existence as a sentient creature, or relations with other hominids, varieties of Homo, or animals. In short, they are denied narratives that link the individual to its past, its future, its mates, its kin, and to remembered events. On the one hand, a protolinguistic adaptation enables some formulation of basic imperatives for survival and reproduction in a harsh setting. On the other hand, while a protolanguage may promote self-awareness, social attribution, crude syntactical reasoning, communicating, and a capacity to store episodic and rudimentary memories, these cannot be framed in terms of beliefs, myths, and elaborated social rationales. A protolanguage simply does not provide the means for constructing elaborate scenarios about basic existential questions.

Because of the simplicity of a protolanguage, psychopathology in hominids would be dominated by global reactions governed by visceral motor responses and associated with physical actions congruent with the perceived stressors or worldly events. What is most important, the descriptive content of situations presumed to evoke behavioral disturbances, hence the form and social meaning of a potential psychopathology, would be devoid of much interpretive content. From an observer's vantage point, an individual's reaction to an irritant would seem to be emotional impulsiveness, regardless of whether it is directly perceived or imagined.

To understand the basis for this claim, recall that the propositions of a protolanguage, that is, the meanings it provides its users, are only simple associations of symbols denoting objects, events, and states of being. Because the self is the center of any predication about the world, the individual is in a position to make sense of how a situation or event in the outside world could influence it, especially in case of potential threats.

Since the individual would have no elaborated constructions to buffer, blunt, or mediate real or conjured-up problems or threats, there would be no such constructions to influence behavior in the long term, either. Consequently, hominids could not be governed by delusions, for this usually requires stories of situations that persist in consciousness and influence actions. At most, disturbed hominids would be troubled by distortions about the immediate, concrete world.

Cognitive Modules and Their Significance for Understanding Psychopathology

Analogues of psychopathology in the earliest hominids and varieties of Homo that might have been manifested in cognitive disturbance, as indicated by faulty use of a protolanguage, bring up the topic of the modularity of the mind. The idea of cognitive modules implies information processing that is domain specific, automatic or mandatory, fast, impenetrable and encapsulated, triggered into action by special forms of sensory input, ontogenetically privileged, and yields outputs of information to other cognitive centers, where it is interpreted. Cognitive modules resemble what others have termed psychological adaptations or mental algorithms, although some of the latter integrate higher levels of social cognition than do cognitive modules, which in many respects are based on perception. There is much debate about how cognitive modules operate in cognition generally and their relationship to consciousness and so-called general intelligence factors (Boyer 1994; Carruthers and Smith 1996; Fodor 1983; Hirschfeld 1996; Karmiloff-Smith 1992; Keil 1989; Mithen 1996a, 1996b; Nelson 1996).

The capacity for language is a commonly cited example of a cognitive module, but others have been described. Interpersonal behavior and social, political expertise as a biological trait is held to reflect a highly developed, complex social intelligence, a pivotally important concept in human evolutionary studies (Byrne 1995; Humphrey 1976), yet this trait appears to function in ways like the purported functioning of cognitive modules (Tooby and Cosmides 1990a, 1990b, 1992). In this interpretation, social intelligence centered on the detection of cheaters is served by a special psychological adaptation or mental organ posited to have been present in emergent form in the last common ancestor and the earliest hominids. Along with a general capability that promotes trial-and-error learning and reinforcement routines in scheduling and organizing behavior, social intelligence is viewed as a specialized mental faculty. Psychologists interested in human cognition, evolution, and ontogenetic development believe that something like a social cognition module programs, regulates, and accounts for behaviors that reflect an innate theory of mind faculty—that is, a capacity to perceive the motives of others and to use this knowledge constructively.

General intelligence factors, cognitive modules, and psychological adaptations are thought to be ecologically directed modes of reasoning, some of which are designed or hard-wired by natural selection. It is assumed that the earliest hominids used such rudimentary information-processing structures that created special forms of expertise for dealing with the environment, for making and using tools, for classifying types of organisms, and inferring the intentions of others. Some of these forms of intelligence or cognitive modules have been named: for example, there is a natural history module and a technical intelligence module. Mental modules are held to embody innate intelligence geared to adaptive behavior for promoting survival and reproduction. Much of the physical cognition that Tomasello and Call (1997) draw attention to as a feature of primates and to some extent mammals can be related to or explained in terms of cognitive modules.

Steven Mithen, whose discipline is cognitive archeology (1996a, 1996b), develops a model of the prehistory of the mind centered on such cognitive modules. He posits that changes in behavior during human evolution can be interpreted as reflecting changes in the way such specialized modules operate in relation to a general intelligence factor. Keep in mind that even chimpanzees, held to be but a small distance in evolutionary time from the Australopithecines, are clever social strategists, easily acquire expertise about the ecology, and have been observed to use tools in natural habitats (Goodall 1986; McGrew 1992). The ease with which they acquire these social and physical behaviors, respectively, and their alleged, arguable special nature when compared to the Cercopithecines, all suggest that cognitive modules played an important role in the behavior of the earliest hominids. Such mental modules are thought to have become better defined, or at least more elaborated cognitively, among the early hominids or ancient humans, but this is debatable.

The topic of an organism's general problem-solving skills, a general intelligence factor of some sort, and that of specialized mental modules featuring special information-processing facilities raises the question of their possible role or influence on the character of psychopathology among the earliest and early hominids, Archaic humans, and, indeed, among anatomically modern humans before the alleged Rubicon of the late Upper Paleolithic. Implied in the concept of special modules is the assumption that the corresponding forms of cognition are linked to or programmed by special circuits of the cortical and subcortical regions of the brain, and in humans, this is of course true not only functionally but also neuroanatomically—for example, involving language, spatial relations, face recognition, and music appreciation. General attention, concentration, and problem solving are currently judged to be global, frontal lobe functions. A module dealing with social intelligence involves access to a neural machinery linked to recognition and expression of emotions, and this suggests involvement of the limbic system (Aggleton 1993; Armstrong, Clarke, and Hill 1987). Technical intelligence or physical cognition would seem to engage sensory mechanisms, along with motor programs associated with regions of the frontal and parietal lobes. What all of this implies is that organisms sustaining localized structural lesions in different regions of the brain are impaired in the skills and behaviors ordinarily regulated by that part of the brain. Just as vascular lesions and head injuries in modern humans are associated with distinct neuropsychological syndromes universal in Homo sapiens, it can be inferred that hominids possessing similar but arguably less elaborated, perhaps more sensation-driven modules would have shown syndromes of psychopathology marked by striking impairment of various specialized skills and behaviors. That the hominids experienced intense competition not only with carnivores but also with other related species means that physical trauma and, in particular, head trauma, were not rare among them.

Much more frequent than traumatic local brain lesions among hominids were deficiencies and impairments of specialized, neuropsychologically mediated cognitive and emotional behaviors resulting from faulty development. Infants and juveniles suffering such functional lesions during critical periods when mental modules are developing and maturing would, of course, have been most vulnerable to such impairments. Individuals subjected to pronounced stress, deprivation, malnutrition, and infection can be expected to have shown impairments in the functioning of different regions of the brain, the organization of which, to be sure, is difficult if not impossible to determine. Consequently, in addition to brain lesions, neurophysiological impairments devolving from purely developmental factors would have been mechanisms for producing special syndromes of psychopathology. Thus, during hominid evolution significant forms of social stress and deprivation, along with physical injuries and disease, impaired social behavior. A general lack of intellectual resourcefulness and self-control, social ineptitude and timidity, poor ability to find shelter and food, track prey, or avoid predators, and neuromuscular clumsiness in using tools all might have been expressions of behavioral pathologies.

For the sake of completeness, it is appropriate to mention that Mithen's speculations (1996a, 1996b) about the evolving modular structure of the hominid brain includes the notion that with the emergence of fully modern humans, information processed by the various specialized modules—for example, linguistic, technical, manipulative, social, interpersonal, natural history—gained access to or became integrated with information processed by a general intelligence faculty. The latter undoubtedly was modified and expanded with the emergence of language as we know it. Whether language constitutes a purely communicative adaptation, and the extent to which it plays a primary cognitive role, is a central problem in the philosophy of mind and of language that cannot be pursued further here (Carruthers and Smith 1996). Suffice it to say that Mithen's formulation implies that sometime during human evolution so-called specialized intelligences were influenced by or integrated with language. In other words, a re-representation in language of material learned and cognized through the specialized modules apparently took place. Mithen argues that the integral connection between forms of information processing linked to the multiple intelligences that one sees in humans had not fully emerged among early hominids and earlier varieties of Homo. If this is true—as suggested by his interpretation of the diverse and seemingly disconnected cognitive achievements of the hominids as interpreted from fossils and other remains—the character of hominid psychopathology may have been manifested as syndromes of discrete deficits of behavior that were strikingly bounded. In other words, accepting the validity of the form of evolving hominid modularity of mind, it is conceivable that structural and functional lesions producing psychopathology could have caused organisms or individuals to display abnormalities only in selected suites of behavior.

A modular interpretation of cognition during human evolution, along with the notion of a supramodal general intelligence factor, enables us to visualize other features of psychopathology. Mithen holds that the products of a module's operation are available only to a general intelligence factor that is supramodal, but that the latter's capacities were poorly developed during evolution. Information processed by the various modules could not be integrated. Only the products, and presumably the mental states and actions that realized them, were interpretable. This implies that physical injury or lesion, perturbation, a general underdevelopment or deficiency, or a dysfunction caused by general physiological or metabolic disturbances could affect the operation of one or several modules independently, producing focal disturbances in behavior. An individual that could not integrate information from different modules would be aware of deficits that could hardly be explained. That is, hominids and early varieties of Homo could not have had much insight about neurocognitively centered varieties of psychopathology. Some of the manifestations of psychopathology during early evolution would thus have been known only to conspecifics. Individuals who displayed focal abnormalities would have lacked an awareness of them because the information-processing machinery that produced them, that is, disturbances in a specialized, isolated module, would not have been influenced by or integrated with information from other modules and, what is more important, minimally or poorly integrated with the functioning of the general intelligence factor. Some neurological lesions in humans are associated with profound denial, for example, neglect of hemiparesis, but, if Mithen is correct, denial behaviors would have been far more prevalent during earlier phases of evolution.

There exists a host of neuropsychiatric disturbances in humans caused by focal brain lesions that seem related to cognitive modularity. They involve disorders of knowing (agnosias) and cognitively and linguistically formulated actions (apraxias), of which the victim has no understanding and sometimes seems unaware. Some of these disturbances arise because the language system of the brain is disconnected from cerebral centers where special motor and sensory functions are represented, and hence the individual is unable to formulate or understand the aberrant behavior stemming from the lesioned center. Other disturbances, however, are observed in the absence of any such apparent disconnection. It is possible that in hominids with protolanguage or no linguistic competence at all, but who showed cognitive modularity as Mithen describes this, unusual forms of psychopathology involving glaring or bizarre denial may have been prevalent. There is little doubt that language provides a form of representation and cognition that profoundly shapes the content and structure of psychopathology generally. During phases of evolution when language was less evolved yet modularity more so (à la Mithen) or, conversely, may have been even less well articulated, psychopathologies incorporating neurocognitive disturbances would have manifested in behavior difficult to predict.

It is perhaps more realistic to propose simply that early hominids would have perceived manifestations of psychopathology in the self in a fragmented, deficient, perhaps impersonal way. In other words, they would have lacked personal insight about the manifestations, symptoms, or signs of psychopathology. Lest this supposition be thought to reflect a naive neuroanatomical locationism, recall that many, perhaps most, instances of psychopathology would have been produced by the hardships and strains of meeting environmental challenges and conflicts in social relations. It is thus possible that individuals with manifestations of psychopathology caused directly by social relations and social worries about the external world would also, given Mithen's reasoning, have shown some degree of fragmentary or deficient insight as well.

Clearly, early hominid forms of behavioral pathology were costly to the afflicted organisms or individuals. They could compromise behavior, adaptation, and welfare, and produce suffering. The harsh reality, of course, is that certain varieties of psychopathology would have not allowed survival in the wild, however tamed it may have become as a consequence of protocultural developments and/or periods of ecological stability. Besides causing a failure to thrive, all varieties of psychopathology would have constituted grist for the mill of natural selection: it would be their impact on the basic ability to subsist, compete, mate, reproduce, and care for offspring by which the effects of such pathologies would be socially calibrated. One immediate, short-term cost or penalty of behavioral pathology, in other words, was elimination because of failure to thrive, falling prey to predators, or vulnerability to disease and injury. Another one was limited success in the enterprise of food gathering and hunting, social competition, and ultimately reproduction, which in the long run could entail eventual termination or extinction.

At some point in the evolution of culture and cognition, a social awareness of anomalies or disturbances of health and well-being, including psychopathology, supervened. A shift between, or at least a modification of, strict evolutionary compared to clinical, or cultural, moral, ways of dealing with these crises took place. Obviously, aspects of cognition and culture covered in this chapter—for example, penetration into strict life chaining routines, transformation of mimetic/early mythic forms of cognition, a protolanguage, and appropriation of cognitive module functions by a general intelligence factor—would have significantly affected how psychopathologies were interpreted, handled, and dealt with, as well as how they were produced and fashioned.

Comment

Psychopathology refers to a complex set of conditions that have been described in modern times for disturbances of behavior among enculturated human beings. To visualize its purely psychological and social behavior characteristics, leaving aside symptoms and signs involving bodily pain and dysfunction, aspects of language, cognition, and systems of meaning involving symbols or culture must be given consideration.

The sphere of behavior associated with these traits emerged and became elaborated during human biological evolution. Controversy exists as to whether human language and cognition constituted singular unique developments; came late and as a mere add on to cruder forms of organized social life involving purely gestures, signals, and/or calling systems; or whether such capacities for symbolization stretched over hundreds of thousands of years and conformed to a progressive and gradual development. Regardless of which interpretation or model one adopts regarding the evolution of language, cognition, and culture, a topic covered in much greater detail in later chapters, one is confronted with the problem of visualizing the changes that did take place, for it is indisputable that these vaunted capacities of humans have a natural history of some sort.

In this chapter four models and schemas about the evolution of language, cognition, and culture of varying degrees of generality and specificity have been reviewed. Each provides a slant on or window to how social and psychological features of psychopathology were configured and enacted during evolution. Even if one accepts tentatively the usefulness and validity of the respective models, it can safely be said that none is fully satisfactory and each leads to conjectures that are incomplete, raise as many questions as solutions to the conundrum of animal varieties of psychopathology, and can also be faulted and are hence contestable. The models, of course, are not mutually exclusive.

Richards's model is complete, thorough, highly abstract and mechanical, but in some ways it merely provides an elaborated description of salient differences of the behavior of primates compared to humans, leaving much of the substantive details one would want to visualize up to the imagination. Donald's schema is more detailed and imaginative and provides a fuller description of what social and psychological behavior could have been like during human evolution, but it leaves unanswered details of what comprises mimetic behavior and its transformation into mythic culture, in many ways the blood and guts of the story of human biological evolution. Bickerton's proposition about protolanguage is heuristically useful since it provides a discrete handle in terms of which one can begin to probe the content of cognition and culture of hominids and thus their psychopathology; but whether it provides more than a simple, descriptive bridge connecting with language as we know it, skirting over complex issues of grammar, syntax, and semantics, is arguable. Finally, Mithen's picture of the prehistory of the mind creatively uses the idea of modularity and its relation to a general intelligence factor, but in the process it simplifies the subtlety and integrated character of primate cognition, leaves unexplained the commonalities between mammalian, primate, and human cognition, and raises unanswerable questions about the structure of psychopathologies associated with focal brain lesions and possible fragmentation of symptoms and signs resulting from general systemic and more functional brain disturbances.

However, each of the four schemas reviewed can be used to help sort out and deepen understanding of the characteristics of psychopathology during human evolution. Many of the structural aspects of its psychological and social manifestations are rendered more graphic and understandable by linking together the insights that each schema emphasizes. This provides a framework one can then use to develop a fuller picture of parameters traditionally thought of as cultural, which are covered in the following chapter.

- From the standpoint of understanding their social and psychological behavior and especially their psychopathology, the most striking characteristic of hominids was their capacity for language, cognition, and culture.
- The effects of language, cognition, and culture on behavior influence characteristics of psychopathology in important ways. Hence, it is desirable to have models of these capacities in hominids so as to visualize how psychopathology changed during evolution.
- Since Richards's physiomorphic learning scheme outlines changes from strict behavior chains related to survival and reproduction to the later emergence of self-awareness, social intention, and culture, it can be used to picture the evolution of psychopathology.
- Because Donald's cognitive psychology scheme posits that hominids moved from a mimetic, gestural to a linguistically based mythic culture, it also provides a way of visualizing psychopathology during phases of human biological evolution.
- Bickerton's scheme regarding the evolution of language, particularly

his conception of protolanguage, provides a useful heuristic that can be used to formulate changes in the content and structure of psychopathology during human biological evolution.
- Mithen studies human evolution from the standpoint of cognitive archeology. His scheme involving the interplay between cognitive modules and a general intelligence factor can be used to describe structural properties of psychopathology during evolution.
- The four formulations discussed leave questions unanswered and raise quandaries. Nonetheless, they are instructive and also complementary, and they provide a more vivid picture of what and how varieties of psychopathology were configured during evolution.

Chapter 11

The Impact of Meaning Systems on Psychopathology

EARLIER CHAPTERS EXAMINED the role of culture in human evolution from different points of view. One is that of biologists, paleoanthropologists, and comparative psychologists who adopt a broad conception of culture that includes general characteristics of cognition and behavior. This position allows one to place hominids on a trajectory that stretches from the last common ancestor to modern humans. Observations of primatologists and experiments on higher primates provide background information about the possible behavior and cognition of hominids, who are seen as following a more or less continuous line of development in which progressive, more or less quantitative gains in capacities favored the elaboration of culture. In this view, culture and early forms of communication are viewed as characteristics that changed slowly and led continuously to the cultural, artistic explosion associated with the transition to the Upper Paleolithic Era in Europe 40,000 to 50,000 years ago.

Another approach concentrates on the qualitatively *different* attributes of human culture and language (Pinker 1994). These are seen as traits of a new species of modern humans that emerged in Africa around 100,000 or so years ago. One theory holds that populations of modern humans slowly moved out of Africa and replaced Archaic human populations throughout the globe. A special capacity for culture and language has been attributed to people of the Middle Stone Age period in sub-Saharan Africa long before they moved out of Africa and colonized the world. Possibly, this species acquired genetically based changes in the structure and organization of the brain. Whether late varieties of Homo erectus and Archaic humans showed language and culture or had an actual capacity for this is contested (Clark 1997; Deacon 1997; Hayden 1993). The compelling manifestations of language, culture and art that appear dramatically in the archeological record in Europe around 40,000 years ago are explained as due to changes in social organization and ecology that facilitated and made possible a new way of life (I. Davidson 1991; R. G. Klein 1995). A point of contention here is whether *culture* and also language should be used to explain characteristics of the way of life of earlier varieties of species Homo.

At issue, then, is how best to characterize human culture in general, especially how to explain the changes associated with the Upper Paleolithic Era. A particularly controversial question concerns the importance of a capacity for symbolic reference as compared to symbolic culture. Symbolic reference enables one to refer to phenomena by means of arbitrary conventional symbols whose meanings are known to a particular community. This capability is generally held to have prefigured the development of symbolic culture. Researchers differ as to when culture as symbolic reference can be ascribed to hominids and how it differs from symbolic culture. Symbolic culture embodies an integrated system of symbols and their meanings. In this instance, symbolic phenomena, such as objects, beings, happenings, and explanatory models, are not isolated from one another but are instead integrated within an overarching system of symbols. Interpretations of phenomena and social life depend on a conventional system for representation agreed upon by members of a community, for it provides for a shared way of life. Chase puts the distinction as follows: "This situation, where not only are symbols in common use, but where symbolism goes beyond reference and where all actions and all things are caught up in a web of symbolic meanings, is what I refer to when I use the terms 'symbolic culture'" (1999, 37). Not all researchers subscribe to the clear distinction between these two concepts, although most appear comfortable with a definition of culture as symbolic and see symbolic culture as authorizing the creative, artistic developments linked to the transition to the Upper Paleolithic Era in Europe.

A final question regarding the evolution of culture involves the selective influences that ushered in this presumably new way of life. This topic is receiving increasing attention. One problem that bears on matters discussed above is whether a capacity for culture, and indeed language, is a naturally designed feature of the brain selected for its special function or is simply a secondary byproduct of other evolutionary changes in brain function merely exapted for culture and language. Another problem is how a capacity that promotes cooperation and group-level phenomena came to be selected for in the first place. In strict Darwinian terms, individuals who cheat on social expectations and agreements, or those who exploit the advantages of group living but never pay anything back for its benefits and fruits, will win out over altruists, and yet culture and language require conventions and artificially created representations that depend on mutual agreement, sharing, and reciprocity. How could the social contracts required by symbolic culture ever get established if individuals were better off undermining such contracts by exploiting them and gaining advantage over others?

Ritual is held to have played a pivotal role in this process. A ritual entails the establishment of group conventions about how to interpret and carry out a socially important group activity. The emergence of ritual, a quintessentially group-level phenomenon and a central attribute of culture, has been linked to the emergence of a special means of communication or language among insiders who promote and uphold the ritual. Each individual sharing the ritual or culture is seen as benefiting in a Darwinian sense from established social bonds and contracts. A scenario currently emphasized in this process is the role of female coalitions against males involving the exchange of sexual access for high-energy foods, a strategy that became especially crucial for females during later phases of evolution. In this scenario, the loss of visible estrus and relying on menstruation as a sign of fertility promoted practices of sham menstruation via rituals of fertility and hunting that included the use of body coloration and artistic decorations of the body (see Dunbar, Knight, and Power 1999).

Culture as a Symbolic System: Evolutionary Explanations

The concept of culture has been used by anthropologists interested in evolutionary questions in two ways. Foley (1987, 1991, 1995a) admonishes against heavy reliance on the concept of trying to understand human evolution generally and the evolution of human behavior in particular. He judges culture to be a polytypic or synthetic concept that joins a variety of attributes of human social behavior but finds that its use begs important questions. What is crucial, in Foley's view, is the process underlying the origins of such attributes as food sharing, hunting, tool making, male parenting, a capacity for language and speech, a capacity for elaborate social learning and symbolizing; this is very different from a global notion of culture that seemingly collapses and blurs the distinctive traits of Homo sapiens. Foley points out that each of these attributes may have evolved separately, under different selective influences. In his words: "Thus, to treat them collectively as 'culture' is to remove the possibility that hominids may, in the past, have possessed only part of their present behavioral repertoire, or that repertoire combined in different ways. There is little advantage in using a term that bestows the advantages of a descriptive shorthand (which the term culture certainly does) if it begs the very question we are asking, thus buying descriptive ease at the expense of analytical precision or evolutionary process" (1991, 27–28). Foley's view is consistent with that of McGrew (1992), who points to the cultural aspects of chimpanzee behavior and also of Dunbar's writings (1988, 1992, 1996a, 1996b), which examine group size and brain enlargement from a phylogenetic standpoint using mathematical models to explain the evolution of various characteristics of modern man. The ideas of these researchers can be taken to imply that during evolution inherited adaptations and mechanisms linked tightly to biological functions, termed earlier as evolutionary residuals, were impacted by systems of meaning that rationalized and clothed fitness imperatives in a symbolic idiom.

Foley's position implies not only that the concept of culture is not necessarily needed to understand the evolution of the human way of life, but also that it may actually obfuscate matters. By integrating diverse aspects of behavior under the banner of social symbols and their meanings, the concept of culture leaves unclear how other characteristics of social life besides symbolic communication contributed to its shaping and structuration during evolution. Works by Merlin Donald (1991) and Steven Mithen (1996a, 1996b) address Foley's concerns in a manner that still permits a cognitively centered and integrated conception of the evolution of culture.

Foley's approach complements that of the evolutionary psychologists John Tooby and Leda Cosmides (see Barkow, Cosmides, and Tooby 1992). Their misgivings about a concept of culture are based on a critique of the Standard Social Science Model (SSSM) that has reigned in anthropology, psychology, and sociology. SSSM holds that the character of human social behavior has everything to do with the system of symbols that infants find in the social environment—society, culture—and that give content and structure to the adult mind as a result of development or enculturation. They conceive of SSSM as positing general-purpose mechanisms like classical and operant conditioning, social learning, and trial-and-error induction. Instead, they posit an Integrated Causal Model (ICM) pertaining to human cultural behavior, which they believe connects the social sciences to the natural sciences. ICM, in contrast to SSSM, concerns the inherited, innate, evolved mental modules or psychological adaptations that have been designed in the environment of evolutionary adaptedness by natural selection. Tooby and Cosmides enumerate the following points that must be taken into consideration in any exposition of character and significance of human social behavior:

> a. the human mind consists of a set of evolved information-processing mechanisms instanciated in the human nervous system; b. these mechanisms, and the developmental programs that produce them, are adaptations, produced by natural selection over evolutionary time in ancestral environments; c. many of these mechanisms are functionally specialized to produce behavior that solves particular adaptive problems, such as mate selection, language acquisition, family relations, and cooperation [they later include sexual jealousy mechanisms, signals for communicating mother-infant emotions, social contract algorithms]; d. to be functionally specialized, many of these mechanisms must be richly structured

in a content-specific way; e. content-specific information-processing mechanisms generate some of the particular content of human culture, including certain behaviors, artifacts and linguistically transmitted representations; f. the cultural content generated by these and other mechanisms is then present to be adopted or modified by psychological mechanisms situated in other members of the population; g. this sets up epidemiological and historical population-level processes; and h. these processes are located in particular ecological, economic, demographic and intragroup social contexts or environments. (1992, 24)

The formulations of Foley and Tooby and Cosmides about human evolution and what is otherwise handled as cultural and symbolic, together suggest that hominids inherited information-processing mechanisms from their predecessors and that they developed, added to, refined, and specialized these mechanisms as they attempted to solve the problems posed by changing paleoecologies. What social scientists gloss as *culture* incorporates in a less than satisfactory way an array of mechanisms and adaptations that have a phylogenetic history and were progressively designed during evolution, and these should be the targets of analysis.

The more descriptive and biologically rooted view of culture may be compared to a holistic, integrated, and qualitative view. In this instance, anthropologists and evolutionary psychologists and biologists single out key milestones or transition points at which aspects of social organization and personal experience came together to produce a qualitatively different symbolic mode of adaptation. Such developments as food gathering by females, hunting by males, food sharing, tool making, and male parenting—these and others being traits of the parceling approach—are examples of potentially significant sparks or prime movers that reorganized life ways by producing a series of reactions that led to the cultural way of life. Kinzey (1987) critically discusses these issues. The most widely held view of this singular, integrated, and qualitative change in the mode of adaptation that led to the emergence of culture is of course that involving language. (See Noble and Davidson 1996.)

A highly provocative account of the synchronic, qualitative approach to the emergence of culture is provided by Chris Knight (1991, 1996), a cultural anthropologist heavily influenced by Marxian social theory as well as sociobiology. To him, the origin of culture constitutes a major breakthrough in hominid evolution, conceptualized as providing the construction of a truly unique symbolic niche. Knight views culture as a system of symbols and symbolic meanings that provided the basis for organized social life in human communities. Knight contrasts human culture with the kind of learning and behavior observed in monkeys and is particularly concerned with the possibility of their having a culture: "Why, then, can we not speak of vervet monkeys as having 'culture'? Part of the reason is that such learning-dependent skills are peripheral to the political determinance of structure. However great their survival value, they are marginal to the maintenance of social-structural continuity from one generation to the next" (1991, 11). Knight draws liberally from the ideas of Dawkins about memes, the analogues of genes, which are held to constitute units of cultural information at the root of cultural continuity: "Symbolic culture ... has its basis in the immortality of whole sets of extremely complex memes—culture—constituting instructions shaped not just by behavioral interaction between organisms and their environments, and derived not only from the genetically based phylogenetic conservatism of the species, but shaped also through the relationship between these and a highly specific, rich and accumulating fund of collective wisdom or tradition materialized in technology, design, language, art, ritual, kinship and so on. . . . What were the conditions which had to be established to enable such complex memic patterns to be preserved? Central to my argument is politics. There could be no memic immortality in the absence of the essentially political capacity to establish agreements, rules and contracts" (Knight 1991, 12–13).

Knight (1991), Noble and Davidson (1996), and Chase and Dibble (1987, 1992) ascribe great importance to symbolic phenomena—meanings, rituals, language, myths—in the creation of the human adaptive mode. They draw attention to the integrated character of this form of adaptation and emphasize that during prehistory it was an integrated transformation involving how individuals could picture and formulate their world, how they constructed and encapsulated this view in a specific, all-encompassing narrative account and a political ideology that provided the conditions for humans' explosion of creativity. Knight offers a complex account of how female synchronization of menses and related behaviors, including offering sexual favors to males in exchange for food, created the conditions for the evolution of culture. Knight and collaborators later modified this theory of the evolution of culture and even, it seems, of language (Knight 1996; Knight, Power, and Watts 1995; Power and Watts 1996). Their theory synthesizes information about ecology, sociobiology, social and cultural anthropology, evolutionary psychology, cognition, social organization, and cultural materialism to argue for the importance of culture as an integrative concept central to understanding the emergence of a qualitatively different, symbolically and politically integrated form of human behavior.

Scholars who hold a partible, descriptive, and adaptationist, in a biological sense, view of culture play down the privileged role of the concept of culture in understanding the seemingly synchronized, symbolically orchestrated, origins of human behavior. They view the behavior of the hominids as essentially continuous, that is, analogous and perhaps homologous, with those of the great apes and anatomically modern humans. They regard the different social ecological conditions that characterized different phases of evolution as having provided selective factors shaping the content and form of normal, adaptive behavior in all its variety.

In this light, psychopathology during earlier phases of human evolution would be understood as equally divided and differentiated when compared to psychopathology in fully culturalized moderns. Each new refinement of a particular psychological adaptation, each new behavioral routine, and each new milestone of social organization and cognition that might constitute culture would be associated with changes in corresponding types of behavioral pathology, and all would be seen as grounded in a shared biology and a shared environment of evolutionary adaptedness. In short, behavioral normality or adaptation, and consequently behavioral pathology, would be interpreted within a framework of comparative socioecology. This implies that such basic humanoid attributes as social bonding, sexual mating, agonistic competition, the making and using of tools, socially cooperative foraging, scavenging, hunting, and an acquired sense of group identity or history can be seen as adding features to and complexifying the forms of behavioral pathology shown by members of hominid and early Homo groups. It is how pathology interferes with and is enacted in the context of existing paleoecologic and changing contingencies that would be emphasized.

The partible, diacritical, and biologically based view gives comparatively less importance to features of social life devolving from the ideology, political economy, and systems of social contracts of the culture, such as totemic myths, rituals of renewal and affirmation, moral and ethical imperatives, and political organization. These essential attributes of symbolic culture, then, would be less important in explaining aspects of social behavior and, by implication, in explaining how psychopathology would be configured and enacted. On the other hand, more emphasis would be ascribed to more molecular, purely adaptive routines linked to ecology, habitat, food procurement and processing, mate selection, sexual possessiveness, and social organization.

By contrast, a perspective on the origins of human behavior and social organization like that of Knight (1991), which places high significance on the concept of culture, would emphasize such symbolic factors, seeing culture as a transforming force. It is the transformation to a political and structurally anchored arrangement of social life of a group, as embodied in rituals and myths and sanctioned by rules and contracts, that gives members of a culture a distinct identity and history. Its roots reside in the qualitatively new modes of cognition displayed by human neonates and infants in interactions with their caretakers, types of communicative exchanges that realize cultural learning as reviewed earlier (Tomasello and Call 1997). The establishment of such a newly evolved cultural mode of social organization is an emergent psychological property of identity and experience, a change in how members of a group see and relate to one another. A sense of corporate identity with associated narratives and knowledge formulas about how life should be lived emerges, affecting the quality of social solidarity among the age and sex groups. What group members think and feel for one another, what they believe they owe to their group mates and is owed to them, what they see as the essential social rationales of community life, what they view as the life well spent or badly carried out, what they expect in the way of help and comfort in coping with the exigencies of a harsh existence, and related symbolically constituted tenets and narratives are all integral in shaping and configuring group life and the lives of individual members. Stated differently, what group members view as normal behavior depends on an evolved cultural rationale—that is, on the content and meaning of the conventions and social rules that govern how they conduct their lives. By extension, what group members mean by social abnormality or deviance, which an external observer might call psychopathology, and how they handle such behavior, all would bear the stamp of the processes and developments that culminated in their cultural rationale. In this second perspective, then, the requirements of adaptation tied to the climate, geography, and ecology of the habitat are less important in determining what is normal, as compared to pathological, than are the meanings of the cultural symbols that give corporate life social, moral, political, and spiritual significance.

The advantage of relying on the partible, differentiated perspective for understanding the archeology and prehistory of culture is that one can use it to explain how different aspects of cognition, social organization, and types of relationships (see Hinde 1987) influence and shape the expression of psychopathology. In other words, if the concept of psychopathology is to be applied to hominids, early Homos, and Archaic humans, it should be linked to and grounded in the variegated spheres and routines of social life of groups attempting to survive in their ecological community. The exigencies compelled by various behavioral ecologies create stages and standards and conventions for measuring relative fitness and reproductive success and hence elements of psychopathology. These standards can be understood only in a specific, locally anchored, and relatively differentiated way. A disadvantage of this view is that it fails to account for how group members interpret behavior in light of their cultural understanding and political structure of their world. This may bleach out significant facets of social life linked to the cultural interpretation of normality, as compared to pathology or deviance. Conversely, an advantage of the second, synchronic, symbolically orchestrated perspective is that, provided one has a realistic model for conceptualizing the origin of culture, it becomes easier to conjure up how normal social behavior is understood and how it preserves social organization. This means that one can better appreciate how deviance and pathology are likely to have been understood and dealt with as social and cultural realities during this breakthrough to a human mode of life.

Yet any model that might be developed to explain cultural emergence is necessarily provisional and theoretically tenuous. If one's model of cultural emergence invokes, as an example, the solidarity among adult females in their sexual strike against male dominance and the role of menstruation in all of this, as Knight (1991) does, then one is compelled to interpret behavioral normality and pathology or deviance in terms relevant to this type of symbolic accounting and political contracting. A related disadvantage is the tendency to overemphasize purely symbolic factors at the expense of concrete social and ecological factors that condition and contextualize adaptive and maladaptive behavior.

The advantages of both perspectives can be realized if one adopts a general, abstract frame of reference. The two perspectives are complementary. A comparative social ecological perspective allows one to incorporate contingent facts about evolutionary residuals of behavior, social structure, and subsistence activities in an analysis of normal, as compared to abnormal, behavior, while the symbolic one allows one to incorporate ideological, political, and meaningful narratives of the group that determine how they become social and cultural objects for the group.

Culture, Language, a Sense of Reality, and Psychopathology

For the psychiatrist or psychologist, reaching a clinical diagnosis of psychopathology involves evaluating an individual's sense of reality. The assumption is that while social background influences how individuals experience themselves and the world, there is nevertheless a real, objective world out there, independent of the observer; and that its salient features, including its objects and modes of operation, are mirrored in culturally shared perceptions, a sense of reality about how the world works. The whole tradition of natural theories of mind and prepared forms of cognition is consistent with principles of mental state testing and the clinical assessment of psychopathology (Byrne 1995; Sims 1988).

Manifestations of a pathological sense of reality involve false perceptions such as hallucinations and distorted ways of explaining one's actions and experiencing social boundaries, such as experiences of being controlled, ideas of being specially referenced by agents or happenings in the world, and false beliefs or delusions (Oltmanns and Maher 1988). To evaluate an individual's sense of reality for possible pathology, the clinician must take into account the individual's cultural background—for example, assumptions about the self; interpretations of the needs, intentions, and expectations of others; and the bases of interpersonal relations. Only against an understanding of the patient's view of the behavioral environment—the social universe, the meanings of objects in it, notions of selfhood and personhood—and knowledge of the patient's system of symbols and their meanings can a clinician identify distortions interpretable as psychopathology. Distortions of reality testing—for example, hallucinations, delusions—are taken as criteria of psychosis, roughly translated as a breakdown in an individual's sense of reality. Psychosis can also be understood in terms of breakdowns in social behavior, although ultimately the person's reasons or intentions for behaving in a particular manner are invariably invoked, and this often entails a consideration of the ability to test reality.

Mental state assessment and the science of descriptive clinical psychopathology have a long history in psychiatry (Berrios 1996; Berrios and Porter 1995). This science posits universalistic criteria about personal experience and judgment that allow the psychiatrist to reach conclusions about possible pathology. However, one may contest the basic tenets of this science from a cultural standpoint (Fábrega 1994a, 1994b). Cultural assumptions influence a sense of reality in innumerable ways. Nonetheless, the concepts of psychosis and of distortions in the sense of reality, as defined culturally in different communities, are useful in clinical diagnosis.

Given the importance of subjective experience, the clinician's task is to determine whether the patient's sense of reality is distorted. Obviously, language provides the critical device or instrument whereby a clinician, knowing the culture of the individual being evaluated, makes this determination. A clinical assessment of a deaf person is extremely difficult unless both participants use a conventional sign system such as American Sign Language or an experienced signer serves as translator. Even so, assessing the mental state of deaf and mute individuals who cannot read and write is very difficult. Evaluating a sense of reality necessitates a shared system of communication that elicits ways of thinking and feeling, the patient's subjective point of view about self and the world, and this is precisely what is lacking in the deaf mute.

Language is obviously crucial to an individual's sense of reality. One's point of view is formulated linguistically. Most hallucinations are auditory, reported as consisting of voices and verbal utterances of agents that have no basis in the common understanding of reality shared by others in the culture. Delusions are generally understood to be false beliefs or propositions framed in language. Delusions reflect distortions of self, of boundaries, of a sense of volition and personal control, and of interpersonal relations—all of which depend on an individual's perspective toward the self and the world, and a capacity to represent the self and the social world in linguistic and narrative form. Because shared language and culture are central to articulating and formulating a sense of reality, perhaps the basic pillar of the science of psychopathology, it is impossible to fully understand psychopathology in individuals who lack this capacity.

Tracing the evolutionary origins of psychopathology implies a conceptual continuity between the traditional clinical perspective, which links culture and language in testing the sense of reality, and related issues in varieties of hominids, Homo, and Archaic humans who are in the process of acquiring human forms of experience. How these earlier species perceived and experienced their world, and how they represented it, must be considered when conceiving possible scenarios of psychopathology during human evolution. There is no logical inconsistency in equating a modern human being showing psychopathology and an Australopithecine or early variety of Homo whose behavior one might hypothetically evaluate from a clinical standpoint. This correspondence might be demonstrated in at least three ways: (1) both individuals harbor profound distortions in their sense of reality that are reflected in negative deviations in the adaptability and adaptedness of their social and or ecologically oriented behavior; (2) they show no distortions of thinking and feeling, that is, no disturbances in mental state, but manifest impaired or diminished adaptability and adaptation in their natural setting; and (3) they show neither an abnormal mental state nor obvious negative deviations in adaptability and adaptation. Another logical possibility is that when properly assessed as to what is normal for their group or species, they show pathological deviations as in (1), but these are not reflected in the adaptability of their behavior, which otherwise appears normal. However, this interesting exception would pertain only to members of a culture in which psychological concerns were important and whose intellectual and emotional resources provided them with ways of dealing with circumscribed mental problems or simply isolating them. It is largely a theoretical possibility, since distress and impairment are usually central to the clinical enterprise, including diagnosis.

In the scenarios presented, an observer should be able to equate forms of mental state functioning elicited through interviews, or inferred from observation and knowledge of behavior and communication typical of the species, and reach comparable diagnoses. There is no logical contradiction in equating a psychotic, socially maladapted person who harbors delusions, experiences hallucinations, and is governed by a distorted sense of boundaries, control, and volition with an Australopithecine or early variety of Homo whose abnormal mental state or information-processing capabilities are reflected in such abnormal behavior as motor stereotypies, vigilant searches for nonexistent objects, displays of aggression and violence with minimal provocation, and persistent self-clutching, biting, or clinging.

Yet one may ask: In an individual lacking a language capacity, what does an abnormal mental state or distorted sense of reality consist of, and how is one to understand it? What would constitute an analogue (or possibly a homologue) of a delusion, hallucination, or an abnormal conception of social relations and personal control in such an individual? My view is that delusions would be inferred from persistent, faulty, nonsensical, or ill-guided activities involving the pursuit of food, choice of nesting sites, attempts at bonding or mating, selection of materials for constructing tools, and responses to environmental hazards or potential predators. Savage-Rumbaugh et al. (1996) attempt to diagnose intentional, symbolic aspects of the ecological behavior of bonobo chimpanzees. Whether an individual's mode of awareness and consequent sense of reality is framed in mimesis, some mental language or mentalese, or language proper, one finds the same underlying symbolic content and structure of psychopathology. An observer seeking to diagnose psychopathology using something like a sense of reality or mental state must be able to communicate with the individual by means of a shared system of communication. However, an observer could reach a diagnosis by interpreting social and ecological aspects of behavior. In this case, one is required to observe, follow, and compare an individual's actions to the behavior and responses of group mates, then interpret their rationality in terms of posited motives of the organism (Savage-Rumbaugh et al. 1996).

A factor influencing the character of psychopathology during human evolution would have been the emergence and elaboration of a sense of personal awareness made possible by culture and a fully evolved capacity for language (Carruthers and Smith 1996). Increasing linguistic competence enabled hominids to more elaborately label, conceptualize, and classify phenomena, including physical and social objects, one's mental state and those of others, and events in the outside world of the environment and the inside world of the group. As concepts of things were joined to those of actions and intentions, early hominids could better communicate with themselves and others about activities related to subsistence, strategic moves in the habitat, such as for better access to food resources and safe living spaces, and about ongoing social relations in the group. Matters pertaining to the self, others, the group as a whole, and the natural history of the habitat and its inhabitants—all of which were central to adaptation, personal well-being and survival, and group stability—were rendered conceivable by means of more elaborate language.

The Emergence of Culture and Psychopathology: The Role of Social Emotions

The emergence of culture viewed as the creation of a symbolic niche can be conceptualized as being caused by psychological changes that took place during evolution in response to specific social imperatives. Of particular importance would be emotions involving responses to social relationships essential to maintaining social organization and order. What I term social emotions, pride and shame being prime examples, are central. These are emotions related to the individual's social status within a group. Such a formulation of the role of emotions in the emergence of culture highlights potentially significant aspects of the importance of psychopathology during human evolution.

Basic or primary emotions—anger, fear, joy or satisfaction, sadness—are understood as basic response patterns of organisms, including humans, that are rooted in innate, phylogenetically old neurophysiological and behavioral routines. Secondary, derived emotions, such as pride or shame, are viewed as cognitively more complex, or as mixtures of primary ones (Ekman and Davidson 1994; Ortony and Turner 1990; Plutchik 1980). In both cases, emotional routines are held to have been designed by natural selection and are common to Homo sapiens, hence the claim that emotional phenomena are universal. Each of the primary emotions is conceptualized as composed of a pattern of complex reactions. These are set in motion by stimuli pertaining to relevant events and circumstances in the environment that affect overall fitness. Primary emotions are expressed in visceral responses, action tendencies or impulses, facial expressions, and overt motor-social actions involving the body, all designed to cope with the environment. They are viewed as continuous in an evolutionary sense with emotional phenomena of other mammals, especially primates. While a narrow view of emotions sees them as purely reaction profiles of the individual in relation to important environmental stimuli, a broad view associates these reactions with underlying motivational and coping mechanisms that involve cognitive appraisal and deliberate actions for adaptively handling high-stakes encounters with the environment (Folkman and Lazarus 1991; Lazarus 1991).

Emotions may be explained in evolutionary terms, and the basic emotions can be traced phylogenetically. In the conduct of basic life routines, such as foraging, mating, avoiding predators, and relating to conspecifics, an individual develops and shows emotions. Some examples would include fear of predators, joy and satisfaction at finding food or in grooming, disgust for ingested toxins, surprise at an unexpected social event, anger at another's refusal to share food or to respond to a mating attempt, and sadness at the loss of a parent or kin. The actual identity of basic emotions is contested (Ortony and Turner 1990), but in all formulations, basic emotions are viewed as fundamental reaction patterns that are universal and phylogenetically represented in the higher primates (Ekman and Davidson 1994; Panksepp 1992; Plutchik 1980). Some researchers hold that specific neurological mechanisms and circuits exist for each of the basic emotions (Panksepp 1991, 1992, 1995; Damasio 1994, 1999).

In social groups, relationships that involve cooperation and competition are closely tied to emotions, since social organization and order, coordinated activities, and resolution of conflicts are essential to group stability, and therefore to individual welfare. In social mammals, particularly nonhuman primates, dominance, submission, and social status and rank are important to an individual's overall behavior and success. An individual's social position in a group not only explains its relations with conspecifics but is also a marker or predictor of its relative fitness. High rank and status generally mean greater access to resources such as food and mates, directly contributing to survival and reproduction. Hence the emotions that pertain to an individual's group standing and help maintain group stability affect how a group functions across time.

In his analysis of social relations among residents of Dusun Baguk, a Malay fishing village, Daniel Fessler (1995) gives central importance to the secondary emotions of pride and shame. These emotions, of course, are common to peoples around the globe and have been analyzed in particular with respect to social organization and order. Irving Hallowell (1960), in a classic formulation of self, society, and culture from an evolutionary point of view, links the capacity for a "self-other" distinction—now conceived as a capacity for mental or social attribution and/or as a theory or model of mind—to the development of a normative orientation in group relations—in other words, to the whole question of moral feelings, orientations, and behavior. Hallowell argues that during the transition from protoculture to culture many features of human cultural behavior, such as notions of property, accomplishment, social responsibility, reciprocity, and obligation, were crucially linked through the ego's capacity to self-objectify and self-differentiate (his terminology) and experience distinct types of emotions in relation to these early social developments. Hallowell's formulation, then, implies that social emotions, such as pride and shame, are fundamentally important in the emergence of culture. Fessler's formulation is especially appropriate because in discussing these particular emotions he gives more explicit attention to evolutionary psychological considerations. The discussion that follows modifies and elaborates upon Fessler's ideas.

Secondary emotions such as pride and shame were no doubt implicated in developments that led to the emergence of culture. Among groups of the last common ancestor, we can assume that an individual who achieved high rank in its group experienced something like pride, satisfaction, and mastery; on the other side, one relegated to a low status would feel something resembling shame, dissatisfaction, and a sense of failure. In other words, much as the primatologists who study chimpanzee behavior conjecture, attaining dominance and high rank must have been associated with pride, whereas being relegated to an inferior position would have brought shame. These secondary or derived social emotions were attributes of an individual's perception of how it stood in relation to others in the context of social relations (recall Kummer's observation).

As culture and cognition evolved, experiences and functions associated with these basic social emotions were elaborated and extended. The development of a capacity for social and mental attribution and, consequently, the capacity to compare and analyze one's social interactions, in terms of one's standing compared to another, made it possible for individuals to experience pride and shame as well as their social complements, contempt, pity, admiration, and envy. A symbolic culture provides a framework for such emotions.

Among individuals with a capacity for self-conceptualization and social attribution, whether these are seen as a protolanguage (Bickerton 1990, 1995; Donald 1991, 1993a, 1993b) or are seen purely from a normative and psychological standpoint (Fessler 1995; Hallowell 1960), pride and shame are central to the emergence and elaboration of symbolic culture. Thus far, these emotions are associated with attaining social rank or perceiving that a particular group mate displays admiration or envy and contempt or pity. However, if we extend the idea of relative social standing to include conforming, or failing to conform, and being commended, or discredited, in the performance of one's social roles—that is, in meeting social obligations—such a definition covers what it means to be incorporated in a shared system of symbols and their meanings and to be governed by the assumed perspective of a generalized other (as per Kummer).

As cognition evolved, then, pride and shame, the emotions linked to dominance and submission, came into play in social interactions generally. Symbolic culture is defined as a group's conventional way of constructing rules, standards of behavior, responsibilities associated with social roles, and accomplishments that merit approval. Consequently, pride and shame play a crucial role in mediating and constituting symbolic culture. More specifically, they were somehow implicated in the internalization of norms, standards, expectations, rules, and moral-ethical obligations, all of which were incorporated in the symbol system that is culture. In social groups made up of evolved organisms capable of self-conceptualization and mental attribution, coping with emotions such as pride and shame was a selective factor in the emergence and evolution of culture.

In a broad frame of reference, pride and shame may be regarded as cardinal features of the psychological experience of individuals whose evolution has given them a cognitive capacity for symbolic culture. Furthermore, the capacity to follow rules and to conform to social standards of behavior may be regarded as a trait naturally selected for during evolution as a consequence of having to cope with these secondary emotions. The functions and implications of these emotions, in other words, were, and continue to be, fundamental to the process of enculturation. At the individual level, such a capacity or trait enabled social cooperation, conformity, and mutualistic behavior. Experiencing pride and avoiding shame were positively reinforcing psychological capacities that contributed to the maintenance of social order, since they promoted social mutualism. These emotions and their correlates may be viewed as the social glue that fostered social organization and order by constructing a cultural way of life. The symbolic niche, in other words, grew out of the internalization of these social emotions and how they contributed to psychological well-being and social order.

This formulation is consistent with any number of scenarios about the selective advantages to the individual of sharing ritual and language. It also complements the ideas of Donald and Bickerton, and Richards's physiomorphic model of the evolutionary process (1989a, 1989b). Recall that Richards emphasized that culture and language constituted traits of hominids and Homo that were selected for because they mitigated the social fragmentation accompanying the increased behavioral variability after individuals were freed from the instinctive restrictions of life routine chaining and dependence on environmental cues for survival. The physiomorphic model, then, requires the biological advantages of culture and language to preserve social order at a time when evolutionary developments had broken apart the tightly constrained patterns of life routines related to fitness. Consequently, Richards's formulation is highly consistent with that which attributes an important role to the social emotions of pride and shame. Indeed, the ideas of Knight and associates (Knight 1991, 1996; Knight, Power and Watts 1995) pertaining to female alliances and contracts with respect to sexual payments to males for meat secured in the hunt—all of this in the context of the timing of menstruation—are not inconsistent with these proposals about emotions. It was in terms of pride and shame that female solidarity was established and contractual gender relations were worked out to promote social bonds that fostered not only group harmony but also effective parenting and better male-female relations in general.

Oliver Goodenough (1995) relates law, the dynamics of information systems, and culture to biological phenomena and especially biological evolution. He offers an interesting thesis about the emergence of culture that complements and supports the idea of the role of social emotions. Goodenough's basic postulates are these: (1) An ability to initiate behavior automatically—that is, without any experiential learning, given appropriate conditions—was an early hard-wired, innate trait of simple life forms naturally selected for. (2) The trait, transmitted via genetic selection, contributed to survival, replication, and evolution of more complex life forms. (3) The ability to make adaptive behavioral choices enhanced the fitness of the system and the organisms that possessed it, producing genetically determined mechanisms for "the ability to learn, identify and be guided by recurring patterns of benefit and danger." A mechanism for checking or monitoring the success of learned behaviors would have constituted a component of such a system. (4) A capacity for imitation bestows advantages upon a learning system and is selected for. (5) Imitation, combined with a simple conditioned learning mechanism, reinforcement, is also a selective capacity, and the disadvantages of or problems with cycles of reinforcing mutual mistakes was abetted by a checking mechanism for evaluating learned imitated behavior. (6) Imitation is particularly powerful in highly social species, and in hominids, no less than among chimpanzees and the last common ancestor, the imitative urge leads to mutual mimicry, which, when combined with learning, leads to standardization and perpetuation. (7) Convergence of imitated behaviors and/or social rules is particularly powerful and selective in the case of conventions about social interaction: "Advanced imitation within a highly social species [is] a plausible culture generator." (8) A "self-sustaining template" for imitation translates back into behavior, and "this process gives rise to a non-genetically-perpetuated set of [controlling] behavior conventions." Thus "we can characterize culture elements as parasitic information systems, colonizing the minds of a succession of hosts." (9) Capacities for learning culture—patterns and habits of transmissible behavior—are generally beneficial, selected for, and eventually "will manifest themselves in an increasingly strong set of compulsions to conform to the expectations of the received model: compulsions which we identify with concepts of morality—right and wrong" (Goodenough 1995, 291–295).

According to Goodenough, then, individuals are guided by internalized, largely unarticulated models of acceptable and unacceptable behavior, although articulated ones complement and reinforce them. Qualifiers such as "right," "wrong," "just," "unjust," "fair," "unfair" record at the emotional level reactions generated by the internalized model of behavioral conventions that constitute culture. By positing a psychological imperative pertaining to behavior, Goodenough joins cultural anthropologists who emphasize the necessary link between self-conceptualization, normative orientations, and social emotions like pride and shame. In short, Goodenough's abstract and general formulation supports those of Fessler, Hallowell, Bickerton, and Donald. Beginning with fundamental biological axioms pertaining to the initiation of behavior, imitation, and learning, taking into account characteristics of social species, and tying all of this to the evolutionary logic or momentum of systems of information to standardize and regulate behavior, he presents a realistic scenario describing the emergence of culture. The human mind is the host of such cultural systems pertaining to behavior, while social emotions like pride and shame are the noise they create in organisms and which organisms must cope with. The effects or correlates of social emotions are transformed into an amalgam or glue that binds organisms to the information system. The units of information of these systems or cultures, termed *memes* by Dawkins (1976), show variation, are learned and passed on, are thus selected for, can evolve, and essentially act as human information parasites.

Thus far I have argued that social emotions, pride and shame in particular, are integrally associated with and closely monitor an individual's social performance and standing in a community. Such emotions, which occur in relation to encounters that bear on an individual's social rank and which can be equated with phylogenetically older relationships of dominance and submission, promote social cooperation and group stability and may have been selective forces during human evolution. Such emotions thus played a central role in the emergence of cultural rules and norms. Now let us address the question of whether—and, if so, how—the psychological changes and social developments described in these scenarios affected not only the character but also the function of psychopathology during human evolution.

These scenarios about culture underscore the pivotal role of psychological, emotional, and cognitive factors in expanding and even dominating the inner life of individuals. The social emotions, of which pride and shame are prototypical, underscore the social basis of mind. The universal drive or motivation for status and mastery among the last common ancestor, and the drives of the nonhuman primates, are transformed by and incorporate the myriad rules, standards, and social concerns of the group. In a basic sense, group relations, group stability or instability, and group identity, quintessentially social phenomena, are mirrored in psychological phenomena. Factors that maintain social organization and order become coextensive with those that maintain emotional well-being. An individual's group membership, in short, is represented in personal well-being, satisfaction, and credibility. Symbolic culture as realized in rules that bind and regulate is identified with notions of identity and self-worth. An individual's social standing and feelings of personal worth are incorporated into reflexive iterations of how one should behave and what is proscribed. Whether group mates admire or scorn the individual is expressed in displays of approval or disapproval to keep individuals behaviorally on track. Maintaining social dignity or at least avoiding a sense of shame and guilt come to dominate mentation and emotional well-being. In the same fashion, the information system instills the need for behavioral conformity in its colonized hosts, that is culture-bearers, thereby generating worry, doubt, distress, and a general preoccupation with conformity to standards and conventions. Goodenough (1995), as an example, describes cultures and information systems as containing traits that favor "conservatism," create "conformity anxiety" at the thought of deviation, instill condemnation of deviant acts, and evolve what he terms "immune systems" that are sensitive to excessive deviation and are designed to maintain a tolerable level of variation and curb excesses deleterious to the culture considered as human information parasites ("mind viruses").

Clearly, in this scenario, the nexus of social emotions and the system of symbolic information that is culture powerfully influence the character of psychological experience and regulate behavior. While negative feedback loops modulate a balance between positive and negative social emotions, conditions are created for emotional dysregulation in the form of escalated, positive feedback loops designed to either punish or enhance individuals both subjectively and in the eyes of group mates. In one respect, then, social emotions, and the system of symbolic information or culture that promotes them, play a dominant role in personal experience, particularly in maintaining a sense of well-being, the sense of being a worthy member of the group. This emotional domain plays both a primary role in regulating behavior and maintaining a sense of inner worth and a potentially significant secondary role. Social emotions can be recruited into cognitive realignments triggered by perturbation of other systems within the individual and in forms of activity removed from the social arena. In other words, vulnerabilities about social standing and worries about earning the admiration and approval of others are directly created as a consequence of social perturbations involving specific group mates to whom notions of obligations, rules, and standards are explicitly applicable. In addition, concerns unrelated to social standing per se, such as the threat of predators, a difficult habitat, and interspecific competition, can be registered in emotionally dysregulated sequences that end up being represented and expressed cognitively in notions of diminished self-worth and social standing.

The scenario about the role of social emotions and the power of information systems more generally in the emergence and expansion of symbolic culture underscores the fact that calculations or miscalculations about one's social status are central to how individuals think and feel about themselves. The tenor of group relations is echoed in psychological experience—notably, pride and shame. Vicissitudes affecting the social emotions can bring about a vulnerability to psychopathology by creating the emotional dysregulations that can affect visceral-somatic processes, undermine motivational drives, and lead to maladaptive coping. The manifestations of psychopathology, specifically, show how such maladaptive social and psychological behavior is played out. Worries, doubts, and ruminations about self-worth, the emotional markers of psychological wellbeing with respect to membership in a community, can be elicited by all sorts of perturbations that affect the information-processing systems of higher organisms. For example, perturbations in visceral-somatic systems—from malnutrition or infection, for example—and those caused by external events and changes—a curtailment of the food or water supply—that do not directly affect social standing can be transformed into secondary distress in the form of socially negative cognitions and distressing emotions and can disrupt behavior.

How does this scenario about the emergence of social emotions and symbolic culture help one to better understand the character and significance of psychopathology during evolution? Here it is fruitful to bring to mind the important role played by social emotions in the formulations of clinical psychiatrists and psychologists regarding the evolutionary bases of psychopathology (Gilbert 1992a, 1992b, 1998; Nesse 1990, 1998; Stevens and Price 1996). While these clinicians imply that the emotions and syndromes of psychopathology have a phylogenetic history, they avoid discussing how issues involving the evolution of language, cognition and culture affect the character of psychopathology. I emphasize that as a consequence of these psychological and cultural transformations, psychopathology becomes an important vehicle for working out an individual's identity, coming to symbolize its preoccupation with a sense of membership in or separation from the group. Specific syndromes will certainly continue to impact on survival and reproduction, as per the evolutionary residuals. Solving basic biological and ecological problems will continue to operate as an important theme in the cause and manifestation of psychopathology, that is, the effects of harmful dysfunctions. In the present context of evolution, however, responsibilities and obligations will come to acquire a hold on psychological experience, since failing to meet them, now that they are internalized directives, generates distress that can escalate and cause disorganization. In a sense, instances of psychopathology become a stage on which problems of social identity are played out in psychologically painful scenarios that can spread and undermine the functioning of visceral somatic rhythms and patterns. The vital importance of social wholesomeness and social relations, and particularly the misery and agony brought by the perception or recognition of failure, are registered in the form of patterns of emotional regulation and dysregulation manifested as psychopathology. The level, intensity, and quality of social emotions and psychopathology become, as it were, sensitive barometers of whether group functions and social relations are suitably regulated and maintained. This creates a highly vulnerable psychological terrain, if not a quagmire, made up of delicate regulators of social relations, pride and shame, and responses to envy, contempt, admiration, and a sense of inferiority. These and related social emotions, and their realization in psychopathology, become a common way of displaying the relative psychological well-being of the individual and the social well-being of the group it belongs to. Alternatively, since disturbed social emotions are represented directly or indirectly, causally or consequentially, in psychopathology, it can come to symbolize an individual's relative social standing and status in its immediate group. In short, the emergence of symbolic culture, the psychological dominion of information systems as infections of mind viruses, the associated social emotions they produce, and the role that all of these factors play in producing and regulating behavior carry with them processes and mechanisms that produce, perpetuate, and heighten the social significance of psychopathology.

It goes without saying that altered states of consciousness and dissociation figure importantly in this formulation about the character of psychopathology among the moderns. Experiences of trance and possession can also draw on social emotions, beliefs, and associated systems of information, and these complicated patterns can generate compelling narratives of cultural significance for the group. An earlier chapter notes the diverse interpretations that can be given to condensed, fused, global, and relatively acute and short behavioral crises of psychopathology in hominids and early varieties of Homo. In the absence of compelling markers of sickness and suffering, actors needed to construct new narratives so as to negate natural retaliative responses and to appropriate federated, affiliative ones before psychopathology could receive appropriate healing. The cultural resources I have described provide the material that make this possible.

- Scientists use the concept of culture in different ways in order to explain developments during human evolution. Its relation to the concept of a language is complex.
- A criticism of the concept of culture is that it has little explanatory power. It lumps behaviors that need to be examined separately in relation to evolution. It also downplays mechanisms underlying behavior and their relationship to specific biological functions.
- To many scholars the concept of culture does have explanatory power. It can be used to refer to the distinct symbolic world and contractual political structure that evolved with modern humans, which helps explain characteristics of the archeological record.
- The point of emergence of culture is a basic disagreement among scholars. Some purport that culture describes even early hominids, whereas others argue that it applies only to a capacity that evolved much later, during the Upper Paleolithic Era.
- Looking at the functional aspects of hominid behavior, the two views of culture can be integrated provided emphasis is given more to their cognitive implications and less to exactly when and how long it took for the change in cognition to occur.
- Evolutionary residuals gear behavior to survival, reproduction, and optimal life history parameters, and they describe one facet of hominid psychopathology. However, the way in which the residuals are configured and enacted in a cultural scenario is also significant.
- Systems of symbols and their meanings, in association with language, opened up, deepened, and endowed psychopathology with a narrative that connects to the worldview and cultural traditions of hominids.
- Clinicians who seek to explain the evolutionary roots and functions of psychopathology view emotions as highly important aspects of the manifestations of psychopathology.
- Secondary social emotions like pride and shame help explain how the emergence of culture changed the character of psychopathology during evolution.

Chapter 12

Dissociation, Psychopathology, and Evolution

THE POWER TO DEVELOP states of awareness and consciousness that differ qualitatively from ordinary, everyday experiences and are relatively discontinuous from them most probably constitutes a human universal. Obvious examples are dreams and states of relaxation in which one's attention is narrowed and focused away from everyday happenings. Less obvious and mundane types of altered states of consciousness have been studied in cultural anthropology and the clinical sciences. Some examples are trance, possession by external agents, hypnosis, and episodes of fugue, which involve sustained, more or less automatic, yet seemingly willed, intentional behavior, during which an individual loses a sense of time, in some instances followed by amnesia, or merely experiences intense concentration, self-absorption, and narrowed consciousness. Altered states of consciousness may induce complex experiences with a more or less distinctive psychobiology (Ludwig 1968, 1983). They can be broadly regarded as a potential component of almost all types of psychiatric disorders and also many forms of psychopathology as defined in this book. Paradoxically, they are also widely experienced as normal behavior.

From a group standpoint, altered states of consciousness serve to organize and intensify behavior, to increase suggestibility and compliance with a leader or important personage, and to coordinate and promote constructive group activities. Altered states of consciousness are used to prepare for a special mission or journey, for a battle against enemies, for a hunting foray, or simply to enhance ritual activities during a ceremonial occasion marking an important event. For the individual, altered states of consciousness are accompanied by changes in sensory perception and in emotion and cognition in conformity with prevailing expectations and symbols. Altered states of consciousness such as trance and possession can also be conceived as socially and culturally constructed behavioral routines or syndromes, for they follow a standard course and are given important meanings by witnesses that fit in logically with the enacted behaviors, especially if the individual becomes completely detached, self-absorbed, and later unable to recall the experience. Regardless of circumstances, the significance of and the role played by altered states of consciousness like trance and possession are based on the system of symbols and their meanings or culture. All forms of altered states of consciousness can be described as generally normal occurrences and subsumed under the relatively neutral category of dissociation. This can be defined as a state of consciousness during which an individual may experience a splitting or division in the content of awareness, with certain elements centrally engaging and shut out or dissociated from the rest of what ordinarily engages the individual and constitutes reality (Hilgard 1986). While not all individuals in all cultures can show such dramatic syndromes of dissociation as trance, possession, and hypnosis, many can be taught to have these experiences, and almost all human beings have experienced brief episodes of normal dissociation.

Scientific and Cultural Aspects of Dissociation

The concept of dissociation as used by psychiatrists, psychologists, and anthropologists refers to a mental capacity. Like psychopathology, then, it is a scientific concept. Dissociation is said to produce a mental state characterized by a splitting of the contents of consciousness. Some of the altered states of consciousness just mentioned are states of dissociation. During a state of dissociation, one's awareness and recall differ from ordinary awareness and recall. Stress and especially trauma evoke an array of psychological and physiological symptoms. Here I will emphasize the social and psychological behavior related to states of dissociation, and especially their cultural appropriation.

In a sustained state of dissociation, which can last minutes, hours, and allegedly days and even weeks, an individual can engage in social and psychological behavior that is culturally meaningful. In other words, the individual has experiences and performs actions whose significance is symbolized by the behavior enacted, be it social or verbal, its meaning inferred by group mates or reported by the individual. Thus, during a state of dissociation the individual seems to enter an altered mode of existence, its meaning determined by what the person says and does or what an audience infers on the basis of their knowledge of the situation. During such states, individuals may or may not answer when called by name and recognize their social characteristics. However, because they are not connected to or are unable to recall aspects of ordinary life, they are, strictly speaking, not themselves. Their identity during states of dissociation is culturally negotiated.

Individuals obviously retain a physical, biological identity while in a dissociated state. However, especially while in a sustained state of dissociation, they can be said to lose their social identity. For example, they may not recall their real names and social characteristics, in which case a condition of functional or hysterical amnesia is said to exist, and would consequently feel disconnected from their real roots in society and behave in a confused or muddled way. Alternatively, while in a state of dissociation, individuals may adopt new names and move into a different social space altogether, in which case they can be said to acquire a new social identity. A change in social identity as a consequence of dissociation can last a variable amount of time and produces a compartmentalization of experience and behavior (Putnam 1995). Again, this new identity is established by surrounding circumstances and interpretations of the behavior by self or others. By definition, the change of social identity seen in protracted states of dissociation is not willful or consciously planned; while fictive in the sense of not conforming to the identity conferred by society, it is real to the person as long as the state of dissociation lasts. In other words, experiences and actions during states of dissociation are not contrived. A dissociation of content, what individuals selectively hold to, and a dissociation of context, what they are not able to bring to mind, account for the observed changes in self-awareness and experience of self (Butler et al. 1996). During states of dissociation the contents of one's awareness and experience are split off from ordinary identity or self and produce an automatization and compartmentalization of behavior. (See Hilgard 1986; Putnam 1995; Spiegel and Spiegel 1978.)

To recapitulate, a capacity for dissociation is a human universal (Brown 1991) and is related to, overlaps with, or underlies several personality traits normally distributed in human populations. Ludwig (1983) contends that the origins of a capacity for dissociation lie in the evolutionary past. The behavior of animals in response to stress, so-called animal defensive reactions like freezing, tonic immobility, aggression, and avoidance of attack, are assumed to be phylogenetic precursors of dissociation as a human mental capacity and explain many psychobiological responses to stress and trauma (Nijenhuis, Vanderlinden, and Spinhoven 1998; Nijenhuis et al. 1998). During the biological phases of human evolution, when organisms often faced trauma and stress, they presumably experienced responses analogous and probably homologous to those of mental dissociation. Both psychopathology and mental dissociation, then, may have been correlated during protocultural phases of human evolution. While a mental capacity for dissociation has not been described for the great apes, the intense activities of chimpanzees under traumatic conditions, cannibalism and violence, for example, suggest that some may be in states of dissociation. The pancultural distribution of dissociation, its universality among human beings, suggests that this mental capacity might be innate. It must be emphasized that dissociation is a normal mental capacity and that an individual's behavior while dissociated is not necessarily pathological. Social circumstances and cultural interpretations can appropriate and thus normalize such behavior. However, while dissociation is not necessarily pathological, dissociation and psychopathology can overlap.

With the evolution of language, culture, and cognition, the capacity for dissociation grew increasingly complex. In other words, among modern humans the forms of experience and awareness, the symbolic content of the behavior produced, and the social and psychological meaning of behavior during states of dissociation show the influence of symbolic culture. Such states acquired meanings that drew on the group's accumulated store of beliefs, myths, rituals, and general knowledge.

While in a state of dissociation, individuals may participate in social activities and have psychological experiences. States of dissociation can occur naturally while persons are under stress or distracted, while daydreaming, as a consequence of diseases affecting the central nervous system, or as a consequence of ingesting mind-altering substances. Whether or not they are accorded special cultural significance depends on the characteristics of the group and its history. Certain forms of dissociation, such as being tuned out, breakaway phenomena, daydreaming, highway hypnosis, enthrallment, can be explained away as not unusual. Of course, when the dissociated behavior, or manifestations linked to dissociation through trauma and stress, include aspects of bodily dysfunction, pain, sensory motor abnormalities, and changes in arousal and consciousness, they are likely to be interpreted in medical terms. Symptoms of some illnesses are linked to mental dissociation and psychopathology.

Two categories used by psychiatrists, psychologists, and anthropologists to describe dissociation are trance and possession. Like dissociation, trance and possession are also scientific concepts, but they are based on cultural distinctions about behavior and especially states of dissociation as behavioral scientists understand them. This means that in theory, social and psychological behaviors popularly seen as trance and possession may not be states of dissociation as a scientist would define them. *States of dissociation*, then, is the abstract description for a behavior complex, whereas *trance* and *possession* are given local content and meaning, taking into account how the complex fits into the fabric of cultural life.

Trance and possession as scientific phenomena can thus be equated with folk or cultural categories or concepts about certain types of behavior that involve altered consciousness. During trance, individuals retain their everyday social and psychological identity but are seen as detached or dissociated from themselves. Some persons while in trance are said to exercise special powers or gifts, to be capable of removing themselves—their minds, souls, or even bodies—from their immediate surroundings and visiting or communicating with other agencies. Shamans and diviners characteristically go into a trance while exercising their healing functions or when receiving knowledge of importance to the group (Eliade 1964). If the state of trance is interpreted by the subject or an audience as involving a new identity, then possession is said to occur. The person is thought to be controlled or possessed by another agency made incarnate—a spirit, ghost, demon, or god—and thus enters an altered state. To repeat, the scientifically defined state of possession is to be distinguished from possession as culturally defined. The latter may only resemble a state of dissociation. That is, an individual may be thought to be possessed but fail to manifest scientific evidence of dissociation.

Anthropological Studies of Dissociation, Trance, and Possession

States of dissociation can promote, organize, and characterize group enterprises and significant commemorative events. In situations of discord and conflict, individuals may enter a dissociated state because of events that cause social stress. Ritualized forms of dissociation like trance and possession are found in association with religious and medical practices across the world and are thus institutionalized in the society. Indeed, one can classify societies in terms of whether institutionalized forms of dissociation are predominantly trance, possession, or a combination of the two. Erika Bourguignon (1968, 1979) was a pioneer in a cross-cultural study of states of dissociation as possession and made it a central topic of psychological anthropology.

States of dissociation and phenomena like trance and possession are part of the normal religious practices of many peoples and cannot be in any way equated with psychopathology. Common, everyday varieties of dissociation that accompany ritual and religious practices but are not institutionalized are also normal phenomena. Fundamentally, the institutionalized forms of dissociation like trance and dissociation largely expand upon naturally occurring, normal varieties of dissociation. However, for researchers in cultural and especially psychological anthropology, culturally specific forms of psychopathology are frequently associated with states of dissociation resembling trance and possession. This connection figures prominently in the literature about culture-bound syndromes and cultural psychiatry more generally (Simons and Hughes 1985). So-called hysterical psychosis, for example, reportedly involves states of trance and/or possession.

Although dissociation is not necessarily abnormal behavior, much of what constitutes clinically relevant psychiatric phenomena can be said to include states of dissociation. Moreover, a number of disorders listed in the contemporary diagnostic manual of psychiatry involve episodes of dissociation during which individuals are reported to lose their connectedness to their ordinary selves and identities and display experiences and behavior that belong to other selves. The most widely recognized is multiple personality disorder—a condition that is actually rather controversial. In general, however, while reported experiences of states of dissociation are usually normal behavior found in many societies, the capacity for entering or experiencing such states can be perturbed, exaggerated, and pathologized as a consequence of sustained social stress. This is especially true during early development. Some hold that disoriented or disorganized attachment to the mother or trauma during early phases of life plays an important role in the etiology of dissociative disorders.

Felicitas Goodman (1972, 1988, 1990), who has studied varieties of trance and possession cross-culturally, contends that a capacity to invent or rely on myth and ritual and to enter these states of dissociation has a long history in the human species. She emphasizes the recurrent, stereotypical routines found in syndromes of trance and possession—for example, the public, ceremonial settings in which they take place, their highly emotional and spiritualized context, the rhythmic stimulation involving the body, repetitive use of physical objects like whistles or rattles to intensify the emotional ambience, chanting, hand clapping, dancing and gesturing, making music, adopting specific postures, and often ingesting mind-altering substances. Goodman suggests that systematic posturings of the body, along with emotional and expressive motor activities, are generic devices for focusing and narrowing attention and entering states of dissociation on religious occasions. Goodman emphasizes the panhuman character of states of dissociation like trance and possession, and she suggests that these practices are probably of prehistoric origin, dodging the question of exactly how states of dissociation may have evolved.

Alondra Oubré (1997) extends the ideas of Bourguignon and Goodman by offering a theoretical formulation about the potentially important role of varieties of dissociation during human evolution. (However, this is only part of her overall thesis.) Oubré's formulation is imaginative but rather ambitious and wide ranging. While she draws on ideas about cognition and culture as well as dissociation, they can only with difficulty be equated, and she fails to sustain the argument that dissociation plays a central role in the evolution of brain, language, and culture. She attempts to integrate (1) basic ideas from cultural anthropology about the importance of rituals and myths; (2) general notions about states of dissociation like trance; (3) Goodman's ideas about the psychophysiological consequences of chanting and rhythmic body motions for producing trance; (4) ideas of Rudolph Otto (1923) about "the numinous," which she defines as "an ethically neutral feeling-response or mental state that is *sui generis*, or totally unique and self-derived" (1997, 7); (5) religious, spiritual, and transcendent features of the numinous as discussed in anthropological and popular literature; (6) ideas about unconscious forms of mentation as developed by Freud and Jung; (7) the nature of dreams and the various stages of sleep as measured by the EEG; (8) the use of psychoactive and medicinal plants by animals and man to create states of dissociation and "transcendent" experiences; and (9) ideas about the triune brain as formulated by Paul McLean. Using all this, as well as studies in paleoanthropology, evolutionary biology, and neuropsychology, Oubré argues not only for the possible early origins of trance and the numinous during human evolution, but also for its role in the actual evolution of symbolization, the brain and language, and for culture as a whole.

In short, Oubré is presenting an original scenario about the origins of human culture and cognition in which she ascribes to dissociation and the numinous a fundamental role. This is a provocative addition to the more standard explanations, which emphasize such factors as bipedality, tool use, food gathering, an aquatic mode of life, hunting, the development of female alliances against men in association with menstrual synchrony, and the emergence of human language. Parker (1996), who emphasizes social learning and transmission mechanisms in the development of self-awareness, elaborates an extractive foraging apprenticeship model to explain the emergence of culture and cognition. Although Oubré does not formulate her thesis on the basis of prior efforts along these lines, her thesis complements them.

Obviously, one need not accept all of Oubré's proposals about the central significance of the numinous in human evolution to agree with her (and with Bourguignon, Goodman, and psychobiologists such as Ludwig) that states of dissociation like trance and possession are replete with significance and that they might have continuity in a phylogenetic sense with phenomena pertaining to earlier hominids and varieties of Homo.

Evolutionary Significance of Dissociation

Despite differences in detail, the ideas of Bourguignon, Goodman, and especially Oubré and Ludwig about dissociation compel one to examine such phenomena from an evolutionary point of view. Something as ubiquitous, psychologically engaging, emotionally compelling, group-affirming, and culturally significant is likely to have both a long lineage in human history, possibly even a phylogenetic history that stretches back into human prehistory.

Although the capacity for dissociation, split consciousness, or alternative mental states has not been observed among chimpanzees or lower primates, the roots of such a capacity may lie in an animal's defensive reactions, motor escape behavior, and tonic immobility or convulsive phenomena (Hinde 1974; Ludwig 1983; Nijenhuis, Vanderlinden, and Spinhoven 1998; Nijenhuis et al.

1998; Oubré 1997; Thorpe 1956). What ethologists view as the release of a behavior routine following a stimulus and ritualization, the evolution of postures and movements as signal displays from nonsignaling behavior, together manifest as animals' reflexively enacted routines. They can come about also as a response to sexual stimuli, such as a ritualized sexual routine, or threat, such as aggressive attack or retreat and avoidance. Both phenomena are more or less automatic. Such responses in animals point to the existence of unconscious and more or less innate behavioral programs or routines. Whether analogous behavior in nonhuman primates (and hominids) is similarly genetic or is socially learned would be contested and hard to resolve since they are complementary.

In human beings, a capacity for ritualization and unconscious behavioral release, in the context of evolved forms of language and cognition, could be linked to dissociation and trance. The antecedents of dissociation in animals are thought to reside in defensive reactions in response to stress and trauma, especially predatory attack. Similarly, a phylogenetically rooted propensity to become immobilized and/or dominated or controlled in the face of environmental threats may have been naturally selected for (or exapted, see Gould 1991) in hominids and may constitute an innate, biological potential that members of evolved societies can draw on in appropriate circumstances. Many scholars have linked the human capacity for dissociation to such types of defense (Goodman 1972, 1988, 1990; Hilgard 1986; Ludwig 1968, 1983; Nijenhuis, Vanderlinden, and Spinhoven 1998; Nijenhuis et al. 1998; Oubré 1997). For example, tonic immobility and sham death in animals such as rats, chickens, and rabbits has been linked to dissociation (Ludwig 1968, 1983; Nijenhuis, Vanderlinden, and Spinhoven 1998; Nijenhuis et al. 1998). This behavior has special neuropsychological characteristics that have been equated with dissociation in humans. In fact, behavioral biologists and comparative psychologists also refer to tonic immobility as "animal hypnosis," underscoring its presumed relationship to altered states of consciousness and dissociation in humans. Tonic immobility is held to be an innate, automatic reflex reaction that is adaptive, like a psychological adaptation that animals have evolved for coping with stress.

Researchers have not observed behavior resembling animal hypnosis, tonic immobility, or any other behavioral complex suggesting dissociation in nonhuman primates. Nor have they provided a comprehensive epidemiology of animal hypnosis. Since tonic immobility has been described in animals about whom the topics of social intelligence and mental capacity are not an issue, and since it resembles unconsciousness, it cannot be defined as social behavior with symbolic significance. Tonic immobility could possibly be an isolated, independent phenotype, with a genetic basis and heritability, of some species not continuous in phylogeny. It is also possible that animal hypnosis does reflect a trait or adaptation with a pedigree that can be traced up the phylogenetic tree to the LCA, and then to hominids, in whom it is an embryonic form of the capacity for dissociation in humans but has simply not yet been observed among the higher apes. Finally, there may be no continuity between these forms of animal behavior and hominids, and, if so, states of dissociation like trance and possession were socially invented, learned, and then transmitted within human groups. In other words, they may be explained as products of cultural selection or evolution and not natural selection or genetic evolution. Language and symbolization could thus be sufficient to produce states of dissociation. But this argument can be challenged.

However, as formulated by psychobiologists such as Ludwig who take a comprehensive approach to human behavioral phenomena, states of dissociation, and the capacity to develop them, constitute normal, common, routine concomitants of individual and especially group experience. They are integral to coordinated group activities that have cultural and biological significance to group members. States of dissociation, I have already indicated, are pancultural, seemingly inherited by man, and I conjecture that higher primates must have them in their repertoire. Furthermore, in the descriptions of group behaviors of higher primates it is not difficult to find instances of group activity that appear to be states of dissociation. Primatologists, for example, have described driven, frenzied behavior of chimpanzees in violent confrontations with competing groups and also in the pursuit of monkeys, which they then consume outright, and in group displays of aggression during which they emit loud, intense cries that are not only deafening but also intoxicating and energizing. Clearly, while engaged in these routines, chimpanzees are marginally aware of other social realities and circumstances—social relations, possibilities for gathering plant foods, the sexual availability of potential mates, who the alpha male is—and appear to show the narrowed attention and concentration, focused cognition, and intense emotionality associated with states of dissociation.

The driven, cannibalistic infanticide routines described for chimpanzees by Japanese ethologists are further examples of repetitive, seemingly automatic, and almost ritualistic behavior during which participants appear cognitively and emotionally focused, driven, and seemingly entranced by the activity of killing and eating one of their own, while some observers, such as the mother of the infant, appear disoriented and seeming to experience a traumatic experience of victimization. In these and other group activities, and, one supposes, in individuals in the pursuit of any number of emotionally, psychobiologically driven activities—for example, sexual mating and aggressive displays—or who react to intense stress, one is likely to observe the narrowed consciousness, the intensified attention and concentration, the disregard for or turning off of other behavioral imperatives, and the full absorption and loss of the self that are hallmarks of human states of dissociation.

Some of the behavior observed in chimpanzees appears to resemble an emergent capacity for entering states of dissociation. The states of detachment, isolation, disinterest, self-absorption, and seeming inward preoccupation of elderly deposed alpha male baboons described by Kummer (1995) can also be cited as examples of routines that resemble states of dissociation in nonhuman primates. If one grants that these and related states of seemingly dissociated mental activity are at least analogues of, preadaptations for, or psychophysiological material that was exapted during later evolution leading to states of dissociation as fully elaborated in humans endowed with language and culture, one can argue that such propensities were part of the adaptive pattern of the earliest hominids. In short, states of dissociation have an archeology in human evolution (Gould 1991).

One surmises that with respect to states of dissociation and their cultural realization as trance or possession, one is dealing with phenomena that while possibly innate and rooted in phylogenetically old routines have been culturally elaborated and, indeed, constructed into something quite different from the purely unconscious forms of ritualized action or inaction as described for other mammals. With human states of dissociation, the circumstances that release or lead to the enactment of a program of behavior are a social context governed by higher forms of intentionality associated with the evolution of language and the cognition and forms of experience it makes possible. In short, as already implied, one could hold that a capacity for trance and dissociation per se was not present among the earlier hominids or early varieties of Homo. However, during a later phase of human evolution, such a capacity was realized because culture and language made it possible to create alternative imaginative worlds, and with this came new forms of psychopathology. This happens, according to this argument, because a capacity for dissociation or trance provided opportunities to engage in alternative levels of consciousness and awareness with which to solve problems of living or simply to relieve painful stress as a result of the special psychological effects of systems of symbols and their meanings. In such an argument, the emergence of symbolic reference and especially symbolic culture during later phases of human evolution gave hominids the special ability to elaborate upon the innate capacity for dissociation. Whatever its origin, the power of dissociation played an important role in human evolution and was certainly broadened and enriched by language and culture (Deacon 1997; Ludwig 1968, 1983; Oubré 1997).

When a capacity for trance and dissociation was realized in social settings governed by rituals, symbols, beliefs, myths, and narratives that explained mysterious or painful phenomena, it enabled the construction of models and scripts that could be implemented and played out in attempts to cope with such problems. A capacity for dissociation is also related to concepts such as repression and psychological defense mechanisms, which have been described as enabling forms of self-deception and more effective ways of manipulating others through deceptive means (Nesse and Lloyd 1992). Clearly, if this is true, this makes the expression and meaning of psychopathology more problematic and complex.

Psychological Trauma, Dissociation, and the Origins of Human Psychopathology

Psychopathology can result when vulnerable individuals are exposed to stress. Vulnerability can be a consequence of innate, genetically determined factors or developmental experiences. Psychopathology caused by stress can be acute and brief or more enduring, essentially recurring and chronic, depending on the degree of stress and at what age it occurs. Severe stress, or trauma, physical or psychological, has far-reaching effects on physiology, somatic symptoms, and behavior, including thinking, emotions, and memory (Bremner and Marmar 1998; Butler et al. 1996; Teicher et al. 1993, 1997; van der Kolk 1994). If trauma occurs early in development, it can render individuals vulnerable not only to psychological symptoms affecting thinking, feeling, memory, identity, but also to physiological symptoms such as gastrointestinal, sensory, and motor problems; pain; arousal; level of consciousness; and other ailments and symptoms well into adulthood (Nijenhuis, Vanderlinden, and Spinhoven 1998; Nijenhuis et al. 1998). Many pictures of psychopathology in nonhuman primates reviewed earlier can be interpreted as products of psychological trauma (Schneider 1992a, 1992b; Schneider, Coe, and Lubach 1992; Schneider and Coe 1993). Posttraumatic stress disorder in humans is a well-known effect of trauma and is thought of as psychological; however, the variety of symptoms seen in this and related psychiatric disorders believed to be influenced and partly caused by prolonged and significant stress and traumas such as depression and anxiety are best thought of as complex behavioral and psychophysiological disturbances (Bremner and Marmar 1998; Fábrega 1974, 1975, 1997). These disorders constitute a large proportion of the problems that general practitioners and psychiatrists see today.

One may assume that the character of social life during human evolution, which involved not only coping with natural catastrophes and group conflicts but also recurring competition and warfare with consequent social devastation, caused great psychological trauma. The psychobiological effects of stress and trauma with respect to psychopathology—disturbances in social and psychological behavior, including physiological and bodily symptoms—must have been very significant during human evolution. More broadly, the ubiquity of trauma in groups of primates, and by extension, in groups of early hominids and varieties of species Homo, in light of the wide-ranging effects on fitness and well-being that trauma can induce not only in adults but also during infancy and youth, when the brain is immature and thus vulnerable, must have been a moderating factor in all forms of sickness and health problems as well as psychopathology (Fábrega 1974, 1975, 1997).

One form psychopathology can take, then, involves the changes in behavior produced by psychological trauma, mental dissociation, and altered states of consciousness as these matters have been scientifically defined (Hilgard 1986; Ludwig 1968, 1983; Putnam 1995; Tart 1969). States of mental dissociation, like episodes of psychopathology, are expressed in complex forms of physiological, social, and psychological behavior; and, also like psychopathology, they are subject to folk interpretation. In other words, both are manifested in personal and somatic experience and social activities conditioned by current happenings and given cultural or symbolic meanings. Incidents of psychopathology associated with states of dissociation illustrate the interplay of psychopathology, language, and culture, as noted.

When yoked to an arguably innate capacity for dissociation, psychological traumas can create significant, long-lasting, painful, and debilitating psychopathologies (see Power 1991). However, I wish to consider a different role for dissociation in the origins of psychopathology. Many psychiatric-like syndromes that have been observed in family-level societies, that is, foragers, hunters, and gatherers, include states of dissociation like trance and possession. While one may question the validity of some of these reports (Simons and Hughes 1985), states of dissociation like trance and possession appear to figure very importantly in the construction and meaning of the culture-bound syndromes found in family-level and village-level societies. The ubiquity and overall significance of dissociation phenomena, the distressing, isolating, depersonalizing effects of psychopathology, and its regular appearance in syndromes having psychiatric significance suggest that such phenomena may have played a role in the origins of human psychopathology.

When, how, and why might a capacity for dissociation, specifically trance and possession, have evolved in the line stretching from the last common ancestor to Homo sapiens? It could be maintained that such a capacity is a feature of cultural evolution that emerged sometime during the transition to the Upper Paleolithic Era as part of the explosion in unmistakably cultural, artistic, creative, and expressive activities associated with that period (Davidson 1991; Pfeiffer 1982). Many observers interpret cultural developments of this era with the emergence of a full capacity for symbolization and language. Many of the cave paintings and figurines from this period strongly suggest spiritual or religious functions and resemble artifacts and images associated with shamanism, as found among contemporary foragers. It seems reasonable that states of dissociation and especially trance and possession could have emerged during the cultural flowering of the Upper Paleolithic. They could have played a role in religious practices and shamanism, and the curing of sickness, including special forms of psychopathology. From there, it could be only a short cultural distance to the incorporation of dissociative experiences into the material content of modern human psychopathologies.

On the other hand, one could apply the ideas reviewed earlier about the pancultural anthropology of dissociative phenomena, their possible paleoanthropological roots, and the possible preadaptation for all of this in the neuropsychology of tonic immobility or animal hypnosis to argue for a biologically based evolutionary role of such phenomena in the origins of psychopathology. As mentioned earlier, some researchers who study dissociation in humans have related it to animal hypnosis and tonic immobility (Ludwig 1968, 1983). Dissociation can be observed in very young infants, in various environments of attachment and mothering, which suggests that it could be innate and has a phylogenetic history (see Main and Morgan 1996).

Incidents of psychopathology may be conceptualized as loci of personal distress and behavioral maladaptation found in groups of individuals who are mutually self-aware, capable of theory of mind thinking and mental attribution, share a high degree of social awareness and connectedness, and maintain a reasonably well-organized social system with some behavioral traditions or a protoculture. Since these properties of cognition and culture are arguably consistent with qualities observed in chimpanzees and attributed to the LCA, one can safely ascribe them to earlier varieties of hominids and Homo (Parker and Russon 1996; but see Tomasello and Call 1997). For hominids, then, an incident of psychopathology could have been almost by definition a socially isolating experience, and often socially alienating as well; an experience arising from stress, causing pain and distress, and from which the individual would want to escape. Moreover, an experience that underscores the fragility of social connectedness and the threat of losing it might very well activate a capacity for acquiring coherence and stability and achieving not only emotional relief but also restoration. In other words, an individual capable of entering a state of dissociation might respond to an experience of psychopathology by becoming withdrawn and self-absorbed and experiencing the narrowing and intensification of consciousness that facilitates entry into states of dissociation.

Bourguignon's research underscores the ubiquity of altered states of consciousness cross-culturally, and Goodman establishes the pancultural uniformity of postures, rhythmic stimulation, stereotypical bodily movements, and the entrainment of physiological and neuropsychological processes involved in inducing trance and possession. Their work raises the question of the origins and natural substrata for dissociative phenomena. Oubré's bold thesis brings into focus additional parts of the puzzle of states of dissociation like trance. First, what types of psychological experiences must socially evolved organisms be capable of in order to enter a trance in the first place? Second, do organisms capable of trancing depend on a repository of knowledge, or at least a set of symbols, that, through attention, concentration, and elaboration, they can dissociate and enter a trance? Third, what sort of cultural and social setting makes trance possible? In the present instance, pathological trance seems to arise in situations of psychological distress, social conflict, and social marginality with associated symbols and lore about disability and sickness. Finally, what are the cognitive substrata that underlie states of dissociation and trance-related experiences? And when and how did they emerge during human evolution?

Symbolization in States of Dissociation and Psychopathology

I am assuming that psychopathology and dissociation experiences involving phenomena like trance and possession depend on and elaborate on psychological states marked by symbolization, as classically manifested in protolanguage and language (Deacon 1997; Savage-Rumbaugh and Lewin 1994). As we have seen, during human evolution psychopathology came to clearly entail a capacity for comprehending one's condition in relation to the group; in other words, psychopathology entailed a capacity for symbolization by means of at least protolinguistic representation. Furthermore, even if individuals are merely plunged into states of dissociation without awareness of what is happening, as one can assume is the case in rabbits and chickens, in highly social organisms like hominids and early varieties of Homo, outcomes are likely to be very different. Among the latter, their bonds with group mates and the fact of living in a public setting composed of shared meanings necessarily implies that disturbances of the flow of consciousness and social awareness, that is, trance and even possession-related phenomena, are likely to be noticed, made sense of, and dealt with, even if the group has only a primitive form of protolanguage and symbolization. Moreover, to whatever extent the phenomena are caused by distress occasioned by social conflicts and environmental turmoil, and especially if they are consciously sought for whatever reason—to relieve distress, to elicit support and compassion—to that extent the individual is using a capacity for symbolization through linguistic representation. This reasoning leads to two generalizations. On the one hand, symbolization as realized in language or protolanguage is almost by definition required to meaningfully configure and enact not only psychopathology but also to enter states of dissociation like trance and possession. Conversely, since highly social organisms or individuals are vulnerable to social stress and conflicts, and are also capable of dissociation, these two different behavioral tendencies might be joined as a likely consequence.

Deacon's formulation (1992a, 1992b, 1997) about brain evolution in relation to animal call systems and human language is not inconsistent with these generalizations about the connections linking symbolization, dissociation, and psychopathology. Deacon emphasizes the continuity between animal calls and human language while acknowledging profound neuroanatomical and behavioral differences. He believes that a capacity for symbolization linked to the beginnings of language emerged around 2 million years ago. He has reviewed material involving the comparative organization of the brain of vertebrates, mammals, especially monkeys and apes, and projections about hominid brain organization drawn from studies of fossil remains and anatomically modern humans. Deacon relates changes in the sizes and connectedness of different regions of the cortex, limbic system, and the midbrain, and the possible transformations from the primate emotional calling systems to early and advanced forms of language among varieties of Homo, culminating in modern humans.

In arguing for a possible continuity between primate emotional calling systems and human language, Deacon sees language as having retained its connections with the limbic system and midbrain, but over evolutionary time emotion has modulated it in the service of rational propositional thinking as realized in language and speech. The brain of varieties of Homo, in other words, has undergone changes from patterns of neural connections pertaining to stereotyped, restricted, highly emotionalized forms of expression, on the one hand, to the development of language, a more controlled form of communication given meaning in relation to cultural activities, which uses brain centers involving the tongue and larynx, on the other. Connections between the higher cortical regions, the language circuit, the limbic system, and the midbrain remained operative and functional as language evolved, but during evolution there was a lessening of the size of the limbic-to-midbrain connections, subserving emotional expression, and a heightening of connections with nuclei subserving the production of speech proper. The evolved neural capacity for speech and language, then, retains its connectedness to emotional circuits of the brain, and in certain circumstances a spillover between the two modalities of communication and representation can still occur.

Deacon's thesis about the evolution of the brain and language, then, contains an exposition about the shift and transformation in the forms of mentation, emotion, and behavior that occurred during human evolution. It can also be used as a blueprint for how symbolization and protolinguistic capacities may have figured in two realms of experience and behavior that one must presume also emerged during human evolution, dissociation phenomena, and psychopathology. Earlier varieties of hominids and Homo drew on symbols to express phenomena of salience to the individual and group. Unique experiences pertaining to the character of selfhood, personhood, of the other, and of the group, along with more abstract notions about the character of existence, if not actually caused by symbolization, were unquestionably deepened and rendered more compelling through it. Episodes of psychopathology in vulnerable individuals, a not uncommon, recurring feature of hominid life, was undoubtedly affected by these social and cultural changes. Furthermore, psychopathology draws to itself material that is integral to states of dissociation. Finally, both states of dissociation and psychopathology would have entailed symbolic reference and culture in line with the neuroanatomical and associated psychobiological transformation outlined by Deacon. In states of dissociation and psychopathology during human evolution, one might see an amalgam composed of different and changing mixtures of emotion, motor programs and displays, symbolization, social and psychological behavior, and scenarios composed of narratives having mythic, ritualistic, and existential import. The neuroanatomy for all of this would have been constructed through changes in the connections among neural centers whose size and importance also changed during evolution in a manner not unlike that outlined by Deacon.

Dissociation and the Evolution of Psychopathology: Concluding Comment

The adaptive value of dissociation, in theory, is that it allows a splitting of consciousness and the creation of focused, inner worlds of attention that can be appropriated for functions promoting survival and hence contributing to fitness. A mental capacity that allowed hominids and early varieties of Homo to dissociate from ordinary consciousness and to enter or create an altered state of consciousness governed by limited, intensely focused, and forcefully enacted routines of behavior must have given an individual a competitive advantage in the environment of evolutionary adaptedness.

The ability to realize personal or group-centered goals and objectives would be enhanced by a mental capacity like dissociation. In other words, an individual could pursue those goals in a frame of mind or consciousness dissociated from ordinary, now irrelevant concerns and thus focus on concerns requiring a more demanding resolution, given current imperatives. Intense, sustained concentration could help one avoid predators, pursue prey, win battles over contested territory, and galvanize and focus others' attention. One could enter states of reverie that promoted a sense of self-worth and personal integrity; one could establish intense personal relationships associated with mating and reproduction; one could lose oneself while working on a basic survival task, such as foraging, building a shelter, or tracking an enemy; one could cope with the agony of losing a relative, mate, or child; or one could project oneself into an imaginary scenario to solve an anticipated crisis. Such crises, which were undoubtedly common in the environment of evolutionary adaptedness, would have been eased if connected psychophysiologically to a genetically built-in capacity for mental dissociation.

This argument for the adaptive value of dissociation emphasizes the important relationship that exists between such a capacity and, more generally, aspects of cognition and culture. As indicated earlier, the evolution of symbolic culture and cognition greatly affected the kinds of experience and behavior made possible by states of dissociation. Knowledge, reasoning, concepts, beliefs, myths, and patterns of social and psychological behavior linked to states of dissociation became ever more meaningful from a symbolic standpoint. Spirituality, the numinous, a sense of the supernatural, and religion constitute obvious examples (Goodman 1972, 1988; Oubré 1997). While the origins of religious thinking may flow naturally from innate, fundamental, and essentially almost mechanical ways in which cognition worked as a consequence of natural selection (Boyer 1994), a capacity to enter altered states of consciousness, and to fashion them in symbolic ways that elicited special and strong emotions, gave religious concepts and attitudes considerable psychological and eventually social power. Details pertaining to the biological and cultural evolution of a mental capacity like dissociation are not primary concerns here, nor is the evolution of religion, numinous experience, and dissociation (Goodman 1972, 1988; Oubré 1997). What is of concern is the posited relationship between a capacity for dissociation and the evolution of psychopathology.

Researchers have reasoned that the capacity for dissociation constitutes a trait that is universal and possibly selected for and hence innate. It may have been exapted in relation to the evolution of culture and cognition (Gould 1991; Gould and Vrba 1982). Originally serving other, perhaps more fundamental biologically adaptive functions, the capacity may have been biologically and culturally appropriated for any number of important psychological and social functions. It made possible increasingly elaborated routines of social and psychological behavior that contributed to the emergence of important social institutions, not only religion but also institutions and practices involving healing, forming political alliances, discipline and punishment, and celebratory practices involving creative uses of music and dance. These ritual practices and associated myths enabled individuals to use them creatively, based on their innate capacity for mental dissociation to solve problems.

The capacity for dissociation was also exapted and culturally appropriated for working out and symbolically reconstructing psychological conflicts and strains. In other words, it provided a means for dealing with incidents of psychopathology. By linking psychological correlates of social dilemmas tied to competing demands and pressures—that is, conditions leading to psychopathology in vulnerable individuals—to an innate capacity to enter altered states of awareness and consciousness and use them according to cultural directives and imperatives, individuals found alternative ways of expressing and relieving the stresses and agonies elicited by those dilemmas. Two factors are involved here.

First, social crises and dilemmas created psychological conditions that could, in addition to precipitating psychopathology, bring on states of dissociation. Second, states of dissociation provided inner spaces or psychological arenas in which those stresses tied to psychopathology could be worked out by channeling psychological experience in a positive, conflict-alleviating direction and by producing scenarios of behavior that communicated the distress and played it out in ways that were safe and culturally understandable, and capable of eliciting sympathy and support. In this way, the capacity for mental dissociation, and the experiences and behavior it made possible, was channeled into the amalgam of material that made up incidents of psychopathology. In other words, in vulnerable individuals, the capacity for mental dissociation could be yoked to those forms of psychosomatic and somatopsychic experience and behavior that registered the effects of stress triggered by current events and crises.

The capacity for mental dissociation constitutes a feature of the human phenotype if not genotype that through symbolic culture has been elaborated in significant ways. The general, protean, and essentially adaptive features of dissociation, the many ways in which it could be played out, made it a positive, selected-for trait. We have seen that psychopathology can also be seen as a human trait that is selectively neutral, negative, and/or even positive. Regardless of how the question of the adaptive significance of psychopathology is answered—whether one views it as a pure mechanical debilitation, a vulnerability, defect, or blemish, or a capacity that in some instances can produce benefits—it seems clear that how psychopathology was configured and enacted, and especially its social consequences, were significantly affected when connected to the material that realized the innate capacity for mental dissociation. In other words, the potential social and cultural significance of episodes of psychopathology, and the epistemic, essentialist meaning of psychopathology itself, considered purely in abstract terms, were all modified significantly when its manifestations, that is, signs and symptoms, were mingled or amalgamated with social and psychological behaviors correlated with states of dissociation, altered states of consciousness interpreted as trance and possession.

I have emphasized that states of dissociation when configured and played out have been adaptive in a number of ways and that they often draw on culturally meaningful rituals, myths, beliefs, and traditions. For these reasons, states of dissociation as behavioral phenomena can be said to carry a great deal of potentially positive social value and ethical-moral utility in the cultural economy of human groups during both biological and cultural phases of human evolution. The behaviors that they can valorize and the potential enhancement that they can mobilize on an individual's behalf may be thought of as social capital that a culture offers to its members. Such currencies of value must have benefited vulnerable individuals in the throes of coping with stressors that contributed to the formation of psychopathology, provided its symptoms and signs drew on social and psychological behaviors that constituted states of dissociation and could be labeled and interpreted by group mates in culturally meaningful terms as trance or possession. States of mental dissociation and the behaviors that they authorize, then, may be viewed as offering individuals susceptible to psychopathology potentially exonerative and exculpatory devices that might neutralize the potentially objectionable and divisive effects that symptoms and signs of psychopathology can have because of their link to social conflicts and antagonisms.

- Altered states of consciousness are described and explained scientifically as dissociation, the compartmentalization and potential isolation of the contents of personal experience and behavior. Dissociation is common and natural in communities around the world.
- In cross-cultural studies, states of dissociation such as trance and possession are frequently seen in relation to healing, divination, and religious devotion and have been described as integral to the practices of shamanism.
- Members of a culture may validate states of trance and possession although such states may not involve true altered states of consciousness, that is, dissociation. However, proving mere role-playing compared to actual altered consciousness can be very difficult.
- Dissociation appears to constitute a human universal, and evidence that it may have a phylogenetic pedigree as well as a biological function suggests that hominids could enter states of altered consciousness and derive important benefits from them.
- The emergence of culture and language certainly facilitated entering states of dissociation and also broadened their scope and social significance since beliefs, rituals, and other aspects of symbolic meaning could explain and fashion elaborate scripts and scenarios.
- Dissociated states are not always pathological, yet such states are commonly associated with psychopathology cross-culturally. When dissociation is pathological, the result is a complicated clinical picture, with an explanation rich in interpretive symbols and meanings.
- Because dissociation and psychopathology are both socially dramatic and culturally complex, they enhance each other and hence their frequent association constitutes an important milestone in the evolution of culture, cognition, and psychopathology.
- Psychological trauma greatly expands the resonance between dissociation and psychopathology, and because such traumas were common during human evolution, the prevalence and significance of pathological varieties of dissociations was probably high.

Chapter 13
Psychopathology in Archaic Human Societies

WHAT IS THE relationship between a group's psychopathology and three aspects of its evolved way of life, namely, its cognition, language, and culture? These concepts are complexly interrelated but analytically independent. Understanding their role in the evolution of psychopathology poses an intellectual problem. To phrase it slightly differently: Viewed against the expanse of human evolution, how do characteristics of psychopathology in a particular hominid group differ in relation to characteristics of its evolved cognition, language, and culture?

The Later Phases of Human Evolution

The divergence of the genus Homo from the earliest hominids, the Australopithecines, was a major milestone in human evolution. Assumed to have taken place approximately 2 million years ago, this shift in adaptive pattern is associated with a number of major transformations in morphology, lifestyle, and cognition. With the shift to the more seasonal and open environment of the savanna, where animal resources were abundant, groups grew larger because of a need for better protection against predators. The increased size of groups, which is believed to have made social relations more complex, was facilitated by an increase in neural tissue that allowed more complex forms of cognition (Aiello and Dunbar 1993; Dunbar 1992, Mellars and Gibson 1996).

Archeological research offers no clear evidence suggesting that Homo erectus showed elaborate forms of symbolic behavior. The Acheulean hand axe tradition associated with erectus populations is taken to imply low levels of symbolic communication or language skills, while the stylistic uniformity and absence of innovation of material culture suggest that these artifacts were produced by imitation rather than verbal instruction. There is also no compelling evidence suggesting the construction of living shelters or the consistent use of fire.

Two anatomical characteristics of Homo erectus suggest that vocally based symbolic communication was rudimentary. First, there is no evidence of an expanded spinal canal, suggesting lack of development of control over muscles used for respiration, which are important in sustained vocalization. Homo erectus also appear to have lacked the adolescent growth spurt characteristic of modern humans, and evidence of skeleton growth shows an advance in dental development similar to that found in chimpanzees. This implies that erectus infants and juveniles did not experience a long period of dependency, which is necessary for the transfer and learning of symbolically based cultural knowledge.

Homo erectus probably communicated through incipient forms of vocal language, nevertheless. Expanded group size, more social organization, increased vocal and emotional communication, and cognitive development all placed greater demands on neural tissue, manifested in a larger brain (approximately 1,000 grams). The increase in brain size meant a greater demand for energy because of the comparatively high metabolic rate of the brain tissue. This demand for energy may have led to a greater reliance on animal protein as a source of food, and associated with this change came a smaller gastrointestinal tract and changes in the structure of the mandible and dentition, both of which facilitated a shift to meat eating (Aiello 1996a, 1996b; Foley 1996).

The period between 1.6 million and 300,000 years ago is generally viewed as a period of stasis in the archeological record, owing to the stability and relatively successful adaptation of varieties of the species Homo erectus. About 300,000 years ago, there was an important shift in the degree of encephalization. Archeological records and artistic flake technology provide supporting evidence. Because social groups increased in size, Homo erectus needed more complex forms of social organization and had to develop more sophisticated social intelligence, all of which are assumed to have caused the expansion of the prefrontal cortex (Aiello 1996a, 1996b; Foley 1996; Roberts, Collins and Robbins 1996). Aiello and Dunbar (1993) believe that groups grew larger than the critical size for using social grooming to maintain social order and resolve social conflicts. The need to maintain functional forms of social relations in larger, more complex groups is assumed to have selected for increased linguistic capacity, as well sheer cognitive ability, both of which necessitated larger, more encephalized brains.

Part of this shift, scholars believe, was the ability to communicate by means of ordered lexical units or syntax to refer to abstract concepts and symbols. The increase in brain tissue, fostering changes in cognition and language beyond the level of that of Homo erectus, is associated with a range of major changes in demography and life history. For example, giving birth became more hazardous for females because fetuses developed larger brains, leading to the delivery of more immature fetuses. This slowed the growth rate and extended and delayed the development and maturation of modern humans, whose infants are dependent for a longer period, that is, altriciality, who have longer intervals between births, who first reproduce at a later age, and who live longer (Foley 1996). The period between 300,000 and 200,000 years ago, then, is viewed as critical in the evolution of human social behavior. During this period, groups increased in size, social complexity, and social cognition; and linguistic capacity developed, that is, language acquired syntax and symbolization; there were major changes in life history traits and the capacity to plan and use technology for subsistence. This shift in adaptive strategy led to the emergence of Archaic humans. This group, arguably, may have included Neanderthals, or may have evolved into Neanderthals, and eventually modern humans, Homo sapiens sapiens.

Anatomically modern humans, Homo sapiens, are held to have originated in Africa around 150,000–100,000 years ago and dispersed from there across the globe over the next 60,000 years. Archeological remains reveal no major changes in social behavior during this period. What is striking is the fact of dispersal itself, which is felt to be a consequence of changes in social organization (Foley 1996; Mellars 1996a, 1996b, 1996c; Gamble 1993, 1996). Foley writes:

> It can be argued that the ancestral condition for modern humans consisted of moderately large communities, with coalitions of males linked by kinship, and unrelated females attached to specific males or possibly several males; . . . such groups are likely to have at least an element of inter-generational lineage structuring (patrilineal, given the pattern of male residence). Assuming conditions of net local population growth, two significant characteristics would arise, from this ancestral social organization. . . . The first is that with male kin-bonded groups, communities would be at least partially closed to each other and hostile, resulting in some form of territorial or agonistic behavior between communities; . . . competition for both resources and reproduction would increase. This would lead to demographic fission of communities . . . promoting geographical dispersal and leading to the colonization of new regions . . . [and] would lead to a segmented lineage structure dispersed and dispersing across the region as whole, and thus producing larger-scale networks of cultural groups, and a regional pattern of ethnic differentiation. (1996, 106–107)

In the model proposed by Foley, qualitative changes in the brain and associated cognitive changes did not account for the Upper Paleolithic transition. In his words, "Art, 'symbolic behavior,' blade technology, and so on (such as those found in the European Upper Paleolithic), are not so much manifestations of a radically different form of behavior but of specific demographic and ecological conditions arising from the successful dispersal of the descendants of the middle and early Upper Pleistocene African hominid populations" (1996, 107). Evidence suggesting actual major changes in social behavior, cognition, and culture are found in Europe beginning around 40,000 years ago. These involve blade technology, changes in dwelling sites suggesting complex organization and role differentiation, and a flowering of artistic creativity in the form of decorative figurines and cave painting.

The comparatively small brain, the rudimentary culture and cognition, and the long, relatively stable tenure of Homo erectus (1.8 million to 500,000 years ago) provides a sharp contrast to changes that took place during the later phases of human evolution involving Archaic and modern humans (500,000–100,000 years ago). Along with a larger brain came advances in cognition, language, and culture. This was associated with changes in life history traits, particularly the birth of smaller, more neuromuscularly immature infants and an extended period of dependence. This presumably correlated with greater paternal investment in caring for their offspring to ensure their survival, with consequent changes in family organization generally and relations between the sexes in particular. This transformation of social life brought larger and more dense populations and changes in political and economic arrangements. New patterns of social life are held to have engendered myths and rituals as a means of formalizing relations between the sexes, and for giving meaning to groupwide institutions, or patterns of roles and behavior, for planning and executing collective actions against competitors and advancing the economic, moral, and emotional health of the group.

Basic Concepts

The later phases of human biological evolution are understood to have brought changes in cognition, language, and culture. Because the social and psychological behaviors of hominids were obviously affected as a function of these changes, one must assume that psychopathology, as a subset of such behaviors, was also affected.

Cognition, or thought, is defined as conceptual activity and reasoning ability, a property of virtually all higher life forms and especially primates and hominids. Roughly analogous to intelligence, cognition refers to abilities that enable perceiving, categorizing, reasoning, and drawing inferences, planning, problem solving, and the like. These powers derive from activities and processes of the central nervous system. An important aspect of cognition involves the imaginative faculties of understanding and reasoning, which include creating frames, image schemata, metaphor, metonymy, polysemy, and other figurative ways of drawing meaning from experience (Lakoff 1987; M. Johnson 1987).

Structures of imaginative understanding should be construed not merely as aspects of speech and language but as fundamental aspects of conceptual activity through which an evolved hominid group creates and establishes meaning. Imaginative and figurative aspects of understanding, while conceptual in nature, derive from basic experiences. In other words, the physical body not only locates the organism in space and has a distinctive morphology, but also situates and makes possible a range of experiences about worldly phenomena ordered in terms of the body's shape and motor activities. These would include the apprehension of verticality and distance in space, the duration of time, the purely sensory properties of external objects, sensations of the body, including pleasure, comfort, or pain, and the experience of handling objects and locomotion. These structures of bodily experience provide a material basis for imaginative understandings of the world; in other words, they function as experiential structures of meaning integral to abstract understanding, rationality, and a sense of group identity and historical placement (Johnson 1987).

Thought and cognition, including imaginative aspects of understanding and meaning, are universal properties of what has been termed the modern mind and are assumed to have an evolutionary basis (Donald 1991). They are also integral to what neuroscientists mean by symbolization (Deacon 1997) and are influenced by and in turn influence language and culture.

Language is the innate human capacity that makes verbal communication possible. All natural languages contain a system of rules, or linguistic universals (Aitchison 1996; Pinker 1994). Language embodies rules and rule systems for the formation and use of morphemes, words, syntax, grammar, and the range of related phenomena that enable us to think, reason, and communicate verbally, frame propositions in sentences, and construct narratives about the world. Language is thus integral to the construction of human meaning and understanding generally and is closely related to aspects of thought and cognition, particularly the imaginative and figurative aspects of understanding and meaning.

The definition and meaning of the idea of culture is much contested (see Janicki and Krebs 1998), especially in evolutionary studies. There is controversy about the relationship between culture and biological processes—for example, whether they have independent effects on fitness, how far culture is based on genetic factors, and the interaction between biology and culture with respect to evolution. Although these questions are not examined here, I assume the capacity for culture, like that for language and cognition, is greatly influenced by genes, and that all three are products of human evolution.

In general, the term *culture* is useful when applied to a domain kept analytically separate from the two other concepts used here—language and cognition. However, both language and cognition clearly influence what culture consists of. By *culture* in this context I have in mind something more specialized yet elaborated than the general ecological conception formed by comparative psychologists, for whom *culture* roughly means behavioral traditions and social styles (Heyes and Galef 1996; Russon et al. 1996). Here, I equate culture with a large pool of information evolved by a group. The information pool varies as to scope and breadth, structure and complexity, and, most important, content or meaning. Culture, for those who share it, stipulates and contextualizes the what, where, why, and how associated with thinking about and relating to the behavioral environment (Hallowell 1960). It can be equated with mental lists and inventories, recipes, rules, standards, rationales, beliefs, myths, rituals, the content of programs for coping with the environment, and the social meanings of artifacts produced (Geertz 1973). Culture is a system of symbols and their meanings, a specially demarcated symbolic niche (Chase 1999). The language and thought of a people are components of culture in that both are used to create distinctive types of meanings and understandings that distinguish one group's sense of social reality and social practices from that of another.

On this account, then, members of a hypothetical hominid group embody three analytically separate yet interrelated mental traits or properties. These are the following: (1) a reasoning apparatus that makes conceptualization possible; (2) a language capacity that enables the representation of products of conceptual activity in linguistic form, making possible further refinements of thought and communication by means of words and sentences; (3) a culture or system of social symbols and their meanings that give specific content to experience and social life, all of which are made possible and influenced by the first two, thought and language. The knowledge and understandings shared by group members, their picture of social reality and of the behavioral environment (Hallowell 1960), and their sense of continuity or history are all part of culture, and yet culture is made possible and influenced by thought and language.

Without a capacity for conceptualization and the use of language, it is inconceivable that human beings could have achieved all that is encompassed by culture. One cannot but assume that language, a conceptual apparatus, and a system of social symbols and their meanings, are not only closely related but also intertwined and perhaps indissoluble (Ingold 1996, 147–198). For example, conceptual distinctions infants and toddlers make appear to enable them to mark verbal expressions grammatically. Many concepts are grammaticalized, in that many aspects of our conception of social reality vary in relation to grammar and the lexicon, and figurative and imaginative concepts are realized in terms of culturally meaningful categories. Moreover, even forms of nonlinguistic reasoning can be influenced by language. Thus language and symbols provide the substrata for the creation of social symbols and meanings that make up how members of a group see and imagine social reality (see Lakoff 1987). Despite these interconnections among thought, language, and culture, it is fruitful nonetheless to handle them as analytically independent domains.

Psychopathology, as we have seen, involves a range of phenomena manifested in changes, anomalies, and breakdowns of behavior. It consists of psychological changes—involving reasoning, emotion, beliefs; changes in social behavior—aberrant actions, threatening behaviors, social withdrawal and avoidance, impaired social functioning; and changes in physiological, bodily functions—pain, digestive system changes, disturbances in sexual activity, sleep, appetite, energy, and motivation. Next we need to clarify how psychopathology varied in relation to the character of a group's cognition, language, and culture during the later phases of human evolution.

Effects of Language and Culture on Psychopathology in Archaic Humans

There are four logical possibilities involving language and culture: (1) At some point in prehistory, a simplified, nascent, or incipient language developed, sometimes referred to as a protolanguage, but which need not necessarily be equated with Bickerton's definition, described earlier. We assume that something like a simplified or protolanguage existed during earlier phases of human evolution and that it may have been a feature of Archaic humans. (2) Natural language as we know it today is assumed to have existed since the flowering of the Upper Paleolithic Era and perhaps existed as far back as 100,000 years ago (see Aitchison 1996; Bickerton 1981, 1990; Davidson 1991). (3) During evolution there emerged a rudimentary culture, or protoculture (Hallowell 1960). (4) Culture as we know it in contemporary societies has been associated with modern humans since the Upper Paleolithic Era (Pfeiffer 1982). Two variables, language and culture, each of which has two values (nascent and impoverished vs. mature and elaborated) means that four conditions or types of societies are possible.

Let us examine what psychopathology might have looked like in four hypothetical groups of hominids during different phases of evolution. All known human societies since the Upper Paleolithic Era are endowed with full, mature culture and language as we know them. Psychopathology in this type of society is well covered in the literature of medical anthropology and is not dealt with here. It is theoretically unwarranted to find protolanguage in a fully elaborated culture. Here I will emphasize the two remaining conditions: (1) protolanguage in a protoculture, and (2) language as we know it associated with protoculture. For purposes of exposition, let us assume that a hypothetical group of Archaic humans, roughly equivalent with the Neanderthals, demonstrated each of these two conditions.

Social Life among Archaic Humans and Anatomically Modern Humans

The cultural capacities of Archaic Humans or Neanderthals constitute a dominant puzzle regarding the later phases of human evolution. Later, I will use them as a test case for analyzing changes in the character of psychopathology during this phase of evolution and so it is desirable to elaborate here on their characteristics. Many scholars, as mentioned earlier, are only comfortable ascribing a full capacity for language and culture to modern humans of the Upper Paleolithic (Chase and Dibble 1987, 1992; Davidson 1991; Klein 1995; Noble and Davidson 1996). According to them, a capacity for language and culture explains the archeological material found in Europe, dated to around 40,000, but especially since around 25,000, years ago, which presents unmistakable evidence of artistic production and organized arrangement of social spaces. I have taken the position that for purposes of hypothesizing about psychopathology, and especially in view of material available for primates (and hence LCAs), it is reasonable to consider earlier phases of evolution. The emergence of language and culture is partly a semantic issue. Even if a conservative position is adopted, it involves a process extending over an indeterminate amount of time, and the developments of language and culture during this transition are crucial when compared to exactly when in chronological time the transition is assumed to have taken place.

It is appropriate to use Neanderthals as test cases in this thought experiment, since the status of language and culture in this species, compared to the later anatomically modern humans, is subject to considerable controversy and speculation (Gibson and Ingold 1993; Mellars 1996a, 1996b, 1996c; Mellars and Gibson 1996; Stringer and Gamble 1993; Trinkaus and Shipman 1994). Indeed, using Neanderthals in this essentially heuristic manner is consistent with characteristics described by archeologists and paleoanthropologists about early anatomically modern humans, as well as Neanderthals proper.

Although archeologists continue to make new discoveries (Mellars 1996a, 1996b, 1996c; Stringer and Gamble 1993), Neanderthals are held to have evolved from later forms of Homo erectus and to have peopled Europe and Asia at least 200,000 to 30,000 years ago, when, it is assumed, their population was extinguished or absorbed into that of Homo sapiens sapiens. Archeological remains of the Neanderthals are generally assigned to the Middle Paleolithic Era, although Neanderthals sometimes produced Upper Paleolithic tools. Early anatomically modern humans are believed to have emerged around 150,000 years ago, and in some regions their territory overlapped with that of Neanderthals. Late anatomically modern humans are classically associated with the Upper Paleolithic Era, which began sometime around 40,000–50,000 years ago. Late anatomically modern humans are classified as Homo sapiens sapiens and are associated with the explosion of a creative, artistic, and symbolically rich culture, in high contrast to that of the Neanderthals (Pfeiffer 1982).

While Neanderthals and anatomically modern humans coexisted in some regions and their archeological remains resemble one another in quality, strong differences exist—although this is much debated. Anatomically modern humans of the Upper Paleolithic are assumed to have evolved language and culture equivalent to that found in contemporary nonindustrialized groups. There is much controversy and speculation about Neanderthal language and culture, as well as associated characteristics of brain organization and function. Some paleoanthropologists and psychologists hold that there is a qualitative difference in culture, and especially language, between Neanderthals and anatomically modern humans (Chase and Dibble 1987, 1992; Noble and Davidson 1996). Yet even these scholars admit the possibility of borderline cases and acknowledge the difficulty of determining the span of time during which a special capacity for language and culture emerged.

Other scholars see the difference between Neanderthals and anatomically modern humans as quantitative, with a continuity of function between the two groups. They recognize that late Neanderthals may have had levels of cognition, brain function, and linguistic capacities comparable to those of anatomically modern humans (Foley 1987, 1995a, 1995b; Gamble 1993). The latter school holds that it was essentially culture and social organization, more than linguistic, cognitive, and brain-centered factors, that distinguished the two groups. Such cultural considerations include how groups were physically arranged, their foraging strategies, their relations with adjacent and regional subpopulations, the means of reckoning time and what kinds of planning these made possible, the modes of social communication across space and time, as well as the complexity and diversity of tools and artifacts.

In this context, and with reference to the Neanderthals in particular, Hayden (1993) has nicely summarized information regarding their cultural capacities. Basically, he reviews and analyzes carefully and elaborately archeological data from well before the Upper Paleolithic that suggest cultural and presumably linguistic competencies. "To summarize the Neandertal technological domain, there is excellent evidence for curation, foresight, planning, skilled craftsmanship and more than adequate mental abilities for the manufacture of blades, bone tools and most tool types characteristic of the Upper Paleolithic" (Hayden 1993, 120). A sense of aesthetics, in his estimation, was present well before classic Neanderthals, inferred from the fact that even Acheulean bifaces well exceeded functional requirements. "Thus, rather than trying to explain the emergence of 'art' in the Upper Paleolithic, it may be more realistic to assume that there was a long tradition of art . . . possibly even including representational forms, extending back into the Middle or Lower Paleolithic" (125).

The inconstant, variable, and sometimes contradictory evidence presented by Neanderthal remains regarding artistic, cultural, and presumably linguistic competencies Hayden ascribes to several factors. He suggests that the cultural, linguistic puzzle regarding the evolutionary status of the Neanderthals, uncertainty about what they were like, devolve from regional differences in such things as their ecologies and economies, harshness of living conditions, access to material for artifact construction, need for the deployment of material tools, the degree of productivity and developed technology of the economy and its lower resource extraction potential, the need for greater social mobility, lower population density, and the bias of only relying on unmistakably artistic objects and presumably ritual as a basis of inferring about culture and language.

Hayden proposes that the allegedly special archeological remains associated with the Upper Paleolithic stem from social economic developments and not biological or neuropsychological ones. He suggests that during this phase (especially after 25,000 years ago) there existed a more elaborated, differentiated, and stratified set of communities richly and comfortably supplied with resources and sees the so-called artistic, creative explosion as linked to interpersonal, economic, and social organizational developments (see also Gamble 1986, 1993). The difference between Neanderthals and Europeans of the Upper Paleolithic he ascribes to the differences between structural and organizational features in general compared to complex hunter gatherers. Hayden (in press) has recently elaborated these ideas in relation to the evolution of religion.

Gibson's formulation (1996; see also Gibson and Ingold 1993) is useful in making sense of differences between Neanderthals, which Gibson also terms Archaic humans, and late anatomically modern humans. Gibson emphasizes that the two groups are not distinguished by the absence versus the presence of language, nor specific types of cognitive skills, declarative learning, involving knowledge, as against procedural learning, involving acquisition of skills and routines. Rather, they differ in their use of a more or less common biocultural brain. In such a type of brain, Gibson writes, "The generative interplay between varied learning, cognitive, and technical domains indicates that modern cognitive and neurological capacities may well have been present long before such capacities manifested themselves in the archeological record in the form of advanced artistic and technological accomplishments" (1996, 37). Gibson explains the unique features and genius of the Upper Paleolithic peoples as the result of ecological, cultural, and climatic factors that contributed to an "expanded technologically assisted symbolism" among anatomically modern humans, rather than the emergence of a strictly biological capacity involving cognition and neuropsychology. She writes, "Keys to resolving these issues may reside in new archeological evidence that Neanderthals and AMHs practiced different patterns of environmental exploitation" (1996, 41). Gibson claims, and her formulation seems generally consistent with that of Gamble (1993), "A migratory life style would have provided incentive for the invention of efficient modes of transportation, of cultural means of anticipating future events, and of methods of communicating information and maintaining social ties across the bounds of space and time" (1996, 43).

While Gibson's position seems in sharp contrast to the qualitative view described earlier, she allows that differences in the organization of the prefrontal cortex may have selected for the two styles of social life that she emphasizes. Thus, even if one observes an apparent continuity of traits (the quantitative view) between Archaics or Neanderthals as against anatomically modern humans, qualitative features involving brain organization and function may have contributed to some of the differences.

Thus, because of the very different linguistic capacities and cultural characteristics of Neanderthals and anatomically modern humans, the long period of evolutionary time involved, and their overlap in time and space, it is difficult to capture the subtle character of the social life of these two groups with generic categories like culture and language. The theoretical positions of researchers are not inconsistent with one that holds that there was a continuum with respect to characteristics of social life, specifically culture and language. Such a view supports the hypothesis that there were two groups of Archaics. Keep in mind that the creativity and artistic flowering linked to a more advanced form of language and culture implies greater behavioral plasticity and a capacity to construct programs and create stories, scenarios, and scripts played out in rituals and embodied in myths. Advanced forms of language and culture make possible more elaborate forms of social and psychological behavior, which would include episodes of psychopathology.

The Character of Mentation Associated with a Simplified Language

Unless one believes that language emerged fully formed, perhaps as a by-product of a rare macromutation or a major reorganization of the brain secondary to purely physical factors, one must assume that a capacity for language was naturally selected for and arose gradually. Undoubtedly, gesture played a significant role as precursor. At some phase of evolution, hominids, or, more likely, varieties of Homo, began to use a simplified language or protolanguage (Bickerton 1981, 1990, 1995; Donald 1991). What such a simplified language consisted of is contestable (Pinker 1994; Pinker and Bloom 1990), especially among earlier varieties of Homo. The following general description of a simplified language might be applied to groups of Archaic humans or very early Neanderthals.

A simplified or protolanguage would involve at least two elements. First, a small inventory of lexical items and, second, a limited grammatical system consisting of simple rules of syntax. Obviously, a simplified language bears a relationship to the speaker's culture. It may be assumed to constrain the pool of information that a group shares. Lexicalization provides labels for things, categories, and concepts that aid in coding, remembering, retrieving, and using information (cognitive processing). Only a small number of objects, situations, events, and categories of all of these could be referred to, intellectually processed, and communicated in speech. The second feature, a simple system of grammatical rules, restricts the way lexical items are related to one another. This is ordinarily accomplished by modifiers and connectors that offer interpretations or propositions about actual and potential states of affairs. For example, certain terms would be used to establish time relations and duration of time as these bear on events, motives, causality, and the sequencing of actions as part of implementing a plan. A simple grammar would determine the conceptual depth, intellectual density, and detail of the assertions, interpretations, and narratives evolved to rationalize and symbolically contextualize a shared world. Given the limitations of a simplified language, the symbols and meanings that could be presented as propositions about social life and the world would have to be relatively uncomplicated. Gesture would have prominently complemented language.

This brings up the highly controversial question of the relationship between language and conceptual capacity (see Cromer 1991). As we have seen, there is general consensus that thought is independent of language, that thought is based on basic, universal experiences that depend on such elements as body structure, spatial orientation, and perceptual capacities, and that habitual ways of thought based on these experiences, prelinguistic mind-sets, played critical roles in influencing the evolution of language and the particular character language assumed in various groups (see Aitchison 1996; Johnson 1987; Pinker 1994). Here, one may suppose that conceptual capacities realized as grammaticalized words mirrored thought and enabled communication of specific types of propositions. There is also consensus that once a language capacity emerged, it facilitated and enhanced thought.

Given that language and thought are mutually influential, one must assume that members of a group with a simplified, nascent, impoverished language would also be relatively disadvantaged intellectually, compared to those equipped with mature language. In other words, one expects a reciprocal influence between language and cognition, some complementarity in categorizing, reasoning, and problem solving (McNeill 1987). Both linguistic expression and cognition were no doubt limited during the phases of human evolution marked by a simplified language or a protolanguage.

As indicated earlier, people with a simplified language would have only a limited number of words for objects and concepts with which to reason and communicate about their world (nouns). Besides words for things and concepts, many lexical items would refer to actions (verbs) and would be used to address concrete situations involving protection from predators, acquiring food, and other activities necessary for survival, as construed by the prototype theory of categorization and cognition (see Lakoff 1987). Furthermore, there would be a tendency to share and communicate phenomena relevant to the group as a whole. The emphasis would be on the representations of ongoing activities, which depend on an ability to identify, refer to, and communicate about shared, hence verifiable, experience.

Speech would also focus on natural history features of the environment, since knowledge about the habitat is essential to survival. In addition, some lexical items would refer to persons, motives, intentions, feelings, and reasons for action. Hence nouns and verbs would be crucial in facilitating social exchanges, remembering past experience, and ironing out relationships with kin and group mates. However, abstract concepts, whether subjective feelings or interpretations of social life and the social universe, such as morality, justice, cosmology, or religion, would tend to be few. Even assuming some capacity to conceive of such abstractions, representing them in linguistic form would obviously be rudimentary. In other words, an Archaic human's ability to code, remember, retrieve, and cognitively use abstract concepts about social reality would be severely restricted. To use abstract concepts, speakers should be able to store and retrieve a large number of lexical items and definitions and build upon them.

Along with a restricted number of lexical items, a simplified or protolanguage means a limited grammar with comparatively few syntactic rules for modifying and arranging words into phrases and sentences so as to capture the myriad relations among the speaker's perceptions, meanings, intentions, and communications. Aitchison (1996) provides examples of such rules and their possible evolution, based on a study of pidgin and Creole languages (see also Bickerton 1982). Most speakers of these types of languages are already proficient in a native language, and all of them, of course, have the brains of Homo sapiens sapiens, which limits the applicability of the discussion. Putting aside this caveat, in general, a nascent or simplified grammar restricts the speaker's ability to form propositions about the world, real or imagined, limiting how actions, states, and events can be represented linguistically, communicated and, arguably, thought about. As an example, a capacity for offline thinking—thinking about the past, anticipated events, future possibilities, imaginative thinking by means of metaphor and metonymy, the kind of reasoning made possible by counterfactual claims, or simply abstract conjectures about the world—would be extremely limited and constrained or simply nonexistent if grammar and syntax were restricted. Speakers would simply be unable to represent in an efficient, economic, and structured form situations and conditions requiring imaginative uses of language.

Also limited would be the basis on which a group could share experience and develop a shared identity based on past and future events.

As discussed earlier, propositions involve verbs and arguments. By *argument*, linguists mean "any expression that refers to any participant in an action, state, or event, whether that participant performs an action, undergoes it, is its goal or beneficiary, is the instrument with which it is performed, or plays any other role in such an action, state, or event" (Bickerton 1995, 30). A comparatively small number of rules for arranging concepts into propositions by way of verbs and their arguments obviously forces speakers of a protolanguage toward online thinking and closes off possibilities for offline thinking. In other words, it limits the extent to which individuals can speak, imagine, and communicate about phenomena not in front of them.

Psychopathology as a Function of Thought as Compared to Language

One could argue that the critical element of psychopathology, that which determines its social and psychological characteristics, depends upon an individual's capacity for thought, especially aberrations of thought, and that these capacities and aberrations should be handled differently from those pertaining to language per se. Giving primary emphasis to thought and cognition, and not language, in describing the character of psychopathology seems to assert that thought and language are separate capacities. On logical grounds such a distinction is defensible; it appears to explain how children acquire communicative skills, it seems to make sense of the evolutionary progression, and many cognitive scientists and developmental psychologists make such a distinction (Cromer 1991; Fodor 1983; Lakoff 1987; Pinker 1994). Yet distinguishing between thought and language is difficult. This problem helps clarify the issue of the origins of psychopathology.

Language is an observable trait of an individual or group. Thought is not. Hence, equating psychopathology with thought and claiming that language and thought are distinctly different forms of mental activity runs the risk of arguing that psychopathology really involves happenings taking place in a black box virtually closed to inquiry. In other words, it implies that only aspects of pure cognition deserve attention and that aspects of language should be handled secondarily. Yet we intuitively equate thought with language. We assume, in other words, that individuals think through language, by means of language. But linguists and cognitive scientists offer a number of reasons why this is not the case (Aitchison 1996; Lakoff 1987; Pinker 1994). They argue for the existence of a capacity to engage in thought, or what one may regard as mentalese, the language of thought, which is independent of language if language is a special form of representation and communication with special properties, rules, and functions. Some assume that language serves as an instinct or organ that enables individuals to communicate about and put into verbal form that which is independently thought about (through mentalese) (Pinker 1994). Others, like Lakoff, view language as "based on cognition. The structure of language uses the same devices used to structure cognitive models—images, schemes, which are understood in terms of bodily functioning" (1987, 291). Finally, as indicated earlier, there is the view that "human thoughts . . . run along pre-ordained grooves, which are likely to affect the order of participants in a grammar" (Aitchison 1996, 119). In other words, the original preferences that conditioned how language evolved were probably based on prelinguistic mind-sets.

To make a logical separation between language and thought means that it is theoretically possible that a group of individuals, or hominids, might have been capable of fairly complex reasoning but lacked a correspondingly complex capacity to represent and communicate it by means of language. Arguments about the social intelligence of nonhuman primates, about their cunning and capacity for Machiavellian thought, when examined in light of their relatively impoverished capacity for language (but see Deacon 1997; Savage-Rumbaugh and Lewin 1994), would seem to support the idea of an evolutionary discontinuity between sheer conceptual development and language competence or sophistication—more specifically, about the possibility of a species with reasonably advanced cognition but simple, shallow language (Byrne 1995; Byrne and Whiten 1988; Whiten and Byrne 1997). In other words, primates, especially chimpanzees, who certainly lack a capacity for language in the traditional sense, especially under feral conditions as compared to experiences with human trainers, appear capable of subtle forms of deception and subterfuge in social relations. These observations support the idea of a relatively advanced social intelligence or cognitive capacity independent from language (see Heyes 1998).

It is certainly possible, then, that during some early phase of human evolution, prior to the evolution of a full capacity for language, hominids were capable of an elaborate form of cognition or mentalese but were impoverished in their powers of linguistic representation and communication. However, the lack of variety and creativity reflected in archeological remains does not support this conjecture, although such remains cannot tell us much about what sort of products of symbolization, in the absence of language, might have been degraded and lost to analysis.

If this reasoning is defensible—that early hominids might have been capable of relatively elaborate cognition yet were impoverished in language—we must conclude that much of the mental, emotional, and social behavioral informational content of psychopathology during evolution would have been reflected in social and psychological behavior, but nonlinguistically. Moreover, it could not have been communicated verbally, and one can assume that it might even have exceeded their capacity for nonverbal communication by means of gesture and mimesis (see Donald 1991). Psychopathology, in other words, would have been characterized by disturbances of cognition, emotion, and psychophysiology, but would not have been modulated, or rendered socially visible and negotiable, by means of language as we know it. This scenario stipulates that much of the conceptual basis and perhaps the informational content of psychopathology would have existed in covert form—conceptually influential, responded to, expressed behaviorally, and dealt with largely by means of mentalese or some other protolinguistic conceptual resources, but not concretely manifested in (utterances of) language as we know it. In short, the scenario implies the presence of many conceptual distinctions and elaborations that could influence psychological experience and behavior, and especially the conflicts and turmoils associated with incidents of psychopathology, despite the absence of a language for representing all of this. However, as discussed earlier, mentation in the absence of a full capacity for language would still have been comparatively shallow, limiting the range of distinguishable social influences that could be marked and come to affect individual relationships and hence psychopathology. Moreover, the question naturally arises: How could hominids have communicated all of this to each other? Obviously, what group mates could know and do about an individual's plight during incidents would be restricted. Consequently, for reasons already given, this scenario is unlikely.

How Might the Evolution of Language Have Modified the Character of Psychopathology?

If hominids passed through a phase of evolution marked by reasonably advanced conceptual development but an impoverished linguistic capacity, how exactly did the emergence of language as we know it affect the character of psychopathology? Here it seems appropriate to examine the effects of language on behavior in children. We can borrow the notion of behavioral regulation as used in developmental psychology to visualize the potential role of the emergence of language and associated changes in the evolution of the brain in shaping the character of a group's psychopathology during human evolution.

Behavioral regulation comprises subsidiary spheres of functioning during ontogeny, namely, cognition—here thought of as attention, concentration, memory, executive functions—motivation, and affect. The concepts of temperament and especially difficult temperament (Tarter 1990; Tarter et al., 1990; Windle 1991; Windle and Lerner 1986) refer to additional aspects of disordered behavior observed during maturation and development, when language and related neuropsychological functions are affected by a variety of social environmental and physical stressors. Temperament is crucial to discussions of the origins of adolescent and adult psychopathologies, especially antisocial behavior (Clarke and Boinski 1995; Rutter and Rutter 1993).

Psychopathology involving all of the areas of function subsumed under behavioral regulation and temperament may stem from various causes. Interferences in the realization of language and associated aspects of cognition, viewed broadly as important and controlling influences in the modulation and expression of behavior during ontogeny and development, play a determining role in the character of human psychopathology during both childhood and maturity. Given what we know about this problem area in modern populations, we can assume that the emergence of language probably affected the character of psychopathology during earlier phases of human evolution.

The development of language and associated forms of cognition and emotion during ontogeny is closely related to the organization, regulation, and especially the control of behavior. Initially in the form of spoken commentaries during action, verbalized thought, language experience is used as a means for modulating, planning, and controlling behavior (Vigotsky 1962). Later during development, as such audible guides are internalized and become internalized speech or verbal thought, language becomes an important mediator in the regulation of behavior. These linguistic skills, of course, are not acquired in isolation. They are integral to a range of other influences over behavior and emotional experience and expression that come with maturation. Temperament and executive cognitive functions cover a range of abilities, involving attention, working memory, set-shifting, planning, and the regulation of goal-directed behavior, all of which are associated with the establishment and functioning of prefrontal-subcortical circuits of the brain (Cummings 1993; Luriia 1980; Giancola, Mezzich, and Tarter 1998a, 1998b). These are the neural centers that have been singled out as important in making sense of the behavior of Archaic humans as compared to anatomically modern humans (Deacon 1997; Gibson 1996; Roberts, Collins and Robbins 1996; Whitcombe 1996). Changes in how language capacities and skills are deployed in behavior during development—in other words, their controlling and smoothing functions—take place in conjunction with brain-mediated influences on psychological, social, and emotional behavior.

In children, deficiencies in all of these regulatory functions, whether due to problems in mother-infant bonding and handling, stress-inducing environmental influences, developmental abnormalities caused by the ingestion of substances toxic to the brain, or lesions to prefrontal areas or subcortical nuclei of the brain, contribute to the expression of psychopathology, especially that expressed in aggressive, antisocial, and hyperactive behavior (Barkley 1997; Giancola, Mezzich, and Tarter 1998a, 1998b). Lack of inhibition, hyperactivity, violence, affective lability, irritability, and a tendency to display strong emotions in social conflicts are manifestations of psychopathology in individuals showing disturbances in the regulatory functions associated with maturation of the brain, development of language skills, and the unfolding of temperamental proclivities.

Language and associated forms of cognition, then, have a controlling influence over the character of psychopathology in modern humans. During phases of human evolution when psychological and social behavior was not yet controlled, modulated, and smoothed by language and associated cognition and execution, psychopathology may have been manifested in direct, disinhibited, aggressive, emotionally labile, and socially disruptive behavior. With respect to earlier phases of evolution, this means that psychopathology would have been characterized by behavioral breakdowns in social relations, emotional displays, disregulation and affective dysphoria, hostility and aggression, visceral-somatic dysfunction, and failure to perform essential tasks smoothly. While such psychopathologies may have devolved from social conflicts and/or distortions in the testing and evaluation of social reality, however represented, this behavior could not have been rationalized through a system of beliefs and narratives, since by definition these would have been beyond the Archaic humans' linguistic capacities. Hence, they could scarcely understand or assimilate manifestations of psychopathology by means of inherited cultural rationales that would have been made possible by language, nor could they have called on these rationales to contain such behavior or analyze and possibly resolve such problems.

Two other influences on the character of psychopathology need brief emphasis here. Language made possible a more elaborate way of deception and cheating by making it possible to tell lies. It has been claimed that even among living ancestors of hominids, deceptive strategies are important manifestations of social intelligence (de Waal 1989, 1996; Whiten and Byrne 1997). However, whether deception is intentional is a contested issue (Heyes 1998; Cheney and Seyfarth 1996). Hence, it cannot easily be claimed that language made a qualitative difference with respect to deception and cheating. It simply allowed speakers to more elaborately weave stories and provide (mis)information that could benefit the deceiver. In short, telling lies by means of language enabled an individual to cheat in the competition for valued resources and to undermine the social prestige of competitors through innuendo and gossip. However, the potential cost of such behavior was the loss of credibility if one was discovered, unreciprocated future exchanges, and perhaps shame and guilt, which could undermine a sense of well-being. Clearly, lying, cheating, and an ability to detect these tendencies in human interactions were both under competitive selective influences and in many respects, at least to the extent that genetic determination operates, could have driven the other to greater and greater levels of sophistication (Humphrey 1976).

The evolution of language made possible the construction of elaborate stories, myths, and rituals whereby individuals and groups explained their origins, coped with fears arising from attempts to survive and reproduce, and also mitigated crises among group members. However, this same feature of language could be used by vulnerable individuals to rationalize and justify taking unfair advantage of others by exaggerating and feigning physical sickness and psychopathology. In other words, the resources made possible by language served to create cover stories for vulnerable, stricken individuals, thereby in some ways socially authenticating episodes of sickness and psychopathology. It would also have facilitated the elaboration of states of dissociation.

A related aspect of language involves the creation of ideologies that serve as devices or inventions for concealing inconsistencies in the social fabric (see Dunbar, Knight, and Power 1999). Of course, any attempt to establish a clearcut difference between a cultural pattern, a symbolic theme, a myth, or an ideology could be contested. Despite this caveat, one could define an ideology as a social lie promoted by an individual or a group to promote a particular viewpoint that favors some individuals at the expense of others. An ideology needs to be shared, promoted, and continually updated or reinforced to appear logically consistent and convincing. However, despite individual and group efforts, an ideology can fail to explain hardships and misfortune. When this happens, it loses its power to bind individuals when its promises fail to materialize. In this instance, an ideology can be viewed as a handicap of a culture that can cause dissatisfaction, disappointment, and loss of morale, all of which can be potential causes of psychopathology in vulnerable individuals.

Comment

An earlier section brought up the question of the character of psychopathology in a hypothetical group characterized by relatively advanced conceptual development but simplified, nascent language. This allowed us to sharpen the question of how the evolution of language as a distinct faculty could have operated as a mechanism affecting the expression of psychopathology. When theorizing about the character of psychopathology during earlier phases of human evolution, it is prudent to regard the two realms of cognitive activity considered here, namely a language of thought (or mentalese) and language as we know it, as mutually reinforcing if not indissoluble. Both aspects of mental functioning are bound together in manifestations of psychopathology. Furthermore, even if the scenario about a runaway mentalese, as opposed to language as we know it, were valid, it has little theoretical and practical value for understanding the nature and origins of psychopathology. This is so because understanding the character of psychopathology in a linguistically impoverished but highly intelligent hominid group would not be ascertainable by ordinary approaches to the study of behavior and psychopathology.

One can fictively entertain the possibility of conducting PET studies, MRIs, EEGs, skull X-rays, laboratory examinations of blood and body fluids, or relying on tightly operationalized descriptive protocols pertaining to abnormal behavior that would enable one in theory to measure, define, and classify varieties of psychopathology among prehistoric species—that is, if a truly objective and operational neuropsychiatry were possible. However, to make sense of a body of information about diagnosis, one must place the results of such an evaluation in a social psychological behavioral frame of reference that could be meaningfully related to the realms of reasons, motives, intentions, and social values of the group in question. Without a suitable language with which to formulate the inner meanings of psychopathology, aspects of psychological behavior that are socially problematic, and to relate them to the social world of the group, the information produced by neuropsychiatric diagnostic procedures would be of little value. In short, language not only endows psychopathology with its social and psychological meaning but also is the medium for understanding its dynamics in relation to group happenings. It is thus best to leave aside the matter of the character of psychopathology in an evolutionary scenario of runaway thought, that is, mentalese, with only a nascent or lagging language as both theoretically and empirically problematic. By contrast, it is appropriate to pursue the topic of early hominid forms of psychopathology while handling language and thought as logically independent but empirically closely connected, each influencing the other.

Psychopathology in Archaic Humans with Simplified Language and Protoculture

In this hypothetical view of Archaics, one posits a small inventory of verbal, that is, lexical, units. This would apply to phenomena of all domains, but we focus on the social realm, since it is especially implicated in making sense of psychopathology. Archaics would have available a relatively small number of lexical items for separately identified concepts, including, for example, aspects of social identity, mental attribution, and social behavior, whether of motives, intentions, causal agency, factors linked to behavior and feeling, or states of being and feeling.

Shared cultural knowledge would also be restricted. For example, an inventory of knowledge, the shared pool of information about emotions, behavior, social conflicts, and anticipated worries, as well as pure natural history, would be comparatively small. Cultural impoverishment means a restricted ability to formulate happenings in the world. Restricted language and culture, then, imply limited resources with which to regulate, modulate, and socially rationalize behavior in the event of social crises and conflicts. They imply restrictions on how the stream of behavior and continua about it could be broken apart and thought about; what disturbances in behavior could be identified, labeled, and communicated about; what causes could be attributed to changes in behavior; the number and kind of relations that could be made between behavior and events in the environment, physical or social. The kinds of connections that could be made about all of these phenomena and the motives of and rationales for individual behavior would be limited.

These restrictions on language and culture can be expected to have shaped psychopathology and would account for the presumably restricted, shallow, concrete, and direct way in which psychopathology was expressed in social behavior, how it was conceptualized, and how it affected the social fabric. Since the manifestations of psychopathology—altered consciousness, loss of emotional regulation, disordered reasoning—would be represented in a simplified language, little of what precipitated it could be communicated and shared linguistically. For example, the effects of stress on vulnerable individuals would be directly expressed in perturbations of cognition, sensory motor behavior, and visceral physiology. Consequently, individuals showing psychopathology would manifest disturbances of brain and bodily functioning, but their meanings would have to be inferred by group mates, since language could provide only a limited way of expressing, formulating, and communicating them. However, an observer could easily interpret power conflicts, behavioral crises, and overt behavioral breakdowns and dilapidations tied to intercurrent social traumas or environmental disasters. Such types of disturbances are seen in chimpanzees (de Waal 1982, 1989, 1996).

One can assume that altered states of consciousness or states of dissociation would be an important component of psychopathology during earlier phases of human evolution. They could be caused by vulnerabilities produced by frank psychological traumas during development, or by intolerable levels of stress encountered as a result of social ecological hardships. However, individuals with mental capacity and social intelligence that enabled them to conceive of insoluble crises of a social psychological character, would have been shifted psychologically onto altered states of consciousness or dissociation that would have been played out in terms of the limited resources provided by their simplified language and culture. A linguistically and culturally impoverished group would have only a small capacity for imaginative and abstract understanding and, more specifically, for beliefs, myths, and rituals that could provide a narrative meaning for the experiences that led up to states of dissociation. Just as ordinary manifestations of psychopathology would be relatively concrete, individualized, bounded, and interpretively shallow and transparent, these same factors would also constrain the character and content of states of dissociation. In a context of stress, vulnerable individuals could have entered altered states of consciousness but would have played them out in simple routines that conflicted in an obvious way with ordinary assumptions about social behavior. Psychopathology played out and manifested in states of dissociation would likely be stereotyped behavioral sequences that signified more or less directly the individual's situation or an interpretation of a social crisis.

Although those equipped only with a protolanguage would suffer from certain limitations on cognition, culture, and behavior, these do not indicate ineptitude in other areas of social life, such as activities involving technical skill, communicative functions of the body and gesture, as in dance, and perhaps in aesthetic expression and imaginative and figurative modes of understanding, considered as aspects of cognition separate from language. As indicated, many paleoanthropologists hold that these powers are closely tied to a capacity for language and that speakers limited to protolanguage would lack them. The poverty of evidence suggesting decoration, ornamentation, spatially diversified living arrangements, and cave painting among these groups support this position. On the other hand, early hominids may have had expressive capacities not limited to language as we know it during earlier phases of evolution, but the physical remains indicating such capacities have simply not survived the destructive effects of geologic time. Arguments summarized in Donald (1991) suggest that Archaics were capable of expression by means of music, dance, gesture, and large-scale neuromuscular coordination and spatiotemporal arrangements. This probably included decoration and even forms of art, all possible given a modicum of pure mentalese and capacity for symbolization (Deacon 1997; Watts 1999). The expressive and interpretive depth of these forms of cognition and culture would have been comparatively shallow, but they might have been manifested in, and might have helped make sense of incidents of psychopathology, including states of dissociation.

Psychopathology in Hominids with Fully Developed Language but Impoverished Culture

The second hypothetical scenario entailed Archaics who were equipped with full language but only nascent culture. It is outlined and used here for heuristic purposes. In this scenario, one assumes a full capacity for linguistic representation and communication but something less than a fully realized culture, as it is assumed to have existed among anatomically modern humans of the Upper Paleolithic. In theory, a group meeting these specifications has a more expanded inventory of lexical items by means of which it can label, refer to, classify, describe, and interpret objects and events in the environment. They also could link these lexicalized representations and their concepts syntactically and grammatically. They could also, presumably, communicate all of this in utterances that formed meaningful propositions about the world, including changes in the feelings and actions of group mates. This would be accomplished by using notions of time—present, past, and future—and aspects of the objective world shared by the group and about any other world that can be imagined or conjectured about.

Because of the close interaction between language and thought, linguistic competence implies a capacity for enhanced, more structured, and richer forms of cognition. Conceptual capacities for figurative and imaginative understanding that involve image schemata and metaphorical reasoning, for example, would be potentially affected by language. The proposition that a group that shares a language as we know it has not evolved an elaborate culture implies that language and thought do not have a one-to-one correspondence with culture—that culture encompasses more than what language, and forms of cognition tied to it, ordinarily make possible. This something else, one can hypothesize, involves a pool of information: a system of meaning, an interpretive matrix of rules, standards, values, assumptions, and conventions that construct distinct pictures of social reality, including social relationships, cosmology, and a shared sense of history. This scenario about a hypothetical group of Archaics assumes that language, and a potential for metaphor and imaginative thinking, may have evolved in advance of much of what we mean by culture, and that culture, while it may depend on language for its realization, may require a more developed social life for its full elaboration.

Limits on culture, for example, mean that individuals would have a limited and impoverished picture of the behavioral environment (Hallowell 1960). Such a protoculture would be built on a relatively small inventory of themes and stories underlying beliefs, myths, and rituals, and on a few elementary concepts, categories, and cognitive models for constructing a picture of social reality. Such a people would not yet have been able to fully exploit the potential powers of language and thought for enriching experience and bringing understanding. These limits would also curtail expressions of psychopathology and how it could be interpreted and dealt with by the group. In positing a disjunction between culture and complex language and thought, one is forced to clarify how social life contributes to culture apart from language and thought, and how all of this might possibly influence different facets of psychopathology.

Let us assume that a group had linguistic competence but a social life that was culturally impoverished. What would such a group look like? We can equate cultural complexity with (1) the number and variety of concepts about the behavioral environment, including the physical habitat, the local group, the regional world, and especially the abstract world of cosmology; (2) the number, variety, and subtlety of distinctions in conventions of social behavior, such as those regulating mating, kinship, social obligations and entitlements, and adjudication of conflicts; (3) the shared history of the group; and (4) the richness of the beliefs and narratives that integrate all these ideas. A complex culture, in this scenario, then, refers to how elaborately the behavioral environment is constructed in the minds of group members to explain and rationalize events in the world and how they affect the stability and persistence of the group.

Let us draw on the components of language considered earlier as a means of visualizing a scenario of full language but impoverished culture. Even with a normal capacity for lexical inventiveness, members of the group would have only a few basic concepts about what could be represented, including both concrete and abstract phenomena. The cultural bases for acquisition of a large pool of information would have been lacking. Similarly, while their language might be fully equipped with a grammatical structure, they would be able to use only a few of these structures to form propositions, share knowledge and beliefs, and construct social narratives because of a narrowed experiential compass and a comparatively restricted information pool. In brief, imaginative, abstractly formulated propositions about the physical world, the social group and its history, interpretations of social life, and the composition and operation of the cosmological universe could theoretically be formulated linguistically and conceptually, but in a narrowed experiential compass and a comparatively unadorned, rudimentary manner. The group could draw on only relatively concrete, simple stories to explain their existence.

This scenario might be the result of three possible states of affairs. First, thought resources, that is, conceptual capacity or mentalese, that drive, energize, or feed data into the linguistic operating system could be limited. In other words, the user simply could not intellectually exploit the full inventiveness of the language system. Since it was argued earlier that the language of thought could exist independently of a language capacity per se, a people, while linguistically competent, could conceivably lack the inventiveness and intellectual resources for elaborating a rich lexicon for interpreting the physical world, the social world, and the group's history and universe. Stated baldly, if a people's means of reasoning about the world is limited in some way, such as because of biological factors—as a consequence of developmental constraints or limited learning opportunities—or because of nutritional deficiencies, then important markers of cultural complexity considered earlier would not be fully realized and hence would seem restricted and lacking in intellectual depth.

A second reason for the cultural impoverishment in this hypothetical scenario could be external factors. A harsh physical environment and difficult life conditions could account for the relative poverty of a group's culture. In other words, a group severely stressed by a harsh social and behavioral ecology causing sustained environmental challenges and barely able to subsist could not make full use of a capacity to name, label, create complex mental spaces, form propositions, and think and communicate about characteristics of their world. This would be tangibly manifested in a restricted number of concepts and linguistic terms for handling the world and only a few narratives for interpreting it. Hence, a group living in a physically harsh, undifferentiated world—for example, in a desert or in the Arctic—preoccupied by sheer survival and the threat of disease, predators, or competitors, could be culturally impoverished.

Of course, this is not always true. The cultures of Australian aborigines, of peoples living in the Kalahari Desert of Africa and the Arctic, which are described as relatively simple societies of hunters and gatherers, are proof to the contrary. Despite harsh or barren environments, they all show a rich inventory of beliefs, rituals, social institutions and practices, and advanced systems of concepts and knowledge—that is, an elaborated culture. There is reason to assume that ancient foragers of the Upper Paleolithic Era were no different. But despite these counterexamples, I am considering here the character of social life that in principle could have been manifested among populations of severely environmentally stressed Archaics during difficult phases of human evolution.

A third condition that could account for the scenario in question (full language, impoverished culture) would involve both internal and external factors. Limited capacity for thought, perhaps not biological or genetic in origin, could be the result of dietary insufficiencies or harsh living conditions that were traumatic during development, coupled with restricted opportunities for cultural inventiveness. Alternatively, a group whose environment demanded specialized techniques for survival, or one whose hardships failed to tax and challenge its resources for conceptual richness, or one that existed in a state of relative isolation and in a minimalist social environment might theoretically be constrained in its capacity for cultural elaboration. These are all possibilities for the interaction of language and culture. While the scenario I have introduced is possible in a contingent sense, given special circumstances of environmental change, migration, or period of strong selection, as in an evolutionary bottleneck, it is less likely that it ever characterized a species or subspecies in a sustained way, and it is doubtful it constituted a permanent condition. Nevertheless, it is hypothetically possible that a group of Archaics could have had full language but an impoverished culture.

While I have emphasized aspects of culture linked to categories for conceptual reference, or lexical inventiveness, the same arguments apply to the realization of the grammatical potential or language. In other words, a deficiency in the language of thought due to internal or external factors could also hold back how a group could use grammar and syntax to formulate meanings about the world and develop a full potential for culture.

What might psychopathology consist of in a group of Archaics whose members had language as we know it but an impoverished culture? Let us assume that a group of native speakers were fully articulate in their ability to formulate and express what happened in the world and in the group, using verbal expressions that represented whatever understanding was possible given their impoverished culture and sense of history. In short, while linguistically skilled, they would have only a few story lines for constructing narratives of practical or existential import. These limitations would imply that behavioral breakdowns could be communicated about by the afflicted individual or group mates, but only through a small inventory of concepts and categories. Given the relative restrictiveness and barrenness of social and cultural life, as pictured in this scenario, there would be a low supply of abstract concepts, since they could not draw on the full imaginative potential of language. Theories of causation or the concept of a cosmological world, which could be implicated in the onset or interpretation of an instance of psychopathology, would presumably be lacking.

The role of the emotions in psychopathology would be affected. While emotions are integral to manifestations of psychopathology, intruding into and disrupting normal behavior, a culturally impoverished group would have difficulty in representing them conceptually and translating them into propositions, descriptions, and interpretations of aberrant behavior. For example, they would not have an elaborate vocabulary of emotions, nor would these emotions be formulated in a rich psychological language that contained extended networks of concepts about personality, motives, and states of being. Formulations that referred to otherworldly scenarios and that integrated supernatural agencies in complex behavioral scenarios of blame would be unlikely, as would explanations that implicated group mates as sources of emotional and behavioral disruptions. There would be a relative absence of dense, subtle, and complexly structured idioms pertaining to social life and its conflicts. Similarly, somatic manifestations of psychopathology, undoubtedly correlated with emotional conflicts, would be represented in relatively simple body language of form and function and in an easily interpreted idiom, given the closeness of living conditions. Thus, one would not find complex ethnopsychological, ethnophysiological, and ethnoanatomical notions wrought into the explanations for bodily symptoms of psychopathology.

Ways of handling the chaos and crises occasioned by behavioral breakdowns would be direct, forceful, and practical. They would entail physical restraints, emotional redirection, social separation, emotionally visible and directly expressive forms of persuasion, and efforts to mediate and adjudicate among contending claims contributing to manifestations of psychopathology. A group with a more elaborated culture might appeal to elaborated and shared knowledge structures and social institutions that made use of elaborate categories and principles of obligation, entitlement, responsibility, blame, morality, suffering, and retribution.

Altered states of consciousness, that is, states of dissociation, would be played out in more elaborate scenarios of meaning than would be possible for Archaics with only protolanguage and protoculture. However, in a group with full language but limited culture, these states could not convey rich narratives or elaborate myths of renewal and retribution, since, by definition, the resources for such forms of symbolic construction and scenario building would not have been realized in a group that possessed only a limited range of concepts, categories, cognitive models, and systems of knowledge. Instead, because the semantic powers of language and cultural narratives would be restricted, and because social strains and stressors would be augmented, those who manifested psychopathology might be likely to enter a dissociated state easily, but it would be unelaborated.

Finally, given the difficult social and behavioral environment of a group whose members are linguistically competent but culturally impoverished, group members would have little time and energy to deal with incidents of psychopathology. In short, cultural conditions would restrict the symbolic texture of manifestations of psychopathology and how they were processed as social events, and restrictions on social life would limit the impact and interpretations that members of the group could draw from such episodes.

Comparing the Effects of Protolanguage, Protoculture, and Language on Psychopathology

Let us now compare how psychopathology might have appeared in the two hypothetical groups and scenarios about Archaics. In many ways, psychopathology would be similar in the two scenarios. Compared to societies characterized by language as we know it and ordinary cultural resources, as presumably was found among anatomically modern humans of the Upper Paleolithic Era, both kinds of psychopathologies of our hypothetical Archaics would lack symbolic depth. They would not be given explanations that delved deeply into social complexities involving the group's current status, nor would either group have a rich vocabulary for psychopathological symptoms—for example, motivations underlying behavior and interpretations of bodily disturbances. Both groups would have a relative paucity of abstract systems of categories and symbols. As a result, social interpretations of incidents of psychopathology, involving causes, meanings, and potential consequences of manifestations with respect to group processes, would lack depth. For example, emotions would not be understood in a variegated way, and actions would be interpreted in terms of phenomena closely tied to immediate events.

Because language, unlike protolanguage, facilitates abstraction and enriches the imaginative component of thought, through metaphor and metonymy, for instance, the two groups of Archaics would differ in their ability to articulate and communicate about details of the manifestations of psychopathology, its causes and consequences. Archaics of the second scenario would be far more able to elaborate upon and make sense of manifestations in relation to current events as they affected individuals, emotionally and physically. Because the full resources of language could be deployed as psychopathologies unfolded, it is likely that reactions to and impulsive resolutions of the social conflicts intertwined in incidents of psychopathology would be less acute, abrupt, and intense. Language would provide a richer opportunity for resolving the social and emotional conflicts caused by and embodied in incidents of psychopathology. This would mean, for example, that abstract and imaginative explanations of anxieties, worries, suspicions, rivalries, antagonisms, jealousies, sexual rivalry, despair, and seemingly senseless behavior, all of which could constitute the material content of psychopathology, would be more elaborated in the scenario of full language with impoverished culture, compared to protolanguage in a protoculture. States of dissociation would communicate a more understandable story, if still a shallow one, compared to groups featured by language and rich culture, and it would be easier for group mates to translate a plight communicated in dissociated behavior into the fabric of events and deal with their consequences for the afflicted and members of the group.

In general, individuals endowed with full language would be more able to interpret stress and to use this understanding to resolve crises. Full language makes it easier to express the distorted thinking associated with psychotic states and to make it public, explicit, and subject to rational scrutiny. Indeed, the sheer content of psychotic beliefs would be open to interpretation and deliberation, as would ideas, beliefs, rules, standards, conceptions, and misconceptions about social events, and the character of deviations that were incorporated in manifestations or contributed to the causes of syndromes of psychopathology. All of these indicate that individuals would demonstrate a greater degree of emotional regulation and that group mates would have the ability to redirect, channel, and bind emotions socially. This would tend to promote a more deliberative resolution of crises that would otherwise appear nonsensical and potentially threatening. Finally, because of the availability of a fully developed, shared idiom of communication and experience pertaining to behavior, Archaics of the second scenario would be better able to draw on past experiences of similar incidents, memories of disturbances, or social conflicts through which to make sense of a breakdown. In other words, a richer inventory of knowledge and information could be brought to bear on incidents of psychopathology. This would mean that the configuration and enactment of psychopathology could be more standardized and conventionalized.

- Language involves a special representational and communicational competence based on arbitrary symbols and rules about their combination that enable sharing experiences about the world, but both its exact properties and evolution are complex and contested.
- Cognition refers not only to logic, thinking, and problem solving but also to imaginative ways of modeling and of using figurative devices that assign meaning, represent happenings, and construct scenarios about the natural and supernatural world.
- Culture can be equated with behavioral traditions and their psychological bases. It constitutes a pool of information consisting of distinctive content and meaning on the basis of which a group establishes and shares a symbolic perspective about the world.
- The concepts of language, culture, and cognition are mutually interrelated. They help explain the behavior that emerged during the later phases of human evolution and also to formulate scenarios about psychopathology among Archaic humans.
- The archeological remains of Archaic humans, the Neanderthals, are variable and complex. Their linguistic, cognitive, and cultural competencies are contestable. This allows room for hypotheses about their behavior and psychopathology.
- Neanderthals provide a test case for hypotheses about psychopathology as affected by different degrees of evolution of language and culture, termed simplified (proto) language and simplified (proto) culture.
- By examining Neanderthal archeology, human brain function, the neurobiology of mental illness, and the interplay between language, cognition, and emotion, we can propose hypotheses about the kinds of psychopathology they and Archaic humans manifested.

Part III Recapitulation and Synthesis

Chapter 14

Phases of the Biological Evolution of Psychopathology

VIEWED GENERALLY, psychopathology is a composite of heterogeneous changes in individual social, psychological, and physiological behavior—for example, social relations, cognition, body dysfunction. From an evolutionary standpoint, it is special because, besides causing personal distress to the individual, psychopathological behavior is associated with decreased biological fitness. In other words, an individual showing a particular variety of psychopathology is less able to carry out biological functions affecting survival and reproduction. This concept of psychopathology comes originally from evolution but has been applied to the clinical setting in general terms, and connected to conventions and values characteristic of human society and culture. The concept and the behavior designated as psychopathology are replete with human social and cultural concerns: normality compared to abnormality; sickness compared to well-being; social responsibility, accountability, and social function compared to social impairment or disability. From a strictly clinical standpoint, psychopathology refers to actual signs and symptoms involving cognition and personal experience, as well as behavior itself, sometimes very personal and socially significant behavior.

Despite its seemingly logical tie to things quintessentially human and cultural, however, psychopathology has to be granted a place in the evolutionary process for three reasons: (1) psychopathology in its many forms—mood disorders, schizophrenia—is found in all contemporary cultures, in societies studied by historians, and in those studied by anthropologists; (2) vulnerability to psychopathology often has a genetic basis and high prevalence, which suggests that it played some positive role in biological evolution; and (3) different varieties of psychopathology have been artificially produced in higher primates by means of laboratory technologies; indeed, in clinical science, the manufactured syndromes function as animal models of psychopathology in humans. Forms of psychopathology have also been observed in natural communities of nonhuman primates. This attests to the common biological processes and behaviors humans share with the higher primates, emphasizing a continuity in the mechanisms that produce psychopathology.

Psychopathology clearly has a phylogenetic history. In other words, along with other human traits and phenotypes—cognitive, emotional, behavioral—that have been traced back to and studied in the higher primates, psychopathology itself must also be assumed to have had an evolutionary history. If the natural anomalies and eccentricities of behavior described by ethologists and primatologists are taken to represent examples of psychopathology, then psychopathology could be said to have possibly existed before the time of the pongid-hominid split during evolution. Manufactured psychopathologies in the higher primates support this supposition. However, one could interpret natural and manufactured behavioral anomalies among higher primates simply as very unnatural deviations, the effects of highly unusual circumstances. This would suggest that psychopathology as a natural characteristic of hominids originated sometime between the split between the pongid and hominid lines, about 5 million years ago, on the one hand, and the emergence of anatomically modern humans showing culture and language as we know it between 150,00 and 100,000 years ago, on the other. The animal varieties of psychopathology discussed in chapters 7 and 8 suggest an evolutionary continuity. Even a personality trait as replete with human and indeed mental illness implications as psychopathy—a trait associated with antisocial behavior in humans—has proven not only appropriate for description of chimpanzee behavior but also measurable in communities of chimpanzees (Lilienfeld et al. 1999).

In previous chapters discussion of the origins of psychopathology has involved filling in the gap of information about symbolic manifestations made apparent by studies of evolutionary psychologists and psychiatrists. This chapter formulates and discusses a schema of the ways psychopathology has changed during later phases of human biological evolution. A hypothetical individual's way of handling and processing information in relation to relatively segregated areas of psychobiological functioning is used to make sense of both content and meaning, as well as biologically important functional characteristics of behavior including psychopathology. An individual's information processing with respect to different realms of behavior is used to describe changes presumed to have taken place during phases of evolution.

Information handling related to visceral somatic processes and monitoring of the environment constitutes a basic operating function of organisms. Describing psychopathology during evolution involves visualizing how information handling systems relating to biosocial functions come to incorporate information from a sytem about conceptual and symbolic material. The mental apparatus governing adaptation and adaptiveness is a composite of heterogeneous morphological structures, organs, mechanisms, functions, adaptations, and/or programs of behavior, each of which has its own phylogeny. Nonetheless, segregated systems of information handling in relation to psychobiological function manage to integrate with each other in the production of behavior. Such an assemblage can be perturbed by many influences, resulting in breakdowns of behavior that are registered as varieties of psychopathology.

In discussions about the evolutionary antecedents of mental illness, evolutionary psychiatrists and psychologists naturally draw on concepts and principles of evolutionary biology. Some examples would include psychological adaptations or algorithms, life history theory, differences in mating strategies as expounded by the theory of sexual selection, ideas of resource holding power and hedonic and agonic competition derived from ethology, and the interplay of infrastructural systems of behavior in relation to basic biological functions and goals. It is assumed that the actual situations and circumstances involving behavior described by these and related concepts and modes of reasoning operate as evolutionary residuals in order to promote survival and reproduction.

During phases of the biological evolution of psychopathology, these imperatives continue to dominate. However, I am concerned with explicating the slow emergence of a symbols and meanings mien and complexion to psychopathology. While evolutionary goals and functions continue to empower and shape behavior and psychopathology in particular, it must be appreciated that processes involving the evolution of language, culture, cognition, and a sense of history mold these evolutionary residuals into a framework of meaning, values, political contracts, and beliefs and rituals about cosmology and spirituality. The underlying assumption is that hominid behavior in general and the origins of psychopathology in particular have a phylogenetic and a cultural basis. To attempt to describe phases of the biological evolution of psychopathology is to attempt to interlink ideas of evolutionary biology with ideas of cultural anthropology as expounded by medical anthropologists.

One way in which to signal or mark this change of emphasis is to draw a distinction between sex and gender. Aspects of mating, reproduction, and parenting as embodied in evolutionary residuals are based on the concept of sex—namely, the biological identity of an individual as manifested by body size, strength, genitalia, strategies of mating, and parenting in the life cycle. A parallel development during the later phases of biological evolution is that sexual identity becomes transformed into or blended with gender identity. In other words, ideas, values, beliefs, rituals, and social political contracts that make up the culture or symbolic niche of hominids come to influence and explain why and how the sexes behave as members of a group that expounds, claims, and/or is assigned a distinctive set of rules, obligations, entitlements, and roles. The sexes, in short, come to constitute engendered persons. Varieties of psychopathology also, one has to assume, come to be expressed and played out on a stage where individuals behave and pursue goals not only as dictated by sexual selection theory but also as persons whose psychology and social behavior is molded and shaped by a sense of culture and history.

Centers of Focus in the Evolution of Psychopathology

In abstract terms, psychopathology came to occupy a special segment of the behavior space of hominids during human evolution. In other words, psychopathology emerged and became part of the social fabric of hominid behavior. The idea of a behavior space brings to mind the totality of social and psychological behavior needed to fulfill the basic biological imperatives of survival and reproduction. Behaviors designed to carry out biological functions—competing for social rank, mutualism, finding a mate—were central to members of an early hominid group. During human evolution, then, behaviors that can be defined as psychopathology came to occupy a segment of hominids' behavior space.

I do not attempt to explicitly equate a phase of the evolution of psychopathology with a particular species of hominid, or with a specific milestone in the paleoanthropological and archeological record. Experts in the field differ as to dates, and no important analytical function is served by trying to link phases of the biological evolution of psychopathology to specific dates or species. It is more fruitful to develop a comparative evolutionary profile of psychopathology during human biological evolution. In what follows, the phases of the biological evolution of psychopathology are formulated as analytic or logical types. To describe characteristics of psychopathology during a particular phase of human evolution, I rely on generalizations about what is known about hominids during that phase. This would include knowledge about such things as language, cognition, culture, social organization, and group dynamics as well, of course, as biological functions and behavioral ecology.

Two complementary reference points are used in the following discussion. The first is to handle psychopathology as a trait or set of traits of individuals. Manifestations such as signs and symptoms, specifically, or changes in social and psychological behavior that meet criteria for disturbances, more generally, are both examples of this point of view. The second reference point is that of the audience of group mates of the afflicted individual. In a sense, the audience plays a crucial role in causing and influencing how behavior is handled. The second viewpoint involves the natural history of psychopathology: how incidents are brought about and interpreted and dealt with, and hence how long such incidents are likely to persist. During human evolution, both centers of interest, that of the individual and its audience, underwent changes in social and psychological behavior, and any formulation of the origins of psychopathology has to address both centers of interest.

In a small, nomadic foraging group of anatomically modern humans, the audience for an incident of psychopathology was comparatively small and culturally homogenous. However, in a large, complex society, an incident of psychopathology will be witnessed by a large audience with many different orientations, since the afflicted person is shunted across institutions of society. There will be some dispute over how best to name, describe, interpret, and respond to the psychopathology—is it sickness, antisocial behavior, eccentricity, weakness, or over sensitivity?

During human biological evolution, an incident of psychopathology would likely have elicited a dispute and negotiation, or at least conflicting responses. The disputants would have brought into play increasingly varied meanings and points of view. For example, among the earliest hominids, imperatives tied to survival and reproduction would determine what psychopathology meant, how social and psychological behavior was interpreted and dealt with, and how the individual and an episode of psychopathology fared and was played out, respectively. If a psychopathology directly undermined an individual's power and authority, its competitors would take advantage, while allies, depending on the individual's social rank and kinship status, might offer some support and defense. A lessening of a victim's formidability and resourcefulness would have meant a challenge to or lowering of its social rank. However, if a group had acquired a concept of loyalty and support or a sense of group solidarity, there would be less likelihood of outright exploitation by group mates if an individual were compromised by psychopathology.

During a later phase of evolution, when culture, cognition, and language were more evolved, alternative meanings and interpretations about psychopathology would be influential. A sense of group identity, group values, and traditions about what is expected of individuals might influence what was made of a psychopathology. Similarly, current group pressures and imperatives would be balanced by acquired group perspectives and traditions, and the outcome would determine how psychopathology was interpreted and dealt with. Alliances based on kinship, mating, or marriage would affect how individuals' actions were interpreted, including manifestations of psychopathology. As biological imperatives of compassion and consolation, already emergent in the last common ancestor, as data on chimpanzees suggest (de Waal 1996), get elaborated into notions of suffering, need, and healing, these ideas would come to shape responses to psychopathology.

Using the idea of hominid behavior space, then, two important tasks are as follows: (1) to describe that segment of behavior space occupied by psychopathology—what an individual showing a psychopathology might have manifested, given its level of cognition, culture, and language; and (2) to describe how happenings in that segment of behavioral space where psychopathology would have been found was actually understood and interpreted by group members, given their mental characteristics. Psychopathology represents a breakdown of behavior associated with personal distress. By convention, the individual showing psychopathology and, in a generic sense, all individuals of similar hominid groups who showed psychopathology, occupied a segment of the group's behavioral space. A basic question here is: How did group mates interpret this segment of the behavioral space of hominids? What kinds of understandings, symbols, and/or meanings did they use to understand it? What conventions had jurisdiction in this social space? To what extent were pure evolutionary residuals and their basis in self-interest determinative over evolving cultural and moral concerns?

Manifestations of Psychopathology during the Ethological Phase

During this phase, individual behavior would not be far removed from that of the last common ancestor. Psychopathological symptoms would be evident in disturbances of information handling systems related to biological and ecological monitoring and in visceral-somatic functioning. An individual's ability to function effectively would be grossly disturbed, and disruptions in general bodily functions affecting appetite, motivation, energy level, sexuality, and sleep, would take raw physiological form. Evolutionary residuals of behavior linked to biological imperatives involving survival, sociality, and reproduction would have had a determinative influence on the configuration, enactment, and outcome. In other words, these disruptions of basic functions would have some but not much symbolic meaning. Consequently, individuals would have some awareness of their disturbed state and could appeal to kin and group mates for support.

Remember that psychopathological behavior falls into two general classes. Symptoms and signs can consist of feelings, dispositions, and behaviors that are socially maintaining or syntonic, on the one hand, and socially divisive or dystonic, on the other. Of the first type among hominids, if manifestations were not socially affiliative and affirming, at least they would not defy conventions, disrupt everyday activities, or antagonize group mates. Most individuals who were suffering and aware of it would show signs and symptoms that symbolically communicated their distress. However, some psychopathologies consist of behaviors that repel, threaten, undermine, or oppose others. Examples among hominids would include anger, irritability, easy frustration, suspicion, distrust, gross misperception of social reality (of the identity or intention of group mates), or callous disregard of the needs of others. In general, these manifestations could elicit defensive reactions from others that led to avoidance or retaliation. If individuals who manifested socially divisive symptoms of psychopathology also communicated or if their behavior implied suffering and disability, their messages (willful or implicit) might blunt negative reactions from group mates. This generalization about the social valence of manifestations of psychopathology applies to all phases of evolution, although, depending on level of cognition and culture, how they were manifested and their social and psychological significance varied.

To be sure, survival and the capacity to reproduce would in a general sense be compromised by psychopathology. Indeed, with respect to the afflicted individual, sheer survival was a cardinal consideration. I am here leaving aside the instances when psychopathology may have played an adaptive role, or been selected as the best routine in an otherwise hazardous social ecology. If the psychopathology involved disability, social disruptions, or social inefficiency, it also placed a burden on group mates. The survival of the group might even be threatened. Overt behavioral manifestations could range from agitation, intense irritability, and explosive loss of emotional control to apathy, indifference, withdrawal, timidity, and fearfulness. Alternatively, behavior could reflect confusion, ineffectiveness, social awkwardness, and states of dissociated behavior having little discernible social rationale. The availability of food and water in the habitat, the degree of intergroup competition, the pressure posed by predators, and intragroup relations, including the individual's social rank and alliance network, would all influence how the individual fared.

In mating and parenting, there would be prominent sexual differences in social effectiveness and success among those manifesting psychopathology. Males could attempt to coerce females into mating, showing a disregard for alliance networks and group mating patterns, or they might be unable to mate altogether. Females might manifest disturbances by disregarding established systems of consortship and friendships with males or by avoiding males entirely. Psychopathology could undermine the effectiveness of mating and maternal behavior.

Manifestations of psychopathology would have comparatively little expressive content. In other words, the limited capacity for social attribution and theory of mind functioning and cognition and communicative ability, mainly gesture but some vocalization, would mean that individuals who displayed psychopathology would have a relatively unelaborated awareness of an altered state of being and functioning. Consequently, they would not grasp what their condition meant nor how it affected the group and hence would not know how to communicate with group mates about all of this. For example, while individuals were in some way programmed with a sense of what favors sought, favors given, and favors owed might possibly mean—that is, a sense of reciprocal altruism—at this stage of evolution, such routines and expectations would not likely be explicitly formulated or communicated, either to the self or to others. Distortions and misperceptions would not be easily apparent or expressed by afflicted individuals, and thus they would be deprived of group mates' responses that might be meliorative. Similarly, afflicted individuals could not effectively communicate personal suffering to group mates as a way of seeking consolation and favors, let alone efforts at healing. Finally, because of limits on verbal language, cognition, and culture, afflicted individuals could not use the group's historical knowledge to formulate and seek relief, obtain what they need, or elicit the support and favors they believed were owed to them.

States of dissociation would likely accompany incidents of psychopathology during the ethological phase. As discussed in chapter 12, hominids inherit a capacity for such states, and the high frequency of psychological traumas could well have prepared individuals to enter states of dissociation as a way of coping with traumas and crises. The suffering and sense of helplessness associated with incidents of psychopathology could give rise to passive withdrawal, inactivity, emotional detachment, and tuning out the environment. Or a state of dissociation might be expressed as purposeless actions, wanderings, pacing, and rocking motions. However, other than showing a sheer inability to perform basic functions, an afflicted individual of an early hominid group communicates very little while in a state of dissociation. From a biological standpoint, evolution seems to provide few behavioral programs for handling extreme breakdowns—essentially, the failure of an adaptive mechanism or algorithm. When the individual's capacity to cope is overwhelmed, a defensive or reflexive chaining of coordinated but inappropriate motor responses shutting out the environment might take over.

Group Responses to Psychopathology during the Ethological Phase

In general, responses of group mates would be affected by the overall social tenor and significance of the manifestations of psychopathology. What were seen as socially affirming or at least socially neutral symptoms and signs would elicit responses that maintained social relations with the afflicted individual or at least did not harm its well-being. On the other hand, so-called socially divisive manifestations would elicit negative reactions: hostility, restraint, guardedness, opposition, or active avoidance. A psychopathology associated with relative powerlessness of the individual might invite exploitation, but it also would pose a potential risk of retaliation from the victim's allies and/or to the stability of the group, as mentioned earlier. Responses discussed below should be viewed as conditioned by the general social implications—that is, biological, social, cultural—inherent in manifestations.

Cognition and a sense of culture among group members determined how psychopathology was perceived. The group had accumulated a store of knowledge about the habitat, threats to safety, and the group's history, and their habits and behavioral traditions had been learned or acquired as a function of group living. This information was not categorized in any abstract or semantic sense; at most, group mates would be capable of conceptually referring to communicating about objects, situations, and events pertaining to natural history, spatial and geographic orientation, social behavior, and nearby or distant groups. All of this implies that while members of the audience of a psychopathology could be aware of self and could attribute intentions and reasons to others, including the individual showing psychopathology, their ability to interpret another's behavior would be very limited.

While a theory of mind would be operative in members of an audience of psychopathology, what the theory could articulate by way of categories, feelings, intentions, and states of being would obviously be restricted and shallow. Thus a predominant theme in a response to psychopathology would be more or less rigidly determined by behavior routines tied to evolutionary imperatives, discussed as evolutionary residuals in previous chapters. Members of the audience of a psychopathology would thus press for their own advantage. Psychopathology would be perceived or dealt with as aberration, weakness, ineptitude, timidity, insecurity, or inappropriate bravado. Group mates might exploit the chance to appropriate the individual's food, mate, living space, or social rank. In the event of outside attack, the individual could not be counted on. Because the individual's disturbance could undermine the more or less coordinated responses that group mates were capable of, psychopathology risked their well-being. Group mates might also ignore, ostracize, or banish the individual.

If a kinsman, mate, or close ally were afflicted, and suffering were visible, manifested as fearfulness, despondency, weakness, lassitude, grossly aberrant actions, purposeless movements, or physical pain, others might show restraint and condescension. But survival and reproduction would be utmost considerations. Because of a strict ethological system of social organization, it is unlikely that a structure of knowledge, a system of categories, or a defined idiom about key elements of social reality and behavior would be enacted so as to culturally domesticate an incident of psychopathology.

Responses to states of dissociation as part of a psychopathological incident would be similar. The individual showing dissociated behavior that connoted obvious impairment and ineffectiveness could be exploited by competitors and rivals. On the other hand, since a state of dissociation could be largely devoid of expressive content, group mates might simply ignore an individual who sat immobilized and stared into space, oblivious to the surroundings. A dissociated syndrome of stereotypic posturing, grimacing, or pacing might be received with puzzlement or indifference.

Manifestations of Psychopathology during the Precultural Phase

Incidents of psychopathology during the precultural phase would be marked by perturbations of an individual's systems of information handling related to language, culture, and cognition. This means that basic disturbances in social ecological monitoring and vital physiological functions that dominated incidents of psychopathology during the earlier, ethological phase would also be present in somewhat raw form during this phase. In other words, foraging, mating, parenting, avoiding predators, competing, and alliance networking would all be compromised. Similarly, basic functions involving appetite, energy level, motivation, and sexuality would be visibly impaired. This implies that sheer survival would be a central problem associated with incidents of psychopathology, as during the ethological phase. However, these gross behavioral and bodily manifestations harnessing or undermining evolutionary residuals would now be complemented and modified by the individual's more evolved, albeit still simplified, abilities in cognition, language, and culture.

Individuals characterized by emergent language and a sense of self-awareness and self-governance are able to voluntarily call upon and use information located in special stores, or distributed networks, as programs for voluntary action that are formulated in a framework incorporating the self and the group. Such representations are more abstract reformulations of material furnished purely by sensory and perceptual systems. Consequently, individuals capable of emergent language, self-awareness, and self-governance meet conditions for at least the second order of intentionality as described by Dennett (1987). Not only are they capable of conceiving abstractions like beliefs and desires, but also they can form elementary abstract notions about phenomena, that is, ideas about ideas. In addition, such individuals show the level of representation and evolutionary development that Donald terms mimetic (1991). When fully integrated with emergent language, culture, and a sense of historical consciousness, conditions are met for early versions of Damasio's extended consciousness and autobiographical self (Damasio 1999).

Donald (1991, 1993a, 1993b, 1994, 1995, 1997) conceives of mimesis as a foundational form of intentional cognition and communication that enables an organism to think for itself, as it were. The idea of mimetic cognition and communication, which includes some verbal or vocal language, appears to resemble what Bickerton (1990, 1995) calls a protolanguage, a rudimentary capacity for "off-line thinking possible in *principle*" (1995, 120; emphasis in original). Donald and Bickerton suggest that during earlier phases of human evolution, a capacity for self awareness (or system 1 functions, as formulated here) was integrated with an early capacity for language (or information equated here with an emergent system 2). It should be noted that Donald and Bickerton differ regarding the timing and the nature of early phases of human cognition.

To account for the human capacity to communicate meaningfully by means of special symbols and idioms, one must posit the emergence of specialized systems pertaining to language, and a sense of culture and history, which can feed information to an executor or general intelligence factor that regulates and monitors behavior. With these capacities, which are emergent in hominids, individuals begin to acquire the capacity for what Kendon (1991) calls conceptual reference and what Chase (1999) refers to as symbolic reference, which is related to the property of language as a system of communication that Hockett and Archer (1964) refer to as displacement, the ability to communicate about phenomena displaced in space and time by means of conventional symbols.

Kendon (1991) points to a capacity for conceptual reference as an essential precondition for the development of language. Such a cognitive capacity is consistent with a sense of self-awareness and self-governance. In his view, apes (and by implication, Australopithecines and early varieties of Homo) are able to communicate via vocal signals and grunts, but all of this serves to index actual perceptions or aspects of the environment that the group is immediately engaged in, concerned about, and toward which all are related in an actual line of activity. (This is not unlike Donald's formulation about episodic cognition in apes.) Kendon proposes that a special form of social organization, the ability to establish consistent differentiation and complementarity between the activities of different individuals, is a precondition for language, with its special ability to displace reference from the here and now. Richards (1987, 1989a, 1989b) discusses cognitive aspects of evolution. He describes physiomorphic learning and the transformation of innate life routine chains into environment-driven, followed by motivation-driven, behavior, eventually leading to linguistic coding as a basic formulation of human action. Richards's ideas are consistent with those of Kendon, and all of them pertain to the functions associated with self-awareness, self-governance, and emergent language.

Consequently, a better-defined sense of awareness of self and a greater capacity to interpret mental states and social identity of group mates meant that incidents of psychopathology would be associated with a greater degree of expressive, communicative content. During this phase, social attribution and theory of mind functioning were relatively well established, although individuals lacked conceptual and especially linguistic capacities to be able to elaborately conceptualize and communicate with themselves and group mates about their plight.

In other words, during this phase individuals afflicted with psychopathology would have a definite, albeit conceptually shallow, insight into their signs and symptoms.

Individuals afflicted with certain varieties of psychopathology could know in a limited way that their well-being was compromised, that they were suffering, and that they were unable to carry out basic survival and reproductive tasks effectively. Individuals showing socially divisive, antisocial manifestations would be locked into routines that rationalized aggression or avoidance, arrogated benefits and resources, and/or ignored the needs of others. In both instances, the nature of the physical and social environment would moderate their manifestations, their awareness of this, and their group mates' responses.

Evolution had programmed individuals with the ability to maintain themselves in their habitat, to have to seek food and shelter, avoid predators and fight enemies, seek mates and reproduce, and mutually work out balanced exchanges. An incident of psychopathology would upset and strain the rationale and intelligence or culture carried in this social fabric and economy. Similarly, psychological episodes would be crafted and fashioned in terms of behavioral programs devolving from these evolutionary rationales. During this phase of evolution, hominids would have acquired some insight about, understanding of, and better control over these behavioral imperatives. Specialization of an information-handling system that provided a capacity for self-awareness and mental attribution, its evolutionary maturation, would produce more complex social behaviors. They would use the information they had acquired about themselves and their world, what a preculture delivers, to moderate how evolutionary residuals of behavior were implemented. In other words, while biological imperatives in some way operate even in the midst of a full-blown modern culture, the symbolic resources that culture provides can modify how they are carried out and how others respond and accommodate to them—the audience's response. An incident of psychopathology during this phase, then, would produce more than a pure disruption or upheaval of biological routines; an awareness of what was at stake in an incident would affect how manifestations were expressed.

If knowledge of one's behavioral incapacity and suffering were added to knowledge of self and situation, symptoms of psychopathology, or any serious sickness stemming from disease and injury, for that matter, could cause terror. Afflicted individuals would recognize what it meant to be behaviorally disabled and compromised: they would fear for their safety, about getting enough to eat, and about the future. Emergent functions governing language and a sense of culture and history would account for such behavior.

In short, during the precultural phase, individuals would be aware of danger and risk, would see that psychopathology threatened the safety of the group. Similarly, they might regard bodily and emotional suffering, and social disability or impairment, as matters of life and death. These aspects of symbolism were emergent during this phase and would be projected onto the afflicted hominid's condition of psychopathology. However, what individuals could do with this knowledge or insight would be very limited. They could communicate symbolically using body language, nonverbal, emotional gestures, and rudimentary vocal sounds that carry concrete, here-and-now, relatively conventionalized meanings. By these actions, they could communicate their suffering and need for support, although many manifestations of suffering implicitly carry such meanings in themselves.

A precultural level of mentation and capacity for social experience would not include full-fledged explanations of life and death, of personal experience and well-being compared to suffering and disability, of what social circumstances meant in an abstract sense, of group characteristics and its history as conditioning everyday life, and of a sense of cosmological placement. Thus, individuals showing psychopathology would be unable to connect knowledge of their suffering and disability with larger existential themes and questions. Such abstract cognitions were not possible until a later phase of evolution.

The inability to give psychopathology existential meaning does not mean the individual could not care for itself, however. Indeed, because afflicted individuals now had a more elaborated conception of their social situation and of their disability in relation to it, they could look out for themselves better. Evolution, of course, has designed routines of behavior that by definition promote adaptation. Innate response tendencies present during infancy would be elaborated by trial and error and guided learning. Residuals of the evolutionary process would give them the behavioral imperatives to search for food, water, mates, shelter, and safe conditions. Sickness, injury, or psychopathology—a kind of sickness marked by behavioral disturbances that affected physiological and behavioral functioning—would not obliterate basic survival and reproductive routines.

Short of a crippling disability or sense of suffering and loss of emotional control, afflicted individuals could look after themselves and their dependents, if in a compromised way, even during earlier phases of evolution. What was different at this stage is that individuals had more social awareness of their plight, of what they could expect of others who were also aware of it, and they could communicate this awareness to others in a rudimentary way. There was no sense of what one could call a social role of sickness or psychopathology, and thus no groupwide, shared basis of expectations regarding.an infirmity. Yet, a biological preadaptation for such a social role of sickness was programmed in by evolution and would come under the influence of states of awareness associated with precultural knowledge. Individuals would innately know, as a result of developmental unfoldings or social learning, that because of their disability, their survival was in jeopardy. They would know to communicate their plight, and they would know that at least close kin and allies would offer support and comfort.

Group Responses to Psychopathology during the Precultural Phase

Others would respond to an episode of psychopathology in ways conditioned by their level of evolved culture and cognition. This included a relatively well-established capacity for mental attribution and theory of mind functioning; limited conceptions of personhood and motivation; concrete modes of reference without abstract reasoning power; and a simple system of communication. As noted earlier, characteristics of the social ecology would play a moderating role in what responses predominated and how they were played out.

Instead of taking advantage of the afflicted individual, others might be able to make emotional sense of what was happening from signs, symptoms, and nonverbal gestures, provided the manifestations were not socially divisive. Kin and allies would offer support if the suffering individual seemed severely compromised. If psychopathological manifestations caused tensions and conflicts, others might seek to mediate, persuade, and redirect behavior. Those not allied to the individual might not offer direct support, but neither would they press for immediate advantage and exploit it. If manifestations were overtly divisive and antisocial, they might avoid the individual or even seek retaliation.

During this phase of the evolution, audience responses to psychopathology would vary. First, dictated by strict evolutionary imperatives, some might seek to take advantage of an impaired individual who showed extreme fearfulness, pain, debility, ritualistic motor behavior, timidity, and restraint. However, others might refrain out of compassion. Still another response would be sympathy, but not acted upon; and finally, some would actively support and help the suffering individual. Group members would have developed a variety of responses to others' behavior beyond reactions to sickness because during this stage of precultural functioning they would have acquired a better defined sense of self, a more elaborated capacity for social attribution, and a range of concepts with which to understand their perceptions of self and the situation. Responses would differ in salience depending on the severity of the psychopathological symptoms, environmental pressures, the relation of group mates to the afflicted individual, and whether or not psychopathological behavior posed a threat to the group.

States of dissociation during this phase would not be too different from those of the ethological phase, and responses would be similar. However, because others could now recognize what these states communicated, they could better understand the individual's plight and interpret the dissociated behavior. Likewise, because afflicted individuals could better understand themselves and their situation, they could conceptualize their plight on an elementary basis. While states of dissociation entail automatic and detached or disconnected experiences and behaviors, they are hardly unregulated or unprogrammed. Just as individuals could grasp some understanding of themselves and their situation while suffering from psychopathology, they could also be aware of the meaning inherent in a state of dissociation. Those surrounding an afflicted individual would also be able to decipher the meaning of dissociated behavior.

While group mates would have an intellectual ability to understand disease, pathology, injury, imminent death, or behavioral disturbances, these meanings would not be categorized and differentiated. They did not have abstract notions about religion, cosmology, or medicine that might guide interpretation and action. Nor were there behavior traditions about how to handle misfortune. At this stage of preculture, hominids had limited conceptual and linguistic resources with which to interpret incidents like psychopathology. And of course, there did not exist well-formed ideas and there were many fewer institutions or rituals involving cosmology, personal or group identity, spirituality, moral accounting, medicine, or the implications of human mortality.

Manifestations of Psychopathology in the Protocultural Phase

During the protocultural phase, we assume that individuals had acquired not only a capacity for conceptual reference but also a set of concepts and cognitive and linguistic structures that enabled them to think about their world and their situation, including behavioral breakdowns or psychopathology. One would observe capacities for information handling associated with language, a sense of self-awareness and self-governance, and a sense of culture and history.

Such capacities or systems of information handling are responsible for consciousness, a sense of self, and a sense of intentionality. This includes core consciousness, which provides a sense of self for one moment (now) and one place (here); it is not dependent on conventional memory, working memory, reasoning, or language (Damasio 1999). During the course of evolution, systems responsible for a sense of self-governance, linguistic competence, and culture and history became more influential. Together they generate extended consciousness, a far more complex biological phenomenon with several grades and levels of organization that becomes elaborated during the lifetime of the individual. According to Damasio: "Although I believe extended consciousness is also present in some nonhumans, at simple levels, it only attains its highest reaches in humans. It depends on conventional memory and working memory. When it attains its highest peak, it is also enhanced by language" (Damasio 1999, 16). The two forms of consciousness are associated with a core self and an autobiographical self, respectively. The content of these can be inferred from description of the biology of core and extended consciousness. Related ideas are found in the writings of Dennett (1987), Jackendoff (1987, 1994), and Donald (1993, 1997).

During this phase, individuals would have an elementary capacity to explain the workings of the world and their own circumstances—that is, elementary social knowledge, natural history knowledge—and to remember what they had learned and to share it with the group. In a psychological sense, this information and social knowledge mediated between behavioral imperatives designed by natural selection and the opportunities and challenges posed by the social and physical world. This led to the establishment in emergent form of what is termed a symbolic niche for the individual. A condition of psychopathology could thus be cognized and explained in a more elaborate way than during earlier phases.

The signs and symptoms of an incident of psychopathology would be expressed in a more symbolic idiom during this phase. In other words, they would incorporate aspects of the social circumstances and the individual's present suffering and disability. This would include not only what it means to suffer disturbances in feeling and behavior, but also what caused the manifestations and what could be done about them.

If disease or injury were the cause, the individual could relate the pain, physical impairment, altered physiology, and overall sense of sickness to its behavioral difficulties and emotional state. Tensions caused by frictions with rivals or enemies could bring fear, worry, ruminative foreboding, a sense of vulnerability or being hunted; the death of a partner or close kin would bring a sense of loss and emptiness. In vulnerable individuals, stressors like these would cause grief, bodily dysfunction, and behavioral breakdown—in other words, an episode of psychopathology. Given the degree of personal insight and understanding individuals were capable of during this phase, they could be aware of how their loss of well-being and dis-ease related to their circumstances in the group. How accurately an individual appraised the cause of its disability is not the issue; evolution may have programmed individuals to deny or to be unaware of what really ailed them (Alexander 1987a, 1987b, 1990; Nesse and Lloyd 1992). What is important is that they could have a concept of social stressors and strains and understand their own condition—for example, as in states of dissociation. In other words, an incident of psychopathology would now be more than a simple rent in a social fabric and economy made up of evolutionary residuals, along with an awareness of social circumstances. Rather, it could communicate a far richer and symbolically more nuanced interpretation of the afflicted's condition in relation to its group and situation. However, a condition of psychopathology would not be an abstract category or culture pattern internalized as part of the information making up an individual's protoculture during this phase. Rather, the individual would understand its condition as caused by some social and emotional upheaval and would be aware of its plight. Group mates would recognize and understand it as well.

Let us imagine how states of dissociation might be manifested during this phase. We have no compelling reason to assume that psychological traumas were less common or intense in the lives of hominids than they were before. Individuals of all ages would have been frequently exposed to any number of terrifying experiences, including environmental catastrophes, warfare, attacks by predators, group hostilities with violence, sexual coercion, and outright killings or murder, among countless others. If there is an innate capacity for developing states of dissociation, traumatic events could have caused some vulnerable individuals to enter a trance-like state as a way of tuning out or obtaining psychological distance from traumas and conflicts. What difference would a capacity for protoculture have made in how states of dissociation (as a component of psychopathology) were played out and dealt with?

Traumas or threats from outside the group could have produced a syndrome of automatic behavior designed to protect the individual by psychologically or physically removing it from the situation—sustained inattention, emotional detachment, prolonged flight—or by confusing or frightening an assailant—bizarre compulsive acts, aimless motor rituals, exaggerated swagger. In these situations, the individual's ability to formulate its plight would be symbolically manifested through gesture, body language, emotional communication. In the case of stressors or conflicts from within the group—a threatened loss of social rank, attack by a rival, appropriation of the individual's mate, the death of a family member—the group's shared experiences and a knowledge of the personality of individuals causing a trauma or threat would shape the syndrome of dissociation. Because of limited cognitive and linguistic resources, expressive behavior would be simple, direct, and unelaborated: a syndrome might communicate wounded pride, humiliation, controlled rage, betrayal, or the threat of reprisal.

A sequence of organized and meaningful motoric acts would instantiate a psychopathology. Gestures, facial expressions, mimetic emblems, and brief vocal expressions would carry the afflicted's messages about the distress, frustration, and helplessness caused by the trauma or stressor. During this phase, what passed for culture was a relatively small inventory of unorganized information. A state of dissociation would carry little narrative depth because protoculture had not yet evolved beliefs, myths, rituals, shared assumptions about the group's origins and social practices, the efficacy of supernatural agencies, or conventions and institutions for dealing with social conflicts. However, it would still embody symbolic communication.

The role of systems of meaning in the scenario of a psychopathology has been given emphasis. This is understandable since during this phase of evolution functions linked to systems of information handling pertaining to language and a sense of culture and history begin to influence how a psychopathology is configured and enacted. However, naturally designed evolutionary residuals continue to shape behavior generally and psychopathology in particular. Biological functions and imperatives linked to sheer survival, sociality, reproduction, and the optimal timing and operation of life history parameters continue to provide the basic grammar or structure and social economy that define and calibrate an incident of psychopathology. In a sense, much of the evolving cognitive significance and social and existential content and drama of psychopathology is governed by the power of the biological functions and imperatives embodied in the evolutionary residuals. In other words, one should assume that the emotional momentum and symbolic figuration of an incident of psychopathology is influenced in important ways by its innate biology and fitness implications.

Group Responses to Psychopathology during the Protocultural Phase

During this phase of evolution, the audience of an incident of psychopathology would be made up of individuals holding different points of view about ambiguous or enigmatic situations and events. Because all incidents of psychopathology by definition entail psychological and social behavior, behavior now with more meaning and content, such incidents would give rise to competing interpretations as to its cause, significance, and mode of handling it.

The cognitive and linguistic attributes of individuals endowed with a protoculture would enable them to shape a psychopathology in a symbolic idiom that expressed the victim's plight and what might have caused the psychopathology. Even (psycho)physiological signs and symptoms could be used to communicate an individual's plight and responses to circumstances. In other words, the body would provide an array of subjective symptoms and visible signs that expressed a sense of vulnerability, suffering, or need. Social and psychological symbolic behavior would provide a far more meaningful idiom with which to express and communicate mental states involving reasons, intentions, desires, needs, and reprisals—any or all of which can be part of a psychopathological incident. Such a capacity and its accompanying idiom were obviously shared by members of a group, who could form different points of view about the psychopathology.

Groups of hominids during this phase would more than likely be made up of sets or segments of blood-related males and immigrant female mates. These segments may have been distantly related, or were simply allies and friends bound together by common ecological resources and contingencies. There would be differentials of rank and status based on age, strength, cleverness, or hunting ability. Although strength of kinship and friendship ties would differ among group members, they all would have been locked into a tight arrangement of reciprocal social ties of favors and obligations. In such a setting, an incident of psychopathology would be an alarm call that activated competing allegiances among group mates and between the afflicted individual and the segment to which it belonged. The cause of an incident of psychopathology would more than likely have been known to others, but group mates would assign cause and blame differently. Responses to the plight of the afflicted individual would elicit different degrees of fear, condemnation, sympathy, rejection, and resentment, and group members would differ as to how to resolve the crisis.

Cognitive capacities and linguistic resources associated with this phase of evolution would allow for a simplified symbolic fashioning of an incident of psychopathology. Individuals might be able to communicate their view of what brought on their distress and breakdown, making some kind of statement about their situation, their culpability, or lack thereof, and how they would like their circumstances to change. A state of dissociation could be an important feature of the incident, which would be again interpreted variously, depending on the observer's position in the group. While an incident of psychopathology might carry a relatively straightforward meaning to the victim, a meaning that reflected an overriding perception of circumstances, this would not be the case for the audience of the psychopathology. It would consist of potentially conflicting interpretations. Intercurrent group events and ecological pressures would influence how group mates responded and the social consensus they reached. As always, affined relationships, kinship, alliances, and friendship patterns involving the victim would play an influential role in how the incident played out.

Manifestations of Psychopathology during the Early Cultural Phase

Virtually all of the information-handling systems I have used to model social and psychological behavior and body manifestations of psychopathology would be operative during this phase, although not all in an advanced state. In particular, language and a sense of culture and history would be limited, compared to fully modern humans. Much of the material presented about possible scenarios involving Archaic humans applies to this early cultural phase. Individuals would have acquired a simplified culture and would have gone beyond protolanguage (see Bickerton 1990, 1995). They would have been able to use more words and to shape them into sentences with elementary syntax and grammar. They could conceive of past events and anticipate a future. They would have gone beyond the chunks of information that characterized protoculture and would have gathered a pool of information—for example, about natural history, the habitat, the seasons, group history, and social behavior. This shared information would be rudimentary but would possess structure and coherence, nonetheless.

Even a simplified culture would make sense and hang together, composed of a pool of organized information. The culture would enable individuals to use information to give focus, direction, and purpose to social life beyond mere aspects of survival and reproduction—evolutionary residuals—although the more elaborate system of beliefs and rituals shaping an incident would have reflected the biological implications. The pool of information would consist of concepts and generalizations about experience and social life, including the ability to identify mental states, to discuss personal well-being and health, disease and injury—in other words, a simplified pool of information about medicine. Members of an early cultural group would gather information about the size, composition, and organization of the immediate group and of nearby groups, especially social conventions and standards. In other words, they would have evolved a shared understanding of social behavior, group identity, and intergroup relations.

Finally, in line with the characteristics of cognition formulated by Boyer (1994), we assume that religious ideas and notions about cosmology would also be part of this pool of knowledge. Simplified modes of reasoning about social and physical phenomena based on direct experience would be projected onto the cosmos, creating a form of religion. Religion would be used to explain and influence how misfortunes in the group were configured and enacted, including psychopathology, which would be associated with suffering, incapacity, danger, risk, and social divisiveness. Consequently, religious thinking could become integral to psychopathology as to sickness in general, of course (Boyer 1994).

In the early cultural phase, an individual would possess a body of information that allowed it to interpret its own experiences and distress. It would have some cognitively and linguistically structured understanding of the group as a whole and of the social identity and personalities of group mates. States of personal distress and behavioral breakdown, that is, of psychopathology, would not just produce in some physical sense signs and symptoms that, as before, had a biological valence and could be used as emblems or symbols. Such states would now be part of an ensemble that included verbal references to the past, present, and future. In other words, cultural information and language could be used by the individual to assess its level of well-being, explain the causes of its distress, and project itself forward to other circumstances.

An amalgam of motives, intentions, doubts, suspicions, and/or resentment would now be not just implicit and covert in an incident of psychopathology but also explicit, referenced, and communicated. In the event of associated physiological manifestations, individuals could seek medicinal substances. These and related features of psychopathology could make up an individual's explanatory model about it. When cognized and expressed, they would enable the individual to explain its condition in an idiom that would make sense to group mates in both an immediate context and a long-term, existential framework. As before, but now in a more muted and less transparent way, biological residuals would continue to cause and influence manifestations and calibrate or empower the significance of a syndrome of psychopathology.

During this phase, individuals experiencing states of dissociation, distress, and behavioral breakdowns could now tell a story. To be sure, during all phases of evolution, including even modern humans, states of dissociation can be direct, simple, striking, and unadorned. Such behavior as convulsion-like stirrings, states of detachment with subsequent amnesia, and a confused, aimless staggering forward without seeming purpose or direction cannot be said to tell a story, certainly not a culturally meaningful one. At most, they communicate incapacity and helplessness. However, individuals with a shared body of information could use it, in conjunction with altered states of consciousness, to make a statement about their situation that formed a narrative. The conjunction of psychological trauma or stress, vulnerability to psychopathology or states of dissociation, and a sense of distress and breakdown could come together so as to fashion states of dissociation that were meaningful in this early cultural context.

A simplified culture presumably would allow individuals not only to posit agencies and powers that controlled happenings in the world, as in protoculture, but also to attribute directed influence to such agencies. This is in keeping with emergent, elementary forms of cosmological and religious reasoning, as described by Boyer (1994). On this basis, members of the culture could believe that spirits or deities were capable of entering into the bodies of individuals and, by controlling their actions, could possess them for a time. Even if altered states of consciousness were not manifested as possession, they could express how the individual interpreted its disturbances. In other words, early cultural information could include elementary syndromes of dissociation that told stories and that functioned like emergent culture patterns or models. These models may have been part of how psychopathology was configured and played out during this phase. A shared pool of information that enabled an individual to place itself in the world in some extended sense, to formulate ideas about well-being and about what could disturb it, to reason about cause and effect, and to understand its connection to group mates and supernatural agents could lead to incidents of psychopathology, including states of dissociation that not only were symbolic but also had a narrative content.

Group Responses to Psychopathology in the Early Cultural Phase

The earliest manifestations of a psychopathological incident would be breakdowns of behavior associated with personal distress. They would resemble the physical signs of biological incapacity and would be interpreted in terms of strict evolutionary imperatives. Even a simplified culture would frame these disturbances in a meaningful idiom that told a story about its circumstances and its afflicted condition. One may interpret such an incident as not only symbolically connected to immediate events, as the individual construed them, but also as communicated to members of the audience that in the individual's mind were able to understand things and possibly do something about them.

Since members of the audience would have different social connections to the afflicted individual and different ideas as to the reasons and causes for a breakdown, their interpretations of the incident would also differ. They might accept, deny, reject, blame, disregard, or outwardly oppose the messages expressed by means of the incident. They would argue as to how to resolve differences between their interpretations—for example, the individual was sick, acting bizarrely or crazy, malingering, offering excuses, blaming others for its own failures, was antisocial, dangerous, ruled by external agencies, or simply wishing to freeload.

With the evolution of early culture, there could be a variety of responses to an incident of psychopathology. In short, not only would the behavioral space of a hominid community be taken up with social syndromes of psychopathology, but also the afflicted individual would now be surrounded by those who challenged, sought to change, oppose, or meliorate the situation as expressed in the psychopathological incident. In short, early culture meant that incidents of psychopathology could now elicit responses from an audience that commented, challenged, contested, and attempted to control the behavior.

Family members as well as group members are likely to be incorporated as causes of a victim's plight and consequently their role in how an incident unfolded—that is, what was done about it and how it played out—would differ. Not only could a victim's scenario of psychopathology show greater symbolic diversity and carry polyvalent meanings, but members of the audience would also show diversity in the way they responded. Debate and contestation would punctuate the course of an incident. However, a consensus about its significance would be reached. The culture of the group would provide concepts and explanatory models or theories about psychopathology. Social practices and proto-institutions pertaining to healing, arbitration, expiation, and social rehabilitation could be brought into play to handle the incident.

- The evolutionary basis of human psychopathology and mental illness supports the claim that these conditions had a presence in the life experiences and behavioral activities of hominids.
- During evolution, varieties of psychopathology differed in terms of physiology and behavior; contemporary psychiatric disorders provide

the best models for these. However, for analytical purposes, handling them in a general and unitary way is useful.
- To better visualize the origins of psychopathology, the evolutionary terrain of the hominids was divided into four phases: ethological, precultural, protocultural, and cultural.
- During each of the four phases, psychopathology was influenced by evolving systems of information handling pertaining to visceral functions, ecological monitoring, self-awareness, self-governance, and sense of culture and history.
- Syndromes of psychopathology are integrated wholes produced by harmful perturbations of the mechanisms and systems of information handling underlying behavior. The interplay between these mechanisms and systems differ during the four phases.
- Evolutionary residuals are naturally designed mechanisms and systems of information-handling keyed to basic biological functions of survival and reproduction. How they play out during evolution is an important characteristic of hominid psychopathology.
- During human evolution, self-awareness, self-interest, and the requirements of sociality and mutualism modulate but do not diminish the power or influence that evolutionary residuals have on the configuration and enactment of psychopathology.
- The emergence and then elaboration of verbal language, cognition, culture, and eventually a sense of history are factors that become increasingly evident in the way scenarios of psychopathology are played out during the later phase of the evolution of psychopathology.
- During the four phases of evolution, the manifestations of victims of psychopathology and the responses of group mates to their plight show the changing influence of the various mechanisms and systems of information handling underlying behavior.
- Social practices and emergent social institutions about medicine, morality and justice, religion and spirituality, and existential/cosmological integration, all of which influence psychopathology, become increasingly important during later phases of evolution.
- From global, acute and circumscribed events of unrest, crisis, disorder, behavior breakdown, or simply subtle interferences in the flow of social life, psychopathologies evolve to tell meaningful stories about the status of the self and conditions in the group.

Afterword

Visualizing the Cultural Evolution of Psychopathology

ANATOMICALLY MODERN HUMANS evolved sometime around 100,000 years ago, showed unmistakable evidence of culture, and most probably language as we know it, around 50,000 years ago, and became food producers by domesticating plants and animals and adopting a fully sedentary way of life around 15,000 years ago. This dating is a rough approximation of archeological landmarks and is subject to contestation, but it will serve as a basis for discussion (Diamond 1997).

Prior to the advent of food production and full sedentary existence, members of Homo sapiens lived in small, nomadic groups as hunters and gatherers. Since then, in different geographic regions of the world, humans have lived in societies that have differed in a number of characteristics; for example, modes of subsistence, size and density of population, culture, social stratification, diversification of economic roles, technology, degree of literacy, and political organization, particularly degree of centralization of power. These and related characteristics have been used to array societies on a continuum of social complexity (Johnson and Earle 1991).

The continuum of types of societies stretching back to the period of the emergence of culture and language and up to the present outlines the social evolution of human societies. It has involved changes in social complexity, beginning with hunter-gatherers, progressing through the advent of food production involving tribes and chiefdoms, and finally passing through states and onto empires and nation-states. The sequence describes how societies, viewed as structures involving social practices, roles, and institutions, have evolved during the course of human prehistory and history, producing new, increasingly complex forms or organization serving new and more elaborate functions. Social evolution, however, should not be equated with progress or improvement. Moreover it differs from biological evolution in its selectionist paradigm: the type of behavior or units that are selected, what features of the environment do the selecting, and what functions are affected by selection (Runciman 1989, 1998).

As a result of changes in the structure of societies during social evolution, human behavior has been modified. So, one must assume, has psychopathology. Because they have negative social consequences and, like sickness, give rise to a need for corrective action, incidents of psychopathology acquire a social identity. The language and culture of a society provide labels, a vocabulary of meanings, and explanatory models in terms of which actors communicate and conduct social action, including talking about, explaining, and doing something about incidents of psychopathology.

To study psychopathology in relation to the social evolution of human societies it is desirable to view society and culture as separate processes. The former involves a complex assortment of sociological characteristics, the latter, symbolic information about ideas, values, worldviews, beliefs, and rituals. These are externalized in different forms, distributed across a population, and provide rationales, purposes, and overall meanings. Cultural evolution involves changes in the content and organization of ideas and related symbols that accompany the social evolution of human societies.

In this light, the cultural evolution of psychopathology involves changes in the way incidents of psychopathology have been configured and handled as social and cultural objects during prehistory and recorded history. In seeking to study this phenomenon it is necessary to keep these two viewpoints analytically separated: The first, etic viewpoint is external and sets the analysis in motion. It presupposes what constitutes psychopathology in the first place—for example, anxiety or depression. The second, emic viewpoint is internal and consists of the labels, constructs, understandings, and ways of handling phenomena such as psychopathology—for example, emotional suffering or behavioral breakdowns. Cultural evolution of psychopathology implies that these meanings differ across types of societies as a function of their organization and culture. It involves comparing features of psychopathology as formulated in the external, analyst's model with those of a given society, the intracultural model of psychopathology—for example, of a tribe, chiefdom, or state.

Given these stipulations, the task of explaining the cultural evolution of psychopathology can be broken into two logical steps that can be formulated in the form of questions. First, what are the differences in the viewpoints or models of psychopathology that exist in the types of society that make up an evolutionary continuum? Second, how does one describe and hopefully explain such differences in viewpoints? Answers to these two questions outline what the cultural of evolution of psychopathology entails.

Characterizing the Cultural Evolution of Psychopathology

To study cultural evolution in a manner analogous to biological evolution, the idea of cultures as having component parts that are selected has been useful. One may define a meme (in analogy to gene) as a unit of cultural information in the pool of information that is culture: by definition, a meme is easily learned, imitated, culturally selected, spreads among actors of a society, that is, from brain to brain, and has high survival value (Dawkins 1976). Memes provide definitions, explanations, and instructions to actors about memorable objects, happenings, situations, and concepts of a culture and in that way shape behavior. A set of memes related in content and subject matter may be described as a memeplex (Blackmore 1999). Religion, moral conduct, law and order, social responsibility, knowledge about subsistence practices, and virtually any other area of institutional activity in a society including medicine operate and function in terms of memeplexes.

The culture of a particular society includes a memeplex, or a set of memeplexes, that can be denoted as PSP. It consists of items of cultural information pertaining to psychopathology, such as ideas, understandings, and ways of orienting and dealing with incidents of psychopathology. A culture's memeplex regarding sickness and disease, denoted as EMED—that is, ethnomedical, the cultural context of medicine—provides labels and understanding about sickness and disease. A society's PSP and EMED overlap in content. However, EMED and PSP are not totally equivalent in composition. Some of the memes comprising PSP are not found in EMED and vice versa. The degree of overlap in content of information between PSP and EMED constitutes one parameter that changes during the course of social and cultural evolution.

The creative, organizing, and essentially architectural character of memes with respect to behavior needs to be fully appreciated. In other words, memes provide the building blocks in terms of which individuals as symbolic creatures—that is, those occupying a symbolic niche—orient to the world, make sense of their situation, understand what happens to them, and realize how they behave in specific circumstances. This underscores the essential, social symbolic, character of human behavior. Thus, in the event of an episode of psychopathology or sickness, an individual victim draws on material from his or her culture, specifically, material of PSP and EMED, along with other memeplexes that the culture provides for anomalous conditions of the self, in order to understand, make sense of, and adopt a course of action with respect to the relevant incident.

In this regard, one may acknowledge that in an ultimate sense, just like human biology and genes constitute—that is, produce and shape—human behavior, they also constitute incidents of psychopathology and disease. However, from a social behavioral and symbolic standpoint, culture has an equal role in this causation of behavior as well as psychopathology and sickness, the social expression of disease. It is in terms of PSP and EMED that the individual makes sense of the incident. Memes provide the information in terms of which a victim labels, understands, makes sense of, in some way fashions, and expresses incidents of psychopathology and proceeds with a course of action..

Components of PSP and Cultural Evolution

Because incidents of psychopathology are public social phenomena, they command attention and elicit explanations. The package of cultural information members of a society have about the causes, connotations, consequences, and cures of psychopathology—that is, its PSP—enables understanding of and directs efforts to cope with incidents when they arise. To visualize the cultural evolution of psychopathology is to visualize how PSP of one type of society differs from that of another, and to explain what changes in PSP occurred, and why, between types of society located across an evolutionary continuum. To accomplish this, it is useful to break apart the material that makes up PSP.

The material of PSP describes the social behavioral properties and implications of an incident of psychopathology—for example, degree of social senselessness, somatic/body centeredness, social unacceptability—and these can be used to formulate a social theory about the evolution of psychopathology (Fábrega 1975, 1993b). From the standpoint of individuals, the material of PSP can be reduced to three types. First, units of information that code *harm to self* refer to manifestations of psychopathology that give rise to such things as psychological pain and suffering, bodily discomfort, cognitive incoherence, and an undermining of the victim's ability to function in society. Information that inform about these manifestations are regarded as the most basic attributes of PSP and are the ones that it shares with EMED. Second, *injury to others* refers to manifestations of psychopathology that are socially offensive, disruptive, alienating, coercive, divisive, or aggressive to others. In general, they cause disapproval, avoidance, reactive aggression, censure, punishment, attempts at arbitration and adjudication, banishment, and even execution. The difference in interpretation of injury to self compared to injury to others underscores the fact that manifestations of psychopathology—such as social and psychological behavior generally—are subject to labeling and interpretation. This logically requires information about how to make sense of behavior. Thus, almost by definition, in order to mobilize or draw on the symbolic resources and material of PSP requires bringing into play memes about what behavior means and how one should respond to it. The third category of information comprising PSP is *relative jurisdiction of social control institutions*. This category of information refers to general cultural knowledge that all members of a society learn regarding what thoughts, beliefs, and behavior mean; more specifically, their presumed causes, implications, sphere of relevance, and how they should be interpreted and dealt with given the worldview shared by members of a society.

Changes in the Content of PSP during Cultural Evolution

In all societies many of the units of the injury-to-self subset of information, in particular, those that most resemble aspects of emotional suffering, dissociation, bodily pain and dysfunction, cognitive incoherence, and breakdown of social role functioning, retain their salience in PSP. On the other hand, in more complex societies, new ways of conceptualizing injury to self evolves and items of information that represent this expanded conception of behavior are incorporated in PSP. For example, an individual's beliefs, if they deviate from conventional norms and assumptions, may come to be regarded as attributes of behavior that signal an individual's social breakdown; for example, a person's beliefs lose their referential character and come to be viewed as signs and symptoms of a person's thinking. This shift in meaning and interpretation is more likely to exist in a complex society where members live in settled communities, there is accumulated knowledge about behavior, such knowledge is specialized and is the province of experts, and there exists evolved social practices and institutions for dealing with behavioral problems. The same may be the case for manifestations of psychopathology involving excessive fear of objects, changes in ritual behavior—that is, avoidance or extravagance—a turn toward introversion, and/or persistent suspicion and distrust. In elementary societies, legal, religious, and/or moral labels may predominate in the interpretation of these manifestations and the corresponding nonmedical institutions are drawn in to negotiate the meaning of the behavior and its handling in light of social circumstances and the biography of the person, and, of course, the political economic organization of the society. For example, beliefs in spirits and demons may connote access to the supernatural since (1) they are accorded reality, (2) excessive religiosity may be seen as a virtue, (3) phobic fears may be viewed as a conservative outlook and judicious thinking, ritualistic behavior as saintly, and persistent antisociality as morally repugnant and criminal. Yet these social and psychological behaviors can be hallmarks of psychopathology in the external, analyst's model. It is likely that as systems of cultural knowledge are elaborated in relation to social evolution, the behaviors come to reflect on the person's normality compared to abnormality.

It is assumed that during cultural evolution many behaviors of individuals, previously thought about and handled in terms of their sheer negative effects on or relevance to others, come to be seen as examples that also reflect personal incompetence or deviance, that is, as reflecting injury of self. Similarly, there also takes place a contraction of or shift in the injury-to-others dimension. Associated with this, there comes into play tension and contestation in the way cultural information about normality and abnormality is applied and about the appropriate jurisdiction of formal and informal social institutions in the conferral of meaning and way of handling behavior and the person showing it. In traditional ancient but complex societies, medical knowledge, including knowledge about normal and abnormal behavior, expands, and the workings of the medical institution comes into tension with the workings of nonmedical institutions. The model of psychopathology embodied in the categories of information of PSP shift in composition and content. With the advent of biomedical psychiatry and its diffusion across the globe through international systems of diagnosis, and as part of the momentum of capitalist industrialism, changes in the composition and content of PSP are magnified. More specifically, the spread, imposition, and colonization that biomedical psychiatry produces creates obvious tensions, since the content, assumptions, and implications of the biomedical version of PSP differs and may conflict if not undermine the systems of knowledge and institutions that provide meaning and rationale for local, cultural versions of PSP in the areas, regions, or nations to which it spreads.

References

Aggleton, J. P. 1993. The contribution of the amygdala to normal and abnormal emotional states. *Trends in Neuroscience* 16 (8) [182]: 328–333.

Aiello, L. C. 1996a. Hominine preadaptations for language and cognition. In *Modelling the early human mind*, ed. P. Mellars and K. Gibson, 89–102. Cambridge, U.K.: McDonald Institute for Archaeological Research.

———. 1996b. Terrestriality, bipedalism and the origin of language. In *Evolution of Social Behaviour Patterns in Primates and Man. Proceedings of the British Academy*, ed. W. G. Runciman, J. M. Smith, and R. I. M. Dunbar, 88:269–289. New York: British Academy/Oxford University Press.

Aiello, L.C., and R. I. M. Dunbar. 1993. Neocortex size, group size, and the evolution of language. *Current Anthropology* 34 (2): 184–193.

Aitchison, J. 1996. *The seeds of speech: Language origin and evolution.* Cambridge, U.K.: Cambridge University Press.

Alexander, R. D. 1979. *Darwinism and human affairs.* Seattle: University of Washington Press.

———. 1987a. *The biology of moral systems.* New York: Aldine De Gruyter.

———. 1987b. The evolutionary approach to human behavior: What does the future hold? In *Human reproductive behavior: A Darwinian perspective*, ed. L. L. Betzig, M. Borgerhoff Mulder, and P. W. Turke, 317–341. Cambridge, U.K.: Cambridge University Press.

———. 1989. Evolution of the human psyche. In *The human revolution: Behavioural and biological perspectives on the origins of modern humans*, ed. P. Mellars and C. Stringer, 455–513. Princeton, N.J.: Princeton University Press.

———. 1990. How did humans evolve? Reflections on the uniquely unique species. Museum of Zoology, University of Michigan, Special Publications No. 1, Ann Arbor, Mich.

Andrews, M. W., and L. A. Rosenblum. 1988. Relationship between foraging and affiliative social referencing in primates. In *Ecology and behavior of food-enhanced primate group*, ed. J. E. Fa and C. H. Southwick, 247–268. New York: Alan R. Liss.

Arling, G. L., and H. F. Harlow. 1967. Effects of social deprivation on maternal behavior of rhesus monkeys. *Journal of Comparative and Physiological Psychology* 64 (3): 371–377.

Armstrong, E. 1991. The limbic system and culture: An allometric analysis of the neocortex and limbic nuclei. *Human Nature* 2 (2): 117–136.

Armstrong, E., M. R. Clarke, and E. M. Hill. 1987. Relative size of the anterior thalamic nuclei differentiates anthropoids by social system. *Brain Behavior and Evolution* 30 (5–6): 263–271.

Atkinson, L., and K. J. Zucker. 1997. *Attachment and psychopathology*. New York: Guilford Press.

Bailey, J. M. 1997. Are genetically based individual differences compatible with species-wide adaptations? In *Uniting psychology and biology: Integrative perspectives on human development*, ed. N. L. Segal, G. E. Weisfield, and C. C. Weisfield, 81–100. Washington, D.C.: American Psychological Association.

———. 1998. Can behavior genetics contribute to evolutionary behavioral science? In *Handbook of evolutionary psychology: Ideas, issues and applications*, C. Crawford and D. L. Krebs, 211–233. Mahwah, N.J.: Lawrence Erlbaum Associates.

Bailey, K. G. 1987. *Human paleopsychology: Applications to aggression and pathological processes*. Hillsdale, N.J.: Lawrence Erlbaum Associates.

Baldwin, J. M., H. F. Osborn, C. L. Morgan, E. B. Poulton, F. W. Headley, and H. W. Conn. 1902. *Development and evolution, including psychophysical evolution, evolution by orthoplasy, and the theory of genetic modes*. New York: Macmillan.

Barkley, R. A. 1997. Behavioral inhibition, sustained attention, and executive functions: Constructing a unifying theory of ADHD. *Psychological Bulletin* 121 (1): 65–94.

Barkow, J. H. 1989. *Darwin, sex and status: Biological approaches to mind and culture*. Toronto: University of Toronto Press.

Barkow, J. H., L. Cosmides, and J. Tooby, eds. 1992. *The adapted mind: Evolutionary psychology and the generation of culture*. New York and Oxford: Oxford University Press.

Bateson, P. 1988. The Active Role of Behaviour in Evolution. In *Evolutionary processes and metaphors*, ed. M.-W. Ho and S. W. Fox, 191–207. New York: John Wiley and Sons.

Bateson, P. P. G., and R. A. Hinde, eds. 1976. *Growing points in ethology: Based on a conference sponsored by St. John's College and King's College, Cambridge*. Cambridge, U.K.: Cambridge University Press.

Belsky, J. 1997. Attachment, mating, and parenting: An evolutionary interpretation. *Human Nature* 8 (4): 361–381.

Belsky, J., L. Steinberg, and P. Draper. 1991. Childhood experience, interpersonal development, and reproductive strategy: An evolutionary theory of socialization. *Child Development* 62 (14): 647–670.

Berkson, G. 1970. Defective infants in a feral monkey group. *Folia Primatologica* 12 (4): 284–289.

———. 1973. Social responses to abnormal infant monkeys. *American Journal of Physical Anthropology* 38 (2): 583–586.

———. 1974. Social responses of animals to infants with defects. In *The effect of the infant on its caregiver*, ed. M. Lewis and L. A. Rosenblum, 233–250. New York: John Wiley.

———. 1976. Rejection of abnormal strangers from macaque monkey groups. *Journal of Abnormal Psychology* 86 (6): 659–661.

———. 1977. The social ecology of defects in primates. In *Primate bio-social development: Biological, social and ecological determinants*, ed. S. Chevalier-Skolnikoff and F. E. Poirier, 189–204. New York: Garland Pub.

Berkson, G., and J. D. Becker. 1975b. Facial expressions and social responsiveness of blind monkeys. *Journal of Abnormal Psychology* 84 (5): 519–523.

Berrios, G. E. 1996. *The history of mental symptoms: Descriptive psychopathology since the nineteenth century*. Cambridge, U.K.: Cambridge University Press.

Berrios, G. E., and R. Porter, eds. 1995. *A history of clinical psychiatry: The origin and history of psychiatric disorders*. London: Athlone.

Betzig, L., ed. 1997. *Human nature: A critical reader.* New York: Oxford University Press.

Bickerton, D. 1981. *Roots of language.* Ann Arbor, Mich.: Karoma.

———. 1990. *Language and species.* Chicago: University of Chicago Press.

———. 1995. *Language and human behavior.* Seattle: University of Washington Press.

Binford, L. R. 1985. Human ancestors: Changing views of their behavior. *Journal of Anthropological Archaeology* 4:292–327.

Blackmore, S. 1999. *The meme machine.* New York: Oxford University Press.

Blanchard, R. J., and D. C. Blanchard. 1990. Anti-predator defense as models of animal fear and anxiety. In *Fear and defence,* ed. P. F. Brain, S. Parmigiani, R. J. Blanchard, and D. Mainardi. Ettore Majorana International Life Sciences Series, 8:89–108. Amsterdam, Netherlands: Harwood Academic Publishers.

Boesch, C. 1996. The emergence of cultures among wild chimpanzees. In *Evolution of social behavior patterns in primates and man. Proceedings of the British Academy,* ed. W. G. Runciman, J. M. Smith, and R. I. M. Dunbar, eds., 88:251–268. Oxford: Oxford University Press.

Bogin, B., and B. H. Smith. 1996. Evolution of the human life cycle. *American Journal of Human Biology* 8 (6): 703–716.

Borries, C. 1997. Infanticide in seasonally breeding multimale groups of Hanuman langurs (*Presbytis entellus*) in Ramnagar (South Nepal). *Behavioral Ecology and Sociobiology* 41 (3): 139–150.

Bourguignon, E. 1968. *A cross-cultural study of dissociational states.* Columbus: Ohio State University Research Foundation.

———. 1979. *Psychological anthropology: An introduction to human nature and cultural differences.* New York: Holt, Rinehart and Winston.

Bowlby, J. 1969. *Attachment and loss.* Vol. 1. London: Hogarth Press and Institute of Psychoanalysis.

———. 1973. *Attachment and loss.* Vol. 2. London: Hogarth Press; New York: Basic Books.

———. 1976. Human personality development and ethological light. In *Animal models in human psychobiology,* ed. G. Serban and A. Kling, 27–36. New York: Plenum Press.

Boyd, R., and P. J. Richerson. 1985. *Culture and the evolutionary process.* Chicago: University of Chicago Press.

Boyer, P. 1994. *The naturalness of religious ideas: A cognitive theory of religion.* Berkeley: University of California Press.

Brandon, R. N. 1990. *Adaptation and environment.* Princeton, N.J.: Princeton University Press.

———. 1996. *Concepts and methods in evolutionary biology.* Cambridge, U.K.: Cambridge University Press.

Bremner, J. D., and C. A. Marmar, eds. 1998. *Trauma, memory and dissociation.* Washington, D.C.: American Psychiatric Press.

Bremner, J. D., E. Vermetten, S. M. Southwick, J. H. Krystal, and D. S. Charney. 1998. Trauma, memory, and dissociation: An integrative formulation. In *Trauma, memory and dissociation,* ed. J. D. Bremner and C. R. Marmar, 365–402. Washington, D.C.: American Psychiatric Press.

Broom, D. M., and K. G. Johnson. 1993. *Stress and animal welfare.* New York: Chapman and Hall.

Brown, D. E. 1991. *Human universals.* Philadelphia: Temple University Press.

Buirski, P., et al. 1973. A field study of emotions, dominance, and social behavior in a group of baboons (*Pupio anubis*). *Primates* 14 (1): 67–78.

Buirski, P., and R. Plutchik. 1991. Measurement of deviant behavior in a Gombe chimpanzee: Relation to later behavior. *Primates* 32 (2): 207–211.

Burgess, R. L., and P. Draper. 1989. The explanation of family violence: The role of biological, behavioral, and cultural selection. In *Family violence, crime and justice: A review of research,* ed. L. Ohlin and M. Tonry, 11:59–116. Chicago: University of Chicago Press.

Buss, D. M. 1991. Evolutionary personality psychology. *Annual Review of Psychology* 42:459–491.

———. 1995a. Evolutionary psychology: A new paradigm for psychological science. *Psychological Inquiry* 6 (1): 1–30.

———. 1995b. The future of evolutionary psychology. *Psychological Inquiry* 6 (1): 81–87.

———. 1999. *Evolutionary psychology: The new science of the mind.* Boston: Allyn and Bacon.

Butler, L. D., R. E. F. Duran, P. Jasiukaitis, C. Koopman, D. Spiegel. 1996. Hypnotizability and traumatic experience: A diathesis-stress model of dissociative symptomatology. *American Journal of Psychiatry* 153 (7): 42–63. Festschrift supplement.

Byrne, R. W. 1995. *The thinking ape: Evolutionary origins of intelligence.* Oxford: Oxford University Press.

Byrne, R. W., and A. Whiten. 1988. *Machiavellian intelligence: Social expertise and the evolution of intellect in monkeys, apes and humans.* Oxford: Clarendon Press.

Caine, N. G., and M. Reite. 1983. Infant abuse in captive pig-tailed macaques: Relevance to human child abuse. In *Child abuse: The nonhuman primate data*, ed. M. Reite and N. G. Caine, 19–27. New York: Alan R. Liss.

Capitanio, J. P. 1986. Behavioral pathology. In *Comparative primate biology.* Vol. 2A, *Behavior, conservation, and ecology*, ed. G. Mitchell and J. Erwin, 411–454. New York: A. R. Liss.

Caporael, L. R. Sociality: Coordinating bodies, minds and groups. Unpublished paper.

Carey, G., and D. L. DiLalla. 1994. Personality and psychopathology: Genetic Perspectives. *Journal of Abnormal Psychology* 103 (1): 32–43.

Carruthers, P., and P. K. Smith, eds. 1996. *Theories of theories of mind.* Cambridge, U.K.: Cambridge University Press.

Carter, C. S., I. I. Lederhendler, and B. Kirkpatrick, eds. 1997. *The integrative neurobiology of affiliation.* New York: New York Academy of Sciences.

Castel, R. 1988. *The regulation of madness: The origins of incarceration in France.* Trans. W. D. Halls. Berkeley: University of California Press.

Cates, D. S., B. K. Houston, C. R. Vavak, and M. H. Crawford. 1993. Heritability of hostility-related emotions, attitudes, and behaviors. *Journal of Behavioral Medicine* 16 (3): 237–256.

Chagnon, N. A. and W. Irons, eds. 1979. *Evolutionary biology and human social behavior: An anthropological perspective.* North Scituate, Mass.: Duxbury Press.

Chance, M. R. A. 1988a. *Social fabrics of the mind.* East Sussex, U.K.: Lawrence Erlbaum Associates.

———. 1988b. A Systems Synthesis of Mentality. In *Social fabrics of the mind*, ed. M. R. A. Chance and D. R. Omark, 37–46. East Sussex, U.K.: Lawrence Erlbaum Associates.

Changeux, J. P. 1985. *Neuronal man: The biology of mind.* New York: Pantheon Books.

Chase, P. 1999. Symbolism as reference and symbolism as culture. In *The evolution of culture: An interdisciplinary view*, ed. R. I. M. Dunbar, C. Knight, and C. Power, 34–49. New Brunswick, N.J.: Rutgers University Press.

Chase, P. G., and H. L. Dibble. 1987. Middle paleolithic symbolism: A review of current evidence and interpretations. *Journal of Anthropological Archaeology* 6:263–296.

———. 1992. Scientific archaeology and the origins of symbolism: A reply to Bednarik. *Cambridge Archaeological Journal* 2 (1): 43–51.

Cheney, D. L., and R. M. Seyfarth. 1990. *How monkeys see the world: Inside the mind of another species.* Chicago: University of Chicago Press.

———. 1996. Function and intention in the calls of non-human primates. In *Evolution of social behaviour patterns in primates and man*, ed. W. G. Runciman, J. M. Smith, and R. I. M. Dunbar, 59–76. Oxford: Oxford University Press.

Chisholm, J. S. 1988. Toward a developmental evolutionary ecology of humans. In *Sociobiological perspectives on human development*, ed. K. B. MacDonald, 78–102. New York: Springer-Verlag.
———. 1996. The evolutionary ecology of attachment organization. *Human Nature* 7 (1): 1–38.
———. 1999. *Death, hope, and sex: Steps toward an evolutionary ecology of mind and morality*. Cambridge, U.K.: Cambridge University Press.
Clark, G. A. 1997. A Middle-Upper Paleolithic transition in Europe: An American perspective. *Norwegian Archeology Review* 30 (1): 25–53.
Clarke, A. S., and S. Boinski. 1995. Temperament in nonhuman primates. *American Journal of Primatology* 37 (2): 103–125.
Clarke, A. S., and M. L. Schneider. 1993. Prenatal stress has long-term effects on behavioral responses to stress in juvenile Rhesus monkeys. *Developmental Psychobiology* 26 (5): 293–304.
Clarke, A. S., D. J. Wittwer, D. H. Abbott, and M. L. Schneider. 1994. Long-term effects of prenatal stress on HPA axis activity in juvenile Rhesus monkeys. *Developmental Psychobiology* 27 (5): 257–269.
Cloninger, C. R. 1986. A unified biosocial theory of personality and its role in the development of anxiety states. *Psychiatric Developments* 4 (3): 167–226.
Cohen, J. E. 1993. Review of *The egalitarians—Human and chimpanzee: An anthropological view of social organization*, by M. Power. *American Scholar* 62:154–157.
Conrad, P., and J. W. Schneider, eds. 1980. *Deviance and medicalization*. St. Louis: Mosby.
Coplan, J. D., M. W. Andrews, L. A. Rosenblum, M. J. Owens, J. M. Gorman, and C. B. Nemeroff. 1996. Increased cerebrospinal fluid CRF concentrations in adult nonhuman primates previously exposed to adverse experiences as infants. *Proceedings of National Academy of Sciences USA* 93:1619–1623.
Coplan, J. D., D. Pine, L. A. Papp, L. A. Rosenblum, T. Cooper, and J. M. Gorman. 1995. Noradrenergic/HPA axis uncoupling in panic disorder. *Neuropsychopharmacology* 13 (1): 65–73.
Coplan, J. D., L. A. Rosenblum, S. Friedman, T. B. Bassoff, and J. M. Gorman. 1992. Behavioral effects of oral yohimbine in differentially reared nonhuman primates. *Neuropsychopharmacology* 6 (1): 31–37.
Coplan, J. D., L. A. Rosenblum, and J. M. Gorman. 1995. Primate models of anxiety: Longitudinal perspectives. *Psychiatric Clinics of North America* 18 (4): 727–743.
Cosmides, L. 1989. The logic of social exchange: Has natural selection shaped how humans reason? Studies with the Wason selection task. *Cognition* 31 (3): 187–276.
Cosmides, L., and J. Tooby. 1999. Toward an evolutionary taxonomy of treatable conditions. *Journal of Abnormal Psychology* 108 (3): 453–464.
Costa, P. T., Jr., and T. A. Widiger. 1994a. Introduction: Personality disorders and the five-factor model of personality. In *Personality disorders and the five-factor model of personality*, ed. P. T. Costa Jr. and T. A. Widiger, 1–18. Washington, D.C.: American Psychological Association.
———, eds. 1994b. *Personality disorders and the five-factor model of personality*. Washington, D.C.: American Psychological Association.
———. 1994c. Summary and unresolved issues. In *Personality disorders and the five-factor model of personality*, ed. P. T. Costa Jr., and T. A. Widiger, 319–327. Washington, D.C.: American Psychological Association.
Crawford, C. 1998a. Environments and adaptations: Then and now. In *Handbook of evolutionary psychology: Ideas, issues, and applications*, ed. C. Crawford and D. L. Krebs, 275–302. Mahwah, N.J.: Lawrence Erlbaum Associates.
———. 1998b. The theory of evolution in the study of human behavior: An introduction and overview. In *Handbook of evolutionary psychology: Ideas, issues, and applications*, ed. C. Crawford and D. L. Krebs, 3–41. Mahwah, N.J.: Lawrence Erlbaum Associates.

Crawford, C., and D. L. Krebs, eds. 1998. *Handbook of evolutionary psychology: Ideas, issues and applications*. Mahwah, N.J.: Lawrence Erlbaum.

Cromer, R. F. 1991. *Language and thought in normal and handicapped children*. Oxford: Basil Blackwell.

Crow, T. J. 1991a. The demise of the Kraepelinian binary system and the aetiological unity of the psychoses. In *Negative versus positive schizophrenia*, ed. A. Marneros, N. C. Andreasen, and M. T. Tsuang, 424–440. Berlin: Springer-Verlag.

———. 1991b. Origins of psychosis and "The Descent of Man." *British Journal of Psychiatry* 159 (suppl. 14): 76–82.

———. 1993a. Sexual selection, Machiavellian intelligence and the origins of psychosis. *Lancet* 342 (8871): 594–598.

———. 1993b. Origins of psychosis and the evolution of human language and communication. In *New generation of anti-psychotic drugs: Novel mechanisms of action*, ed. N. Brunello, J. Mendlewicz, and J. Racagni, 39–61. Basel, Switzerland: Karger.

———. 1995. A Darwinian approach to the origins of psychoses. (Review article.) *British Journal of Psychiatry* 167 (1): 12–25.

———. 1997. Aetiology of schizophrenia: An echo of the speciation event. *International Review of Psychiatry* 9 (4): 321–330.

Cummings, J. L. 1993. Frontal-subcortical circuits and human behavior. *Archives of Neurology* 50 (8): 873–880.

Curio, E. 1994. Causal and functional questions: How are they linked? *Animal Behaviour* 47 (5): 999–1021.

Daly, M., and M. Wilson. 1988. Evolutionary social psychology and family homicide. *Science* 242 (4878): 519–524.

Damasio, A. R. 1994. *Descartes' error: Emotion, reason, and the human brain*. New York: G. P. Putnam's Sons, a Grosset/Putnam Book.

———. 1999. *The feeling of what happens: Body and emotion in the making of consciousness*. New York: Harcourt Brace.

Davidson, I. 1991. The archeology of language origins—a review. *Antiquity* 65:39–48.

Dawkins, M. S. 1980. *Animal suffering: The science of animal welfare*. New York: Chapman and Hall.

———. 1990. From an animal's point of view: Motivation, fitness, and animal welfare. *Behavioral and Brain Sciences* 13 (1): 1–61.

———. 1993. *Through our eyes only? The search for animal consciousness*. Oxford: W. H. Freeman, Spektrum.

Dawkins, R. 1976. *The selfish gene*. New York: Oxford University Press.

Deacon, T. W. 1992a. Brain-language coevolution. In *The evolution of human languages*, ed. J. A. Hawkins and M. Gell-Mann. SFI Studies in the Sciences of Complexity, Proc. Vol. 10, 49–83.

———. 1992b. The neural circuitry underlying primate calls and human language. In *Language origin: A multidisciplinary approach*, ed. J. Wind et al., 121–161. The Netherlands: Kluwer Academic Publishers.

———. 1997. *The symbolic species: The co-evolution of language and the brain*. New York: W. W. Norton and Co.

Dennett, D. C. 1987. *The Intentional Stance*. Cambridge, Mass.: MIT Press.

———. 1995. *Darwin's dangerous idea: Evolution and the meanings of life*. New York: Simon and Schuster.

Devereux, G. 1956. Normal and abnormal: The key problem of psychiatric anthropology. In *Some uses of anthropology: Theoretical and applied*, ed. J. B. Casagrande and T. Gladwin. Washington, D.C.: Anthropological Society of Washington.

———. 1980. *Basic problems of ethnopsychiatry*. Trans. B. M. Gulati and G. Devereux. Chicago: University of Chicago Press.

de Waal, F. B. M. 1982. *Chimpanzee politics: Power and sex among apes*. Baltimore: Johns Hopkins University Press.

———. 1989. *Peacemaking among primates*. Cambridge, Mass.: Harvard University Press.
———. 1996. *Good natured: The origins of right and wrong in humans and other animals*. Cambridge, Mass.: Harvard University Press.
de Waal, F. B. M., and F. Aureli. 1997. Conflict resolution and distress alleviation in monkeys and apes. In *The integrative neurobiology of affiliation*, ed. C. Sue Carter, I. Izja Lederhendler, and Brian Kirkpatrick, 317–328. New York: New York Academy of Sciences.
de Waal, F. B. M., H. Uno, L. M. Lutrell, M. Lesleigh, L. F. Meisner, F. Lorraine, and L. A. Jeannotte. 1996. Behavioral retardation in a macaque with autosomal trisomy and aging mother. *American Journal on Mental Retardation* 100 (4): 378–390.
Diamond, J. 1997. *Guns, germs and steel: The fate of human societies*. New York: W. W. Norton.
Dicks, D., R. E. Myers, and A. Kling. 1968. Uncus and amygdala lesions: Effects on social behavior in the free-ranging rhesus monkey. *Science* 165 (888): 69–71.
Digman, J. M. 1994. Historical antecedents of the five-factor model. In *Personality disorders and the five-factor model of personality*, ed. P. T. Costa Jr. and T. A. Widiger, 13–18. Washington, D.C.: American Psychological Association.
Dobkin de Rios, M., and M. Winkelman. 1989. Shamanism and altered states of consciousness: An introduction. *Journal of Psychoactive Drugs* 21 (1): 1–7.
Donald, M. 1991. *Origins of the modern mind: Three stages in the evolution of culture and cognition*. Cambridge, Mass.: Harvard University Press.
———. 1993a. Human cognitive evolution: What we were, what we are becoming. *Social Research* 60 (1): 143–170.
———. 1993b. Precis of origins of the modern mind: Three stages in the evolution of culture and cognition. *Behavioral and Brain Sciences* 16 (4): 737–791.
———. 1994. Representation: Ontogeny and phylogenesis. *Behavioral and Brain Sciences* 17 (4): 714–715.
———. 1995. The neurobiology of human consciousness: An evolutionary approach. *Neuropsychologia* 33 (9): 1087–1102.
———. 1997. The mind from an historical perspective: Human cognitive phylogenesis and the possibility of continuing cognitive evolution. In *The future of the cognitive revolution*, ed. D. M. Johnson and C. E. Emeling, 1–20. New York: Oxford University Press.
———. 1999. Preconditions for the evolution of protolanguages. In *The descent of mind: Psychological perspectives on hominid evolution*, ed. M. C. Corballis, S. E. G. Lea et al., 138–154. New York: Oxford University Press.
Draper, P., and J. Belsky. 1990. Personality development in the evolutionary perspective. *Journal of Personality* 58 (1): 141–161.
Draper, P., and H. Harpending. 1982. Father absence and reproductive strategy: An evolutionary perspective. *Journal of Anthropological Research* 38 (3): 255–273.
———. 1987. Parent's investment and the child's environment. In *Parenting across the life span: Biosocial dimensions*, ed. J. B. Lancaster, 207–235. New York: A. de Gruyter.
———. 1988. A sociobiological perspective on the development of human reproductive strategies. In *Sociobiological perspectives on human development*, ed. K. B. MacDonald, 340–372. New York: Springer-Verlag.
Dunbar, R. I. M. 1988. *Primate social systems*. London: Croom Helm.
———. 1992. Neocortex size as a constraint on group size in primates. *Journal of Human Evolution* 20:469–493.
———. 1996a. Determinants of group size in primates: A general model. In *Evolution of social behaviour patterns in primates and man*, ed. W. G. Runciman, J. M. Smith and R. I. M. Dunbar, 33–57. New York: Oxford University Press.
———. 1996b. On the evolution of language and kinship. In *The archaeology of human ancestry: Power, sex and tradition*, ed. J. Steele and S. Shenman, 380–396. New York: Routledge.

———. 1996c. *Grooming, gossip, and the evolution of language.* Cambridge, Mass.: Harvard University Press.
Dunbar, R. I. M., C. Knight, and C. Power, eds. 1999. *The evolution of culture: An interdisciplinary view.* New Brunswick, N.J.: Rutgers University Press.
Durham, W. H. 1991. *Coevolution: Genes, culture and human diversity.* Stanford, Calif.: Stanford University Press.
Eaves, L. J., H. J. Eysenck, and N. G. Martin. 1989. *Genes, culture and personality: An empirical approach.* San Diego, Calif.: Academic Press.
Ekman, P., and R. J. Davidson, eds. 1994. *The nature of emotion: Fundamental questions.* New York: Oxford University Press.
Eliade, M. 1964. *Shamanism: Archaic techniques of ecstasy.* Trans. W. R. Trask. Princeton: Princeton University Press, Bollingen Series 76.
Ellinwood, E. H., Jr., and O. Duarte-Escalante. 1972. Chronic methamphetamine intoxication in three species of experimental animals. In *Current concepts of amphetamine-abuse: Proceedings of a workshop, Duke University Medical Center, June 5–6, 1970,* ed. E. H. Ellinwood and S. Cohen, 59–68. Washington, D.C.: GPO.
Ellison, G. D., and M. S. Eison. 1983. Continuous amphetamine intoxication: An animal model of the acute psychotic episode. *Psychological Medicine* 13 (4): 751–761.
Endler, J. A. 1986. *Natural selection in the wild.* Princeton, N.J.: Princeton University Press.
Epstein, A. W. 1987. The phylogenetics of fetishism. In *Variant sexuality: Research and theory,* ed. G. D. Wilson, 142–149. London: Croom Helm.
Erickson, P. A., with L. D. Murphy. 1998. *A history of anthropological theory.* Orchard Park, N.Y.: Broadview Press.
Erwin, J., and R. Deni. 1979. Strangers in a strange land: Abnormal behaviors or abnormal environments? In *Captivity and behavior: Primates in breeding colonies, laboratories and zoos,* ed. J. Erwin, T. Maple, and G. Mittchell, 1–29. New York: John Wiley and Sons.
Fábrega, H., Jr. 1974. *Disease and social behavior.* Cambridge, Mass.: MIT Press.
———. 1977. Culture, behavior and the nervous system. *Annual Review of Anthropology* 6:419–455.
———. 1979a. Phylogenetic precursors of psychiatric illness: A theoretical inquiry. *Comprehensive Psychiatry* 20 (3): 275–288.
———. 1979b. Neurobiology, culture and behavior disturbances: An integrated review. *Journal of Nervous and Mental Disease* 168 (8): 467–474.
———. 1981. Cultural programming of brain behavior relations. In *Brain behavior relationships,* ed. J. Merikangas, 1–63. Boston: Lexington Books.
———. 1987. Psychiatric diagnosis: A cultural perspective. *Journal of Nervous and Mental Disease* 175 (7): 383–394.
———. 1989a. Cultural relativism and psychiatric illness. *Journal of Nervous and Mental Disease* 177 (7): 415–425.
———. 1989b. The self and schizophrenia: A cultural perspective. *Schizophrenia Bulletin* 15 (2): 277–290.
———. 1989c. An ethnomedical perspective of Anglo-American psychiatry. *American Journal of Psychiatry* 146 (5): 588–596.
———. 1990a. A plea for a broader ethnomedicine. *Culture Medicine and Psychiatry* 14:129–132.
———. 1990b. Psychiatric stigma in the classical and medieval period. *Comprehensive Psychiatry* 31 (4): 289–306.
———. 1990c. The concept of somatization as a cultural and historical product of western medicine. *Psychological Medicine* 52 (6): 653–672.
———. 1991. The culture and history of psychiatric stigma in early modern and modern western societies. *Comprehensive Psychiatry* 32 (2): 97–119.
———. 1992a. The psychodynamic approach in American psychiatry: A case study in ethnomedicine. *History of Psychiatry* 3, pt. 4 (12): 457–472.

———. 1992b. The role of culture in a theory of psychiatric illness. *Social Science and Medicine* 35 (1): 91–103.

———. 1993a. A cultural analysis of human behavioral breakdowns: An approach to the ontology and epistemology of psychiatric phenomena. *Culture, Medicine and Psychiatry* 17 (1): 99–132.

———. 1993b. Towards a social theory of psychiatric phenomena. *Behavioral Sciences* 38:75–100.

———. 1994a. International systems of diagnosis in psychiatry. *Journal of Nervous and Mental Disease* 182 (5): 256–263.

———. 1994b. Personality disorders as medical entities: A cultural interpretation. *Journal of Personality Disorders* 8 (2): 149–167.

———. 1995. Cultural challenges to the psychiatric enterprise. *Comprehensive Psychiatry* 36 (5): 377–383.

———. 1996. Cultural and historical foundations of psychiatric diagnosis. In *Culture and psychiatric diagnosis: A DSM-IV perspective*, ed. J. E. Mezzich et al., 3–14. Washington, D.C.: American Psychiatric Press.

———. 1997. *Evolution of sickness and healing*. Berkeley: University of California Press.

Fedigan, L. M., and L. Fedigan. 1977. The social development of a handicapped infant in a free-living troop of Japanese monkeys. In *Primate bio-social development: Biological, social and ecological determinants*, ed. S. Chevalier-Skolnikoff and F. E. Poirier, 205–222. New York: Garland Publishing.

Fessler, D. M. T. 1995. *Towards an understanding of the universality of second order emotions*. Paper presented at the biannual meeting of the Society for Psychological Anthropology, San Diego, Calif.

Fisher, R. A. 1958. *The genetical theory of natural selection*. New York: Dover.

Fodor, J. A. 1983. *The modularity of mind: An essay on faculty psychology*. Cambridge, Mass.: MIT Press.

Foley, R. A. 1984a. Putting people into perspective: An introduction to community evolution and ecology. In *Hominid evolution and community ecology: Prehistoric human adaptation in biological perspective*, ed. R. Foley, 1–24. London: Academic Press.

———. 1984b. Early man and the red queen: Tropical African community evolution and hominid adaptation. In *Hominid evolution and community ecology: Prehistoric human adaptation in biological perspective*, ed. R. Foley. London: Academic Press.

———. 1987. *Another unique species: Patterns in human evolutionary ecology*. New York: John Wiley and Sons.

———. 1989. The evolution of hominid social behaviour. In *Comparative socioecology: The behavioural ecology of humans and other mammals*, ed. V. Standen and R. A. Foley, 473–494. Boston: Blackwell Scientific Publications.

———. 1991. How many species of hominid should there be? *Journal of Human Evolution* 20:413–427.

———. 1993. Causes and consequences in human evolution. *Journal of Royal Anthropological Institute*, n.s., 1:67–86.

———. 1995a. *Humans before humanity: An evolutionary perspective*. Cambridge, Mass.: Blackwell.

———. 1995b. The adaptive legacy of human evolution: A search for the environment of evolutionary adaptedness. *Evolutionary Anthropology* 4 (6): 194–203.

———. 1996. An evolutionary and chronological framework for human social behaviour. In *Evolution of social behaviour patterns in primates and man*, ed. W. G. Runciman, J. M. Smith, and R. I. M. Dunbar, 95–117. New York: Oxford University Press.

Foley, R. A., and P. C. Lee. 1989. Finite social space, evolutionary pathways, and reconstructing hominid behavior. *Science* 243 (4893): 901–906.

Folkman, S., and R. S. Lazarus. 1991. Coping and emotion. In *Stress and coping: An anthology*, 3d ed., ed. A. Monat and R. S. Lazarus, 207–227. New York: Columbia University Press.

Fox, R. C. 1988a. Medical evolution. In *Essays in medical sociology: Journeys into the field*, 499–531. New Brunswick: Transaction Books.

———. 1988b. The evolution of medical uncertainty. In *Essays in medical sociology: Journeys into the field*, 533–571. New Brunswick: Transaction Books.

Franzen, E. A., and R. E. Myers. 1973. Neural control of social behavior: Prefrontal and anterior temporal cortex. *Neuropsychologia* 11 (2): 141–157.

Friedman, S. 1994. Effects of oral yohimbine on the performance of a perceptual-motor task in nonhuman primates. *Journal of Anxiety Disorders* 8 (4): 301–310.

Fuster, J. M. 1980. *The prefrontal cortex: Anatomy, physiology, and neuropsychology of the frontal lobe*. New York: Raven Press.

Gallup, G. G. 1982. Self awareness and the emergence of mind in primates. *American Journal of Primatology* 2 (3): 237–248.

Gambill, J. D. 1981. The relevance of sociobiology for mental illness. *Perspectives in Biology and Medicine* 25 (1): 155–165.

Gamble, C. 1986. *The palaeolithic settlement of Europe*. Cambridge, U.K.: Cambridge University Press.

———. 1993. *Timewalkers: The prehistory of global colonization*. Cambridge, Mass.: Harvard University Press.

———. 1996. Making tracks: Hominid networks and the evolution of the social landscape. In *The archaeology of human ancestry: Power, sex and tradition*, ed. J. Steele and S. Shennan, 253–277. New York: Routledge.

Gardner, R. 1982. Mechanisms in manic-depressive disorder: An evolutionary model. *Archives of General Psychiatry* 39 (12): 1436–1441.

———. 1988. Psychiatric syndromes as infrastructure for intra-specific communication. In *Social fabrics of the mind*, ed. M. R. A. Chance and D. R. Omark, 197–225. East Sussex, England: Lawrence Erlbaum Associates.

———. 1995. Sociobiology and its application to psychiatry. In *The comprehensive textbook of psychiatry*, 6th ed., ed. H. I. Kaplan and B. J. Sadock, 365–375. Baltimore, Md.: Williams and Wilkins.

Geary, D. C. 1998. Functional organization of the human mind: Implications for behavioral genetics research. *Human Biology* 70 (2): 185–198.

Geertz, C. 1973. *The interpretation of cultures: Selected essays*. New York: Basic Books.

Geist, V. 1978. *Life strategies, human evolution, environmental design: Toward a biological theory of health*. New York: Springer-Verlag.

Geyer, M. A., and D. L. Braff. 1987. Startle habituation and sensorimotor gating in schizophrenia and related animal models. *Schizophrenia Bulletin* 13 (4): 643–668.

Ghiglieri, M. P. 1987. Sociobiology of the great apes and the hominid ancestor. *Journal of Human Evolution* 16:319–357.

Giancola, P. R., A. C. Mezzich, and R. E. Tarter. 1998a. Disruptive, delinquent and aggressive behavior in female adolescents with a psychoactive substance use disorder: Relation to executive cognitive functioning. *Journal of Studies on Alcohol* 59 (5): 560–567.

———. 1998b. Executive cognitive functioning, temperament, and antisocial behavior in conduct-disordered adolescent females. *Journal of Abnormal Psychology* 107 (4): 629–641.

Gibson, K. R. 1996. The biocultural human brain, seasonal migrations, and the emergence of the Upper Paleolithic. In *Modelling the early human mind*, ed. P. Mellars and K. Gibson, 33–48. Cambridge, U.K.: McDonald Institute for Archaeological Research.

Gibson, K. R., and T. Ingold, eds. 1993. *Tools, language and cognition in human evolution*. Cambridge, U.K.: Cambridge University Press.

Gilbert, P. 1992a. *Depression: The evolution of powerlessness*. New York: Guilford Press.

———. 1992b. *Human nature and suffering*. New York: Guilford Press.

———. 1998. Evolutionary psychopathology: Why isn't the mind designed better than it is? *British Journal of Medical Psychology* 71 (4): 353–373.

Goffman, E. 1959. *The presentation of self in everyday life*. New York: Doubleday.
———. 1963a. *Stigma: Notes on the management of spoiled identity*. Englewood Cliffs, N.J.: Prentice-Hall.
———. 1963b. *Behavior in public places: Notes on the social organization of gatherings*. New York: Free Press.
Goldstein, J. 1987. *Console and classify: The French psychiatric profession in the nineteenth century*. Cambridge, U.K.: Cambridge University Press.
Goodall, J. 1971. *In the shadow of man*. Boston: Houghton Mifflin.
———. 1977. Infant killing and cannibalism in free-living chimpanzees. *Folia Primatologica* 28 (4): 259–289.
———. 1986. *The chimpanzees of Gombe: Patterns of behavior*. Cambridge, Mass.: Harvard University Press, Belknap Press.
———. 1988. *In the shadow of man*. Rev. ed. Boston: Houghton Mifflin.
———. 1990. *Through a window: My thirty years with the chimpanzees of Gombe*. Boston: Houghton Mifflin.
Goodenough, O. R. 1995. Mind viruses: Culture, evolution and the puzzle of altruism. *Social Science Information* 34 (2): 287–320.
Goodenough, W. H. 1990. Evolution of the human capacity for beliefs. *American Anthropologist* 92:597–612.
Goodman, F. D. 1972. *Speaking in tongues: A cross-cultural study of glossolalia*. Chicago: University of Chicago Press.
———. 1988. *Ecstasy, ritual and alternate reality: Religion in a pluralistic world*. Bloomington: Indiana University Press.
———. 1990. *Where spirits ride the wind: Trance journeys and other ecstatic experiences*. Bloomington: Indiana University Press.
Goodwin, B. 1985. Constructional biology. In *Evolution and developmental psychology*, ed. G. Butterworth, J. Rutkowska, and M. Scaife, 45–66. New York: St. Martin's Press.
Goosen, C. 1981. Abnormal behavior patterns in Rhesus monkeys: Symptoms of mental disease? *Biological Psychiatry* 16 (8): 697–716.
Gottesman, I. I., and D. L. Wolfgram. 1991. *Schizophrenia genesis: The origins of madness*. New York: W. H. Freeman and Co.
Gottlieb, G. 1992. *Individual development and evolution*. New York: Oxford University Press.
———. 1995. Some conceptual deficiencies in developmental behavior genetics. *Human Development* 38 (3): 131–141.
———. 1997. *Synthesizing nature-nurture: Prenatal roots of instinctive behavior*. Mahwah, N.J.: Lawrence Erlbaum Associates.
Gould, S. J. 1991. Exaptation: A crucial tool for an evolutionary psychology. *Journal of Social Issues* 47 (3): 43–65.
Gould, S. J., and R. C. Lewontin. 1979. The spandrels of San Marco and the panglossian paradigm: A critique of the adaptationist programme. *Proceedings of the Royal Society of London-Series B: Biological Sciences*, B205 (1161): 581–598.
Gould, S. J., and E. S. Vrba. 1982. Exaptation—A missing term in the science of form. *Paleobiology* 8:4–15.
Gray, J. A. 1987. *The psychology of fear and stress*. 2d ed. Cambridge, U.K.: Cambridge University Press.
Griffin, D. R. 1992. *Animal minds*. Chicago: University of Chicago Press.
Guthrie, S. E. 1993. *Faces in the clouds: A new theory of religion*. New York: Oxford University Press.
Hallowell, A. I. 1960. Self, society, and culture in phylogenetic. In *The evolution of man: Mind, culture, and society*. Vol. 2, *Evolution after Darwin*, ed. S. Tax, 309–371. Chicago: University of Chicago Press.
Hamburg, D. A. 1975. Ancient man in the twentieth century. In *The quest for man*, ed. V. M. Goodall and B. Bender, 27–54. London: Phaidon.

Hamilton, W. D. 1964. The genetical evolution of social behaviour. I, II. *Journal of Theoretical Biology* 7 (1): 1–52.
Harlow, H. F., and M. A. Novak. 1973. Psychopathological perspectives. *Perspectives in Biology and Medicine* 16 (3): 461–478.
Harlow, H. F., and C. Mears. 1979. *The human model: Primate perspectives.* Washington, D.C.: V. H. Winston and Sons.
Harpending, H., and P. Draper. 1988. Antisocial behavior and the other side of cultural evolution. In *Biological contributions to crime causation*, ed. T. E. Moffitt and S. A. Mednick, 293–307. Dordrecht: Martinus Nighoff.
Harris, J. C. 1989. Experimental animal modeling of depression and anxiety. *Psychiatric Clinics of North America* 12 (4): 815–836.
Harrison, G. A., ed. 1993. *Human adaptation.* New York: Oxford University Press.
Hasegawa, T., and M. Hiraiwa. 1980. Social interactions of orphans observed in a free-ranging troop of Japanese monkeys. *Folia Primatologica* 33 (1–2): 129–158.
Hayden, B. 1993. The cultural capacities of Neanderthals: A review and re-evaluation. *Journal of Human Evolution* 24:113–146.
———. In press. *The prehistory of religion.* Washington, D.C.: Smithsonian Institution Press.
Healy, D. 1987. The comparative psychopathology of affective disorders in animals and humans. *Journal of Psychopharmacology* 1 (3): 193–210.
Heilman, K. M., R. T. Watson, and D. Bowers. 1983. Affective disorders associated with hemispheric disease. In *Neuropsychology of human emotion*, ed. K. M. Heilman and P. Satz, 45–64. New York: Guilford Press.
Hénaff, M. 1998. *Claude Lévi-Strauss and the making of structural anthropology.* Trans. M. Baker. Minneapolis: University of Minnesota Press.
Heyes, C. M. 1998. Theory of mind in nonhuman primates. *Behavioral and Brain Sciences* 21 (1): 101–134.
Heyes, C. M., and B. G. Galef Jr., eds. 1996. *Social learning in animals: The roots of culture.* San Diego, Calif.: Academic Press.
Higley, J. D., M. Linnoila, and S. J. Suomi. 1994. Ethological contributions: Introduction and rationale for using nonhuman primates to understand psychiatric disorders. In *Handbook of aggressive and destructive behavior in psychiatric patients*, ed. M. Hersen, R. T. Ammerman, and L. A. Sisson, 17–32. New York: Plenum Press.
Hilgard, E. R. 1986. *Divided consciousness: Multiple controls in human thought and action.* Rev. ed. New York: Wiley.
Hinde, R. A., ed. 1972. *Non-verbal communication.* New York: Cambridge University Press.
———. 1974. *Biological bases of human social behaviour.* New York: McGraw-Hill.
———. 1983. *Primate social relationships: An integrated approach.* Oxford: Blackwell Scientific.
———. 1987. *Individuals, relationships and culture: Links between ethology and the social sciences.* New York: Cambridge University Press.
Hiraiwa, M. 1981. Maternal and alloparental care in a troop of free-ranging Japanese monkeys. *Primates* 22 (3): 309–329.
Hirschfeld, L. A. 1996. *Race in the making: Cognition, culture, and the child's construction of human kinds.* Cambridge, Mass.: MIT Press.
Hockett, C. F., and R. Archer. 1964. The human revolution. *Current Anthropology* 5:135–168.
Hodgson, G. M. 1993. *Economics and evolution: Bringing life back into economics.* Ann Arbor: University of Michigan Press.
Hofer, M. A. 1984. Relationships as regulators. *Psychomatic Medicine* 43 (3): 183–197.
———. 1987. Early social relationships: A psychobiologist's view. *Child Development* 58 (3): 633–647.
———. 1995. An evolutionary perspective on anxiety. In *Anxiety as symptom and signal*, ed. R. A. Glick and S. P. Roose, 17–38. New Haven: Hillsdale, N.J.: Analytic Press.

Holcomb, H. R., III. 1998. Testing evolutionary hypotheses. In *Handbook of evolutionary psychology: Ideas, issues and applications*, ed. C. B. Crawford and D. L. Krebs, 303–334. Mahwah, N.J.: Lawrence Erlbaum Press.

Holloway, R. L. 1981. The Indonesian Homo erectus brain endocasts revisited. *American Journal of Physical Anthropology* 55:43–58.

Hrdy, S. B. 1974. Male-female competition and infanticide among the lanqurs (*Presbytis entellus*) of Abu, Rajasthan. *Folia Primatologica* 22:19–58.

———. 1979. Infanticide among animals: A review, classification and examination of the implications for the reproductive strategies of females. *Ethology and Sociobiology* 1:13–40.

———. 1981. *The woman that never evolved*. Cambridge, Mass.: Harvard University Press.

Hull, D. L. 1988. Interactors versus vehicles. In *The role of behavior in evolution*, ed. H. C. Plotkin, 19–50. Cambridge, Mass.: MIT Press.

Humphrey, N. K. 1976. The social function of intellect. In *Growing points in ethology: Based on a conference sponsored by St. John's College and King's College*, ed. P. P. G. Batson and R. A. Hinde, 303–317. Cambridge, U.K.: Cambridge University Press.

Hyland, M. E 1990. A functional theory of illness. In *Recent trends in theoretical psychology. Vol.2, Proceedings of the third biennial conference of the International Society for Theoretical Psychology, April 17–21, 1989*, ed. W.J. Baker, R. Van Hezewijk, M. Hyland and S. Terwee, 423–429. New York: Springer-Verlag.

Ingold, T., ed. 1996. *Key debates in anthropology*. London and New York: Routledge.

Isaac, B., ed. 1989. *The archaeology of human origins: The papers of Glynn Isaac*. Cambridge, U.K.: Cambridge University Press.

Iversen, S. D. 1987. Is it possible to model psychotic states in animals? *Journal of Psychopharmacology* 1 (3): 154–176.

Jablensky, A. 1988. Methodological issues in psychiatric classification. *British Journal of Psychiatry* 152 (Suppl. 1): 15–20.

Jackendoff, R. S. 1987. *Consciousness and the computational mind*. Chapter 12. Cambridge, Mass.: MIT Press.

———. 1994. *Patterns in the mind: Language and human nature*. New York: Harper Collins, Basic Books.

Janicki, M. G., and D. L. Krebs. 1998. Evolutionary approaches to culture. In *Handbook of evolutionary psychology: Ideas, issues and answers*, ed. C. Crawford and D. L. Krebs, 163–207. Mahwah, N.J.: Lawrence Erlbaum Press.

Johnson, A. W., and T. Earle. 1991. *The evolution of human societies: From foraging group to agrarian state*. Stanford, Calif.: Stanford University Press.

Johnson, M. 1987. *The body in the mind: The bodily basis of meaning, imagination, and reason*. Chicago: University of Chicago Press.

Kalin, N. H., and L. K. Takahashi. 1991. Animal studies implicating a role of corticotropin-releasing hormone in mediating behavior associated with psychopathology. In *Central nervous system peptide mechanisms in stress and depression: Progress in psychiatry*, ed. S. C. Risch, No. 30, 53–72. Washington, D.C.: American Psychiatric Press.

Kandel, E. R. 1983. From metapsychology to molecular biology: Explorations into the nature of anxiety. *American Journal of Psychiatry* 140 (10): 1277–1293.

Kaplan, J. R. 1986. Psychological stress and behavior in nonhuman primates. In *Comparative primate biology. Vol. 2A, Behavior, conservation, and ecology*, ed. G. Mitchell and J. Erwin, 455–492. New York: Alan R. Liss.

Karmiloff-Smith, A. 1992. *Beyond modularity: A developmental perspective on cognitive science*. Cambridge, Mass.: MIT Press.

Keil, F. C. 1989. *Concepts, kinds, and cognitive development*. Cambridge, Mass.: MIT Press.

Kemper, T. D. 1992. *The egalitarians—Human and chimpanzee: An anthropological view of social organization*, by Margaret Power. *American Journal of Sociology* 97 (6): 1757–1759.

Kendon, A. 1991. Some considerations for a theory of language origins. *Man (N. S.)* 26:199–221.

Kennedy, J. S. 1992. *The new anthropomorphism.* Cambridge, U.K.: Cambridge University Press.

King, B. J. 1991. Social information transfer in monkeys, apes, and hominids. *Yearbook of Physical Anthropology* 34 (suppl. 13): 97–115.

King, B. H., C. DeAntonio, J. T. McCracken, S. R. Forness, V. Ackerland. 1994. Psychiatric consultation in severe and profound mental retardation. *American Journal of Psychiatry* 151 (12): 1802–1808.

King, J. E. 1999. Attitudes to animals: Views in animal welfare. In *Animal perspectives,* ed. F. Dolins, 101–113. Cambridge, U.K.: Cambridge University Press.

King, J. E., and A. J. Figueredo. 1997. The five-factor model plus dominance in chimpanzee personality. *Journal of Research in Personality* 31 (2): 257–271.

Kinzey, W. G., ed. 1987. *The evolution of human behavior: Primate models.* Albany: State University of New York Press.

Kirmayer, L. J., and A. Young. 1999. Culture and context in the evolutionary concept of mental disorder. *Journal of Abnormal Psychology* 108 (3): 446–452.

Kittrie, N. N. 1971. *The right to be different: Deviance and enforced therapy.* Baltimore: Johns Hopkins Press.

Klein, D. F. 1978. A proposed definition of mental illness. In R. Spitzer and D. F. Klein, eds., *Critical aspects in psychiatric diagnosis,* 41–71. New York: Raven Press.

———. 1999. Harmful dysfunction, disorder, disease, illness and evolution. *Journal of Abnormal Psychology* 108 (3): 421–429.

Klein, R. G. 1995. Anatomy, behavior and modern human origins. *Journal of World Prehistory* 9 (2): 167–198.

———. 1999. *The human career: Human biological and cultural origins.* 2d ed. Chicago: University of Chicago Press.

Kleinman, A. 1988. *Rethinking psychiatry: From cultural category to personal experience.* New York: Free Press.

Kling, A., J. Lancaster, and J. Benitone. 1970. Amygdalectomy in the free-ranging vervet (Cercopithecus Aethiops). *Journal of Psychiatry Research* 7 (3): 191–199.

Kling, A., and H. D. Steklis. 1976. A neural substrate for affiliative behavior in nonhuman primates. *Brain Behavior and Evolution* 13 (2–3): 216–238.

Knight, C. 1991. *Blood relations: Menstruation and the origins of culture.* New Haven: Yale University Press.

———. 1996. Darwinism and collective representations. In *The archaeology of human ancestry: Power, sex and tradition,* ed. J. Steele and S. Shennan, 331–346. New York: Routledge.

———. 1999. Sex and language as pretend play. In *The evolution of culture: An interdisciplinary view,* ed. R. I. M. Dunbar, C. Knight, and C. Power, 228–247. New Brunswick, N.J.: Rutgers University Press.

Knight, C., R. I. M. Dunbar, and C. Power. 1999. An evolutionary approach to human culture. In *The evolution of culture: An interdisciplinary view,* ed. R. I. M. Dunbar, C. Knight, and C. Power, 1–11. New Brunswick, N.J.: Rutgers University Press.

Knight, C., C. Power, and I. Watts. 1995. The human symbolic revolution: A Darwinian account. *Cambridge Archaeological Journal* 5 (1): 75–114.

Kraemer, G. W. 1985. The primate social environment, brain neurochemical changes and psychopathology. *Trends in Neurosciences* 8 (8): 339–340.

———. 1986. Causes of changes in brain noradrenaline systems and later effects on responses to social stressors in Rhesus monkeys: The cascade hypothesis. *CIBA Foundation Symposium* 123:216–233.

———. 1988. Speculations on the developmental neurobiology of protest and despair. In *An inquiry into schizophrenia and depression: Animal models of psychiatric disorders,* ed. P. Simon, P. Soubrié, and D. Widlöcher, 2:101–139. Basel, Switzerland: Karger.

———. 1992. A psychobiological theory of attachment. *Behavioral and Brain Sciences* 15 (3): 493–541.
Kuhn, S. L. 1995. *Mousterian lithic technology: An ecological perspective*. Princeton, N.J.: Princeton University Press.
Kummer, H. 1971. *Primate societies: Group techniques of ecological adaptation*. Chicago: Aldine Atherton.
———. 1995. *In quest of the sacred baboon: A scientist's journey*. Trans. M. A. Biederman-Thorson. Princeton, N.J.: Princeton University Press.
Lakoff, G. 1987. *Women, fire, and dangerous things: What categories reveal about the mind.* Chicago: University of Chicago Press.
Lane, L. W., and D. J. Luchins. 1988. Evolutionary approaches to psychiatry and problems of method. *Comprehensive Psychiatry* 29 (6): 598–603.
Laudan, L. 1977. *Progress and its problems: Toward a theory in scientific growth*. Berkeley: University of California Press.
Layton, R. 1997. *An introduction to theory in anthropology*. Cambridge, U.K.: Cambridge University Press.
Lazarus, R. S. 1991. *Emotion and adaptation*. New York: Oxford University Press.
Lee, P. C. 1992. Biology and behaviour in human evolution. *Cambridge Archaeological Journal* 1:207–226.
———. 1994. Social structure and evolution. In *Behaviour and evolution*, ed. P. J. B. Slater and T. R. Halliday, 266–303. Cambridge, U.K.: Cambridge University Press.
Lehman, A. 1986. Presidential address, 1985: Face the beast and fear the face: Animal and social fears as prototypes for evolutionary analyses of emotion. *Psychophysiology* 23 (2): 123–145.
———. 1987. The psychophysiology of emotion: An evolutionary-cognitive perspective. *Advances in Psychophysiology* 2:79–127.
Lemert, E. M. 1951. *Social pathology: A systematic approach to the theory of sociopathic behavior*. New York: McGraw-Hill.
———. 1967. *Human deviance, social problems, and social control*. Englewood Cliffs, N.J.: Prentice-Hall.
Lessells, C. M. 1991. The evolution of life histories. In *Behavioral ecology: An evolutionary approach*, 3d ed., ed. J. R. Krebs and N. B. Davies, 32–65. Oxford, England: Blackwell.
Levin, H. S., H. M. Eisenberg, and A. L. Benton, eds. 1991. *Frontal lobe function and dysfunction*. New York: Oxford University Press.
Lieberman, P. 1991. *Uniquely human: The evolution of speech, thought, and selfless behavior*. Cambridge, Mass.: Harvard University Press.
Lilienfeld, S. O., J. Gershon, M. Duke, L. Marino, and F. B. M. de Waal. 1999. A preliminary investigation of the construct of psychopathic personality (psychopathy) in chimpanzees (pan troglodytes). *Journal of Comparative Psychology* 113 (4): 365–375.
Lilienfeld, S. O., and L. Marino. 1995. Mental disorder as a Roschian concept: A critique of Wakefield's "harmful disfunction" analysis. *Journal of Abnormal Psychology* 104 (3): 411–420.
Ludwig, A. M. 1968. Altered states of consciousness. In *Trance and possession states*, ed. R. Prince, 69–95. Montreal: R. M. Buch Memorial Society.
———. 1983. The psychobiological functions of dissociation. *American Journal of Clinical Hypnosis* 26 (2): 93–99.
Lumsden, C. J., and E. O. Wilson. 1981. *Genes, minds and culture*. Cambridge, Mass.: Harvard University Press.
Luriia, A. R. 1980. *Higher cortical functions in man*. New York: Basic Books.
MacDonald, K. B. 1988a. *Social and personality development: An evolutionary synthesis*. New York: Plenum Press.
———. 1988b. *Sociobiological perspectives on human development*. New York: Springer-Verlag.

———. 1991. A perspective on Darwinian psychology: The importance of domain-general mechanisms, plasticity, and individual differences. *Ethnology and Sociobiology* 12:449–480.

———. 1995. Evolution, the five-factor model, and levels of personality. *Journal of Personality* 63 (3): 525–567.

MacLean, P. D. 1970. The triune brain, emotion, and scientific bias. In *The neurosciences: Second study program*, ed. F. O. Schmitt, 336–349. New York: Rockefeller University Press.

———. 1973. A triune concept of the brain and behavior. In *Clarence M. Hincks Memorial Lectures*, ed. T. Boag and D. Campbell, 6–66. Toronto: University of Toronto Press.

———. 1982a. A triangular brief on the evolution of brain and law. *Journal of Social and Biological Structures* 5 (4): 369–379.

———. 1982b. On the origin and progressive evolution of the triune brain. In *Primate brain evolution: Methods and concepts*, ed. E. Armstrong and D. Falk, 291–316. New York: Plenum Press.

Maestripieri, D. 1994. Infant abuse associated with psychosocial stress in a group living pigtail macaque (Macaca nemestrina) mother. *American Journal of Primatology* 32 (1): 41–49.

———. 1998. Parenting styles of abusive mothers in group-living rhesus macaques. *Animal Behavior* 55 (1): 1–11.

———. 1999. Fatal attraction: Interest in infants and infant abuse in rhesus macaques. *American Journal of Physical Anthropology* 110 (1): 17–25.

Main, M., and H. Morgan. 1996. Disorganization and disorientation in infant strange situation behavior: Phenotypic resemblances to dissociative states. In *Handbook of dissociation: Theoretical, empirical and clinical perspectives*, ed. L. Michaelson and W. J. Ray, 107–138. New York: Plenum Press.

Marks, I. M. 1987. *Fears, phobias, and rituals: Panic, anxiety, and their disorders*. New York: Oxford University Press.

Marks, I. M., and R. M. Nesse. 1994. Fear and fitness: An evolutionary analysis of anxiety disorders. *Ethology and Sociobiology* 15:247–261.

Maryanski, A. R. 1993. The elementary forms of the first protohuman society: An ecological/social network approach. *Advances in Human Ecology* 2:215–241.

Matza, D. 1969. *Becoming deviant*. Englewood Cliffs, N.J.: Prentice-Hall.

Mayr, E. 1963. *Animal species and evolution*. Cambridge, Mass.: Harvard University Press.

———. 1982. *The growth of biological thought: Diversity, evolution and inheritance*. Cambridge, Mass.: Harvard University Press.

———. 1988. *Toward a new philosophy of biology: Observations of an evolutionist*. Cambridge, Mass.: Harvard University Press.

———. 1991. *One long argument*. Cambridge, Mass.: Harvard University Press.

McFarland, D. J. 1977. Decision making in animals. *Nature* 269 (5623): 15–21.

———. 1989. *Problems of animal behaviour*. London: Longman.

———. 1993. *Animal behaviour: Psychobiology, ethology, and evolution*. 2d ed. New York: John Wiley and Sons.

McGrew, W. C. 1991. Review of *The egalitarians—human and chimpanzee: An anthropological view of social organization*, by Margaret Power. *Nature* 354 (6351): 324–330.

———. 1992. *Chimpanzee material culture: Implications for human evolution*. Cambridge, U.K.: Cambridge University Press.

McGrew, W. C., L. F. Marchant, and T. Nishida, eds. 1996. *Great ape societies*. Cambridge, U.K.: Cambridge University Press.

McGuire, M. T., S. M. Essock-Vitale, and R. H. Polsky. 1981. Psychiatric disorders in the context of evolutionary biology: An ethological model of behavioral changes associated with psychiatric disorders. *Journal of Nervous and Mental Disease* 169 (11): 687–704.

McGuire, M. T., I. Marks, R. M. Nesse, and A. Troisi. 1992. Evolutionary biology: A basic science for psychiatry? *Acta Psychiatrica Scandinavica* 86 (2): 89–96.
McGuire, M. T., and A. Troisi. 1987. Physiological regulation-disregulation and psychiatric disorders. *Ethology and Sociobiology* 8:9S–12S.
———. 1998. *Darwinian psychiatry*. New York: Oxford University Press.
McKinney, W. T., and W. E. Bunney. 1969. Animal models of depression: I. Review of evidence: Implications for research. *Archives of General Psychiatry* 21 (2): 240–248.
McNeill, D. 1987. *Psycholinguistics: A new approach*. New York: Harper and Row.
Mealey, L. 1995. The sociobiology of sociopathy: An integrated evolutionary model. *Behavioral and Brain Sciences* 18 (3): 523–599.
Mehmet, H., and A. D. Edwards. 1996. Hypoxia, ischaemia, and apoptosis. *Archives of Disease in Childhood Fetal and Neonatal Edition* 75 (2): F73–75.
Mellars, P. 1996a. The emergence of biologically modern populations in Europe: A social and cognitive "revolution"? In *Evolution of social behaviour patterns in primates and man*, ed. W. G. Runciman, J. M. Smith, and R. I. M. Dunbar, *Proceedings of the British Academy* 88:179–202. New York: British Academy/Oxford University Press.
———. 1996b. Symbolism, language, and the Neanderthal mind. In *Modelling the early human mind*, ed. P. Mellars and K. Gibson, 15–32. Cambridge, U.K.: McDonald Institute for Archaeological Research.
———. 1996c. *The Neanderthal legacy: An archaeological perspective from western Europe*. Princeton: Princeton University Press.
Mellars, P., and K. Gibson, eds. 1996. *Modelling the early human mind*. Cambridge, U.K.: McDonald Institute for Archaeological Research.
Mellars, P., and C. Stringer, eds. 1989. *The human revolution: Behavioural and biological perspectives on the origins of modern humans*. Princeton, N.J.: Princeton University Press.
Mezzich, J. E., H. Fábrega Jr., G. A. Coffman, and R. Haley. 1989. DSM III disorders in a large sample of psychiatric patients: Frequency and specificity of diagnoses. *American Journal of Psychiatry* 146 (2): 212–219.
Mezzich, J. E., A. Kleinman, H. Fábrega Jr., and D. L. Parron, eds. 1996. *Culture and psychiatric diagnosis: A DSM-IV perspective*. Washington, D.C.: American Psychiatric Press.
Miller, G. F. 1998. How mate choice shaped human nature: A review of sexual selection and human evolution. In *Handbook of evolutionary psychology: Ideas, issues and applications*, ed. C. Crawford and D. L. Krebs, 87–129. Mahwah, N.J.: Lawrence Erlbaum Associates.
Mitchell, G., and A. S. Clarke. 1984. Contributions of behavioral primatology to veterinary science and comparative medicine. *Advances in Veterinary Science and Comparative Medicine* 28:25–50.
Mithen, S. J. 1996a. *The prehistory of the mind: The cognitive origins of art, religion and science*. New York: Thames and Hudson.
———. 1996b. Social learning and cultural tradition: Interpreting early Paleolithic technology. In *The archaeology of human ancestry: Power, sex and tradition*, ed. J. Steele and S. Shennan, 207–229. New York: Routledge.
———. 1999. Symbolism and the supernatural. In *The evolution of culture: An interdisciplinary view*, ed. R. I. M. Dunbar, C. Knight, and C. Power, 147–169. New Brunswick, N.J.: Rutgers University Press.
Moore, J. 1996. Savanna chimpanzees, referential models and the last common ancestor. In *Great Ape Societies*, ed. W. C. McGrew, L. F. Marchant, and T. Nishida, 275–292. Cambridge, U.K.: Cambridge University Press.
Morris, B. 1987. *Anthropological studies of religion: An introductory text*. Cambridge, U.K.: Cambridge University Press.
Murphy, J. 1976. Psychiatric labeling in cross cultural perspective. *Science* 191: 1019–1028.

Myers, R. E. 1972. Role of prefrontal and anterior temporal cortex in social behavior and affect in monkeys. *Acta Neurobiologiae Experimentalis* 32 (2): 567–579.

Nadler, R. D. 1980. Child abuse: Evidence from nonhuman primates. *Developmental Psychobiology* 13 (5): 507–512.

Nelson, K. 1996. *Language in cognitive development: The emergence of the mediated mind.* New York: Cambridge University Press.

Nesse, R. M. 1984. An evolutionary perspective on psychiatry. *Comprehensive Psychiatry* 25 (6): 575–580.

———. 1990. Evolutionary explanations of emotions. *Human Nature* 1 (3): 261–289.

———. 1998. Emotional disorders in evolutionary perspective. *British Journal of Medical Psychology* 71 (4): 397–415.

———. 2000. Is depression an adaptation? *Archives of General Psychiatry* 57 (January): 14–20.

Nesse, R. M., and A. T. Lloyd. 1992. The evolution of psychodynamic mechanisms. In *The adapted mind: Evolutionary psychology and the generation of culture*, ed. J. H. Barkow, L. Cosmides, and J. Tooby, 601–624. New York: Oxford University Press.

Nesse, R. M., and G. C. Williams. 1994. *Why we get sick: The new science of Darwinian medicine.* New York: Random House, Times Books.

Nijenhuis, E. R. S., P. Spinhoven, J. Vanderlinden, R. van Dyck, and O. van der Hart. 1998. Somatoform dissociative symptoms as related to animal defensive reactions to predatory imminence and injury. *Journal of Abnormal Psychology* 107 (1): 63–73.

Nijenhuis, E. R. S., J. Vanderlinden, and P. Spinhoven. 1998. Animal Defensive Reactions as a Model for Trauma-Induced Dissociative Reactions. *Journal of Traumatic Stress* 11 (2): 243–260.

Nishida, T., ed. 1990. *The chimpanzees of the Mahale mountains: Sexual and life history strategies.* Tokyo: University of Tokyo Press.

Nitecki, M. H., and D. V. Nitecki, eds. 1993. *Evolutionary ethics.* Albany: State University of New York Press.

Noble, W., and I. Davidson. 1996. *Human evolution, language and mind: A psychological and archaeological inquiry.* Cambridge, U.K.: Cambridge University Press.

Norikoshi, K. 1982. One observed case of cannibalism among wild chimpanzees of the Mahale mountains. *Primates* 23 (1): 66–74.

Novak, M. A., and S. J. Suomi. 1988. Psychological well-being of primates in captivity. *American Psychologist* 43 (10): 765–773.

Odling-Smee, F. J. 1983. Multiple levels in evolution: An approach to the nature-nurture issue via "Applied Epistemology." In *Animal models of human behavior: Conceptual, evolutionary, and neurobiological perspectives*, ed. G. C. L. Davey, 135–158. New York: John Wiley and Sons.

———. 1988. Niche constructing phenotypes. In *The role of behavior in evolution*, ed. H. C. Plotkin. Cambridge, Mass.: MIT Press.

———. 1994. Niche construction, evolution and culture. In *Companion encyclopedia of anthropology*, ed. T. Ingold, 162–196. New York: Routledge.

Öhman, A. 1986. Face the beast and fear the face: Animal and social fears as prototypes for evolutionary analyses of emotion. *Psychophysiology* 23 (2): 123–145.

———. 1987. The psychophysiology of emotion: An evolutionary-cognitive perspective. *Advances in Psychophysiology* 2: 79–127.

Oliverio, A., S. Cabib, and S. Puglisi-Allegra. 1992. Nonhuman behavioral models in the genetics of disturbed behavior. *Journal of Psychiatric Research* 26 (4): 367–382.

Oltmanns, T. F., and B. A. Maher, eds. 1988. *Delusional Beliefs.* New York: John Wiley and Sons.

Ortony, A., and T. J. Turner. 1990. What's basic about basic emotions? *Psychological Review* 97 (3): 315–331.

Otto, R. 1923. *The idea of the holy: An inquiry into the non-rational factor in the idea of the divine and its relation to the rational.* Oxford: Oxford University Press.

Oubré, A. Y. 1997. *Instinct and revelation: Reflections on the origins of numinous perception.* The Netherlands: Gordan and Breach Publishers.

Panksepp, J. 1991. Affective neuroscience: A conceptual framework for the neurobiological study of emotions. In *International review of studies on emotion,* ed. K. T. Strongman, 1:59–99. New York: John Wiley and Sons.

———. 1992. A critical role for "affective neuroscience" in resolving what is basic about basic emotions. *Psychological Review* 99 (3): 554–560.

———. 1995. The emotional brain and biological psychiatry. *Advances in Biological Psychiatry* 1:263–286.

Parker, S. T. 1996. Apprenticeship in tool-mediated extractive foraging: The origins of imitation, teaching, and self-awareness in great apes. In *Reaching into thought: The minds of the great apes,* ed. A. E. Russon, K. A. Bard, and S. T. Parker, 348–370. Cambridge, U.K.: Cambridge University Press.

Parker, S. T., R. W. Mitchell, and M. L. Boccia, eds. 1994. *Self-awareness in animals and humans: Developmental perspectives.* Cambridge, U.K.: Cambridge University Press.

Parker, S. T., and A. E. Russon. 1996. On the wild side of culture and cognition in the great apes. In *Reaching into thought: The minds of the great apes,* ed. A. E. Russon, K. A. Bard, and S. T. Parker, 430–450. Cambridge, U.K.: Cambridge University Press.

Pfeiffer, J. E. 1982. *The creative explosion: An inquiry into the origins of art and religion.* New York: Cornell University Press.

Piaget, J. 1980. *Adaptation and intelligence: Organic selection and phenocopy.* Chicago: University of Chicago Press.

Pinker, S. 1994. *The language instinct: How the mind creates language.* New York: William Morrow and Co.

Pinker, S., and P. Bloom. 1990. Natural language and natural selection. *Behavioral and Brain Sciences* 13:707–784.

Plomin, R., and G. E. McClearn. 1993. *Nature, nurture and psychology.* Washington, D.C.: American Psychological Association.

Plotkin, H. C. 1988. *The role of behavior in evolution.* Cambridge, Mass.: MIT Press.

———. 1994. *Darwin machines and the nature of knowledge.* Cambridge, Mass.: Harvard University Press.

Plotkin, H. C., and F. J. Odling-Smee. 1981. A multiple-level model of evolution and its implications for sociobiology. *Behavioral and Brain Sciences* 4 (2): 225–268.

Plutchik, R. 1980. *Emotion: A psychoevolutionary synthesis.* New York: Harper and Row.

Porter, R. 1987a. *Mind-forg'd manacles: A history of madness in England from the restoration to the regency.* London: Athlone Press.

———. 1987b. *A social history of madness: Stories of the insane.* London: Weidenfeld and Nicolson.

Povinelli, D. J. 1993. Reconstructing the evolution of mind. *American Psychologist* 48 (5): 493–509.

Povinelli, D. J., and J. G. H. Cant. 1995. Arboreal clambering and the evolution of self-conception. *Quarterly Review of Biology* 70 (4): 393–421.

Power, C., and I. Watts. 1996. Female strategies and collective behaviour: The archaeology of earliest Homo sapiens sapiens. In *The archaeology of human ancestry: Power, sex and tradition,* ed. J. Steele and S. Shennan, 306–330. New York: Routledge.

Power, M. 1991. *The egalitarians—Human and chimpanzee: An anthropological view of social organization.* New York: Cambridge University Press.

Price, J. 1988. Alternative channels for negotiating asymmetry in social relationships. In *Social fabrics of the mind,* ed. M. R. A. Chance and D. R. Omark, 157–195. Hove: Erlbaum.

Price, J., L. Sloman, R. Gardner Jr., P. Gilbert, and P. Rohde. 1994. The social competition hypothesis of depression. *British Journal of Psychiatry* 164:309–315.

Putnam, F. W. 1995. Development of dissociative disorders. In *Developmental psycho-*

pathology. Vol. 2, *Risk, disorder and adaptation*, ed. D. Cichetti and D. J. Cohen, 581–608. New York: John Wiley and Sons.

Quiatt, D., and J. Itani, eds. 1994. *Hominid culture in primate perspective*. Boulder: University of Colorado Press.

Quiatt, D., and V. Reynolds. 1993. *Primate behaviour: Information, social knowledge, and the evolution of culture*. Cambridge, U.K.: Cambridge University Press.

Raleigh, M. J., and D. Steklis. 1981. Effect of orbitofrontal and temporal neocortical lesions on the affiliative behavior of vervet monkeys (*Cercopithecus aethiops sabaeus*). *Experimental Neurology* 73 (2): 378–389.

Rappaport, R. A. 1999. *Ritual and religion in the making of humanity*. Cambridge, U.K.: Cambridge University Press.

Redmond, D. E., J. W. Maas, A. Kling, C. Graham, and H. Dekirmenjian. 1971. Social behavior of monkeys selectively depleted of monoamines. *Science* 174 (7): 428–431.

Reiss, S., G. W. Levitan, and J. Szyszko. 1982. Emotional disturbance and mental retardation: Diagnostic overshadowing. *American Journal of Mental Deficiency* 86 (6): 567–574.

Renfrew, C., and E. B. W. Zubrow. 1994. *The ancient mind: Elements of cognitive archaeology*. Cambridge, U.K.: Cambridge University Press.

Richards, G. D. 1987. *Human evolution: An introduction for the behavioural sciences*. London: Routledge and Kegan Paul.

———. 1989a. *On psychological language and the physiomorphic basis of human nature*. New York: Routledge.

———. 1989b. Human behavioural evolution: A physiomorphic model. *Current Anthropology* 30 (2): 244–255.

Roberts, A. C., P. Collins, and T. W. Robbins. 1996. The functions of the prefrontal cortex in humans and other animals. In *Modelling the early human mind*, ed. P. Mellars and K. R. Gibson, 67–80. Cambridge, U.K.: McDonald Institute for Archaeological Research.

Rodseth, L., R. W. Wrangham, A. M. Harrigan, and B. B. Smuts. 1991. The human community as a primate society. *Current Anthropology* 32 (3): 221–254.

Runciman, W. G. 1989. Evolution in sociology. In *Evolution and its influences*, ed. A. Grafen, 19–33. Oxford, U.K.: Clarendon Press.

———. 1998. The selectionist paradigm and its implications for sociology. *Sociology* 32:163–188.

Runciman, W. G., J. M. Smith, and R. I. M. Dunbar, eds. 1996. *Evolution of social behaviour patterns in primates and man*. Oxford: Oxford University Press.

Russon, A., K. A. Bard, and S. T. Taylor. 1996. *Reaching into thought: The minds of great apes*. New York: Cambridge University Press.

Rutter, M., and M. Rutter. 1993. *Developing minds: Challenge and continuity across the life span*. New York: Basic Books.

Sadler, J. Z. 1999. Horsefeathers: A commentary on "Evolutionary versus prototype analyses of the concept of disorder." *Journal of Abnormal Psychology* 108 (3): 433–437.

Saku, M. 1992. Relevance of ethology to psychiatry. *Japanese Journal of Psychiatry and Neurology* 46 (3): 631–639.

Sapolsky, R. M. 1987. Stress, social status, and reproductive physiology in free-living baboons. In *Psychobiology of reproductive behavior: An evolutionary perspective*, ed. D. Crews, 291–322. Englewood Cliffs, N.J.: Prentice Hall

———. 1990a. Stress in the wild. *Scientific American* 262 (1):106–113.

———. 1990b. Adrenocortical function, social rank, and personality among wild baboons. *Biological Psychiatry* 28 (10): 862–878.

———. 1998. *Why zebras don't get ulcers: A guide to stress, stress-related diseases, and coping*. New York: W. H. Freeman and Co.

Sapolsky, R. M., and J. C. Ray. 1989. Styles of dominance and their endocrine correlates among wild olive baboons (*Papio Anubis*). *American Journal of Primatology* 18 (1): 1–13.

Savage-Rumbaugh, E. S., and R. Lewin. 1994. *Kanzi: The ape at the brink of the human mind*. New York: John Wiley and Sons.

Savage-Rumbaugh, E. S., S. L. Williams, T. Furuichi, and T. Kano. 1996. Language perceived: Paniscus branches out. In *Great ape societies*, ed. W. C. McGrew, L. F. Marchant, and T. Nishida, 173–184. Cambridge, U.K.: Cambridge University Press.

Schneider, M. L. 1992a. The effect of mild stress during pregnancy on birthweight and neuromotor maturation in Rhesus monkey infants (*macaca mulatta*). *Infant Behavior and Development* 15 (4): 389–403.

———. 1992b. Prenatal stress exposure alters postnatal behavioral expression under conditions of novelty challenge in Rhesus monkey infants. *Developmental Psychobiology* 25 (7): 529–540.

Schneider, M. L., C. L. Coe, and G. R. Lubach. 1992. Endocrine activation mimics the adverse effects of prenatal stress on the neuromotor development of the infant primate. *Developmental Psychobiology* 25 (6): 427–439.

Schneider, M. L., and C. L. Coe. 1993. Repeated social stress during pregnancy impairs neuromotor development of the primate infant. *Journal of Developmental and Behavioral Pediatrics* 14 (2): 81–87.

Scull, A. T. 1989. *Social order/mental disorder: Anglo-American psychiatry in historical perspective*. Berkeley: University of California Press.

———. 1993. *The most solitary of afflictions: Madness and society in Britain, 1700–1900*. New Haven: Yale University Press.

Searle, J. R. 1995. *Construction of social reality*. New York: Free Press.

Segerstrale, U., and P. Molnar. 1997. Nonverbal communication: Crossing the boundary between culture and nature. In *Nonverbal communication: Where nature meets culture*, ed. U. C. O. Segerstrale and P. Molnar, 1–21. Mahwah, N.J.: Lawrence Erlbaum Associates.

Serban, G., and A. Kling, eds. 1976. *Animal models in human psychobiology*. New York: Plenum Press.

Simons, R. C., and C. C. Hughes, eds. 1985. *The culture-bound syndromes: Folk illnesses of psychiatric and anthropological interest*. Boston: D. Reidel.

Sims, A. C. P. 1988. *Symptoms in the mind: An introduction to descriptive psychopathology*. Philadelphia: Baillière Tindall.

Smith, C. U. M. 1993. Evolutionary biology and psychiatry. *British Journal of Psychiatry* 162:149–153.

Smith, E. A., and B. Winterhalder, eds. 1992. *Evolutionary ecology and human behavior*. New York: Aldine de Gruyter.

Smuts, B. B. 1985. *Sex and Friendship in Baboons*. New York: Aldine de Gruyter.

———. 1992. Male aggression against women: An evolutionary perspective. *Human Nature* 3 (1): 1–44.

———. 1995. The evolutionary origins of patriarchy. *Human Nature* 6 (1): 1–32.

Sober, E. 1984. *The nature of selection: Evolutionary theory in philosophical focus*. Chicago: University of Chicago Press.

Sovner, R. 1986. Limiting factors in the use of DSM-III criteria with mentally ill/mentally retarded persons. *Psychopharmacology Bulletin* 22 (4): 1055–1059.

Spiegel, H. X., and D. Spiegel. 1978. *Trance and treatment: Clinical uses of hypnosis*. New York: Basic Books.

Squire, L. R., and S. Zola-Morgan. 1991. The medial temporal lobe memory system. *Science* 253 (5026): 1380–1386.

Standen, V., and R. A. Foley, eds. 1989. *Comparative socioecology: The behavioural ecology of humans and other mammals*. Oxford: Blackwell Scientific.

Stanley, S. M. 1996. *Children of the Ice Age: How a global catastrophe allowed humans to evolve.* New York: Harmony Books.

Stearns, S. C. 1992. *The evolution of life histories.* New York: Oxford University Press.

Steele, J., and S. Shennan, eds. 1996. *The archaeology of human ancestry: Power, sex and tradition.* New York: Routledge.

Stent, G. S., ed. 1980. *Morality as a biological phenomenon: The presuppositions of sociobiological research.* Berkeley: University of California Press.

Stevens, A., and J. Price. 1996. *Evolutionary psychiatry: A new beginning.* New York: Routledge.

Stiner, M. C. 1994. *Honor among thieves: A zooarchaeological study of Neanderthal ecology.* Princeton, N.J.: Princeton University Press.

Stringer, C., and C. Gamble. 1993. *In search of the Neanderthals: Solving the puzzle of human origins.* New York: Thames and Hudson.

Struhsaker, T. T., and L. Leland. 1985. Infanticide in a patrilineal society of red Colobus monkeys. *Zeitschrift Tierpsychologie* 69 (2): 89–132.

Strum, S. C. 1987. *Almost human: A journey into the world of baboons.* New York: W. W. Norton.

Strum, S. S., and B. Latour. 1987. Redefining the social link: From baboons to humans. *Social Science Information* 26 (4): 783–802.

Suomi, S. J. 1978. Maternal behavior by socially incompetent monkeys: Neglect and abuse of offspring. *Journal of Pediatric Psychology* 3 (1): 28–34.

———. 1996. Biological, maternal, and life style interactions with the psychosocial environment: Primate models. In *Environmental and nonenvironmental determinants of the east-west life expectancy gap in Europe,* ed. S. Kelly and M. Bobak (NATO ARW Conference Series), 133–142. Amsterdam: Kluwer Academic Publishers.

Symons, D. 1979. *The evolution of human sexuality.* New York: Oxford University Press.

Szasz, T. S. 1961. *The myth of mental illness: Foundations of a theory of personal conduct.* New York: Harper and Row.

Takahata, Y. 1985. Adult male chimpanzees kill and eat a male newborn infant: Newly observed intragroup infanticide and cannibalism in Mahale National Park, Tanzania. *Folia Primatologica* 44 (3–4): 161–170.

Tart, C. T., ed. 1969. *Altered states of consciousness: A book of readings.* New York: John Wiley and Sons.

Tarter, R. 1990. Evaluation and treatment of adolescent substance abuse. A decision tree method. *American Journal of Drug and Alcohol Abuse* 16 (1–2): 1–46.

Tarter, R. E., M. Kabene, E. A. Escallier, S. B. Laird, and T. Jacob. 1990. Temperament deviation and risk for alcoholism. *Alcoholism: Clinical and Experimental Research* 14 (3): 380–382.

Teicher, M. H., C. A. Glod, J. Surrey, and C. Swett Jr. 1993. Early childhood abuse and limbic system ratings in adult psychiatric outpatients. *Journal of Neuropsychiatry and Clinical Neurosciences* 5 (3): 301–306.

Teicher, M. H., Y. Ito, C. A. Glod, S. L. Andersen et. al. 1997. Preliminary evidence for abnormal cortical development in physically and sexually abused children using EEG coherence and MRI. In *Psychobiology of posttraumatic stress disorder. Annals of the New York Academy of Sciences,* ed. R. Yehuda and A. C. McFarlane (821): 160–175. New York: New York Academy of Sciences.

Thorpe, W. H. 1956. *Learning and instinct in animals.* London: Methuen.

Tinbergen, N. 1951. *The study of instinct.* Oxford, U.K.: Oxford University Press.

———. 1952. Derived activities: Their causation, biological significance, origin, and emancipation during evolution. *Quarterly Review of Biology* 27:1–32.

Tomasello, M. 1992. The social bases of language acquisition. *Social Development* 1 (1): 67–87.

Tomasello, M., and J. Call. 1997. *Primate cognition.* New York: Oxford University Press.

Tomasello, M., A. C. Kruger, and H. H. Ratner. 1993. Cultural learning. *Behavioral and Brain Sciences* 16:495–552.
Tooby, J., and L. Cosmides. 1990a. On the universality of human nature and the uniqueness of the individual: The role of genetics and adaptation. *Journal of Personality* 58 (1): 17–67.
———. 1990b. The past explains the present: Emotional adaptations and the structure of ancestral environments. *Ethology and Sociobiology* 11 (4–5): 375–424.
———. 1992. The psychological foundations of culture. In *The adapted mind: Evolutionary psychology and the generation of culture*, ed. J. H. Barkow, L. Cosmides, and J. Tooby, 19–136. New York: Oxford University Press.
Tooby, J., and I. DeVore. 1987. The reconstruction of hominid evolution through strategic modelling. In *The evolution of human behavior: Primate models*, ed. W. G. Kinzey, 183–237. Albany: State University of New York Press.
Toth, N. 1985. Archaeological evidence for preferential right-handedness in the lower and middle Pleistocene, and its possible implications. *Journal of Human Evolution* 14:607–614.
Trinkaus, E. 1995. Neanderthal mortality patterns. *Journal of Archeological Science* 22: 121–142.
Trinkaus, E., and P. Shipman. 1994. *The Neanderthals: Of skeletons, scientists, and scandal*. New York: Random House, Vintage Books.
Trivers, R. L. 1971. The evolution of reciprocal altruism. *Quarterly Review of Biology* 46:35–57.
———. 1985. *Social evolution*. Menlo Park, Calif.: Benjamin/Cummings Publishing Co.
Troisi, A., F. Aureli, P. Piovesan, and R. F. D'Amato. 1989. Severity of early separation and later abusive mothering in monkeys: What is the pathogenic threshold? *Journal of Child Psychology and Psychiatry and Allied Disciplines* 30 (2): 277–284.
Troisi, A., and R. F. D'Amato. 1983. Is monkey maternal abuse of offspring aggressive behavior? *Aggressive Behavior* 9 (2): 167–173.
———. 1984. Ambivalence in monkey mothering: Infant abuse combined with maternal possessiveness. *Journal of Nervous and Mental Disease* 172 (2): 105–108.
———. 1991. Anxiety in the pathogenesis of primate infant abuse: A pharmacological study. *Psychopharmacology* 103 (4): 571–572.
Troisi, A., R. F. D'Amato, R. Fuccillo, and S. Scucchi. 1982. Infant abuse by a wild-born group-living Japanese Macaque mother. *Journal of Abnormal Psychology* 91 (6): 451–456.
Troisi, A., and M. T. McGuire. 1990. Deception in somatizing disorders. In *Psychiatry: A world perspective*, ed. C. N. Stefanis, A. D. Rabavilas and C. R. Soldatos, 3:973–978. Amsterdam: Elsevier Science.
———. 1998. Evolution and mental health. In *Encyclopedia of mental health*, ed. H. S. Freedman, 2:173–181. San Diego, Calif.: Academic Press.
Valzelli, L. 1981. *Psychobiology of aggression and violence*. New York: Raven Press.
van der Kolk, B. A. 1994. The body keeps the score: Memory and the evolving psychobiology of posttraumatic stress. *Harvard Reviews of Psychiatry* 1 (5): 253–265.
van Praag, H. M., S.-L. Brown, G. M. Asnis, R. S. Kahn, M. L. Korn, J. M. Harkavy-Friedman, and S. Wetzler. 1991. Beyond serotonin: A multiaminergic perspective on abnormal behavior. In *The role of serotonin in psychiatric disorders*, ed. S.-L. Brown and H. M. van Praag, 302–332. New York: Brunner/Mazel Publishers.
Vigotsky, L. S. 1962. *Thought and language*. 1934. Cambridge, Mass.: MIT Press.
Waddington, C. H. 1959. Canalization of development and genetic assimilation of acquired characters. *Nature* 183:1654–1655.
———. 1961. Genetic assimilation. *Advances in Genetics* 10:257–293.
Wakefield, J. C. 1992a. Disorder as harmful dysfunction: A conceptual critique of DSM-III-R's definition of mental disorder. *Psychological Review* 99 (2): 232–247.

———. 1992b. The concept of mental disorder: On the boundary between biological facts and social values. *American Psychologist* 47 (3): 373–388.

———. 1999a. Evolutionary versus prototype analyses of the concept of disorder. *Journal of Abnormal Psychology* 108 (3): 374–399.

———. 1999b. Mental illness as a black box essentialist concept. *Journal of Abnormal Psychology* 108 (3): 465–472.

Washburn, S. L., ed. 1961. *Social life of early man*. Chicago: Aldine.

Watts, D. P. 1989. Infanticide in mountain gorillas: New cases and a reconsideration of the evidence. *Ethology* 81 (1): 1–18.

Watts, I. 1999. The origins of symbolic culture. 1999. In *The evolution of culture: An interdisciplinary view*, ed. R. I. M. Dunbar, C. Knight, and C. Power, 113–146. New Brunswick, N.J.: Rutgers University Press.

Weingart, P., S. D. Mittchell, P. J. Richerson, and S. Maasen. 1997. *Human by nature: Between biology and the social sciences*. Mahwah, N.J.: Lawrence Erlbaum Associates.

Weiss, J. M. 1980. Learned helplessness, physiological change and learned inactivity. Part 3: Commentary. *Behavior, Research and Therapy* 18:475–512.

Weiss, J. M., and P. G. Simson. 1986. Depression in an animal model: Focus on the locus ceruleus. In *Antidepressants and receptor function*, ed. R. Porter, G. Bock, and S. Clark, 191–215. New York: John Wiley and Sons.

Wenegrat, B. 1984. *Sociobiology and mental disorder: A new view*. Menlo Park, Calif.: Addison-Wesley.

———. 1990. *Sociobiological psychiatry: Normal behavior and psychopathology*. Lexington, Mass.: Lexington Books.

Whitcombe, E. 1996. The anatomical foundations of cognition: Suggestions for a reinterpretation. In *Modelling the early human mind*, ed. P. Mellars and K. R. Gibson, 81–88. Cambridge, U.K.: McDonald Institute for Archaeological Research.

White, F. 1993. Review of *The egalitarians—Human and chimpanzee: An anthropological view of social organization*, by Margaret Power. *Biological Anthropology* 95:165–166.

White, N. F., ed. 1974. *Ethology and psychiatry*. Buffalo: University of Toronto Press.

Whiten, A., and R. W. Byrne, eds. 1997. *Machiavellian intelligence II: Extensions and evaluations*. Cambridge, U.K.: Cambridge University Press.

Williams, G. C. 1966. *Adaptation and natural selection: A critique of some current evolutionary thought*. Princeton: Princeton University Press.

Wilson, D. R. 1992. Ideas in theoretical biology: Evolutionary epidemiology. *Acta Biotheoretica* 40:87–89.

———. 1993. Evolutionary epidemiology: Darwinian theory in service of medicine and psychiatry. *Acta Biotheoretica* 41:205–218.

Wilson, E. O. 1975. *Sociobiology: The new synthesis*. Cambridge, Mass.: Harvard University Press, Belknap Press.

Windle, M. 1991. The difficult temperament in adolescence: Associations with substance use, family support, and problem behaviors. *Journal of Clinical Psychology* 47 (2): 310–315.

Windle, M., and R. Lerner. 1986. Reassessing the dimensions of temperamental individuality across the life span: The revised Dimensions of Temperament Survey (DOTS-R). *Journal of Adolescent Research* 1 (2): 213–229.

Wrangham, R. W. 1979. On the evolution of ape social systems. *Social Science Information* 18 (3): 335–368.

———. 1986. The significance of African apes for reconstructing human social evolution. In *The evolution of human behavior: Primate models*, ed. W. G. Kinzey, 51–71. Albany: State University of New York Press.

Wrangham, R. W., W. C. McGrew, F. B. M. de Waal, and P. G. Heltne, eds. 1984. *Chimpanzee politics*. Cambridge, Mass.: Harvard University Press.

———. 1994. *Chimpanzee cultures*. Cambridge, Mass.: Harvard University Press.

Wrangham, R. W., and D. Peterson. 1996. *Demonic males: Apes and the origins of human violence*. Boston: Houghton Mifflin.
Zihlman, A. L. 1997. Natural history of apes: Life-history features in females and males. In *The evolving female: A life-history perspective*, ed. M. E. Morbeck, A. Galloway, and A. L. Zihlman, 86–103. Princeton, N.J.: Princeton University Press.
Zorumski, C. F. 1988. The relevance of developmental and genetic studies in animals to the neurobiology of psychiatric disorders. *Psychiatric Developments* 6 (3): 227–240.

Index

abnormality vs. normality, 14–17
ABP (abusive behavior patterns), 165–166
abstract thought, 71, 256, 308, 314, 317, 325, 336, 338–340, 354
abuse of infants, 164–168, 197
adaptation, 11–12, 16, 27–28, 35, 74, 84, 89, 279, 315, 347; and dissociation, 309–312; human, 10, 23, 278; psychological, 60, 68; and psychopathology, 85–86, 91–94, 113–117, 122–124, 130, 133–135, 172. See also psychopathology and adaptation
adaptiveness, 12, 112
adolescence, 49–50, 59
Africa, 217–218, 273, 315, 316, 337
aggression and hostility, 84, 147, 301; among hominids, 60–61, 233–236, 352, 356; among nonhuman primates, 146–148, 172, 175–177. See also infanticide; infants, abuse of; sexual coercion and assault
Aiello, A. C., 314
Aitchison, J., 325
altered states of consciousness, 70, 294–295. See also dissociation; trance; possession by external agents
altriciality, 49–50, 315
altruism, reciprocal, 34, 45–46, 68, 198
anatomically modern humans, 46, 321, 323, 329; early, 320–321; late, 315–316, 320–322. See also Homo sapiens

anatomy. See brain; life cycle; life history
ancestralization, 132
anger, 60
animal calls, 307–308
animal hypnosis, 301–302, 306
animal suffering, 11–12
animal welfare, 11–12
anthropomorphism, 12, 15, 16
antisocial behavior, 83
anxiety disorders, 81–82, 167, 180, 304
Archaic and early anatomically modern humans, 71, 220–221, 247, 280, 282–283, 315–316, 320–323, 333, 339–340. See also Homo, erectus; Homo, habilis; Homo, lineage, early; Neanderthals
Archaic humans, 99, 319–320, 329–330; and morality, 69, 70, 72, 125–126; and myths and rituals, 69–70, 71; with protoculture and mature language, 334–339, 340, 363–366; with protoculture and protolanguage, 332–334, 339–340; and psychopathology, 69–71, 332–335, 336–341, 363–366
Archer, R., 355
Arctic, 337
art, 334. See also culture; Upper Paleolithic Era, culture of
Asia, 320
attachment, 27–31, 34, 53, 306
Australian aborigines, 337
Australopithecines, 59–60, 62, 217–219, 283, 313

401

autobiographical self, 354, 359
automatic behavior, 361
awareness: in chimpanzees, 203–208, 306; in hominids, 62, 246, 247–248, 249, 250, 269–270, 284. *See also* self-awareness

baboons, behavior of, 148–157; in all-female groups, 152; homosexual, 149; infanticide, 153; in old age, 155–156, 303; and personality differences, 156–157; and rank, 149–152; sexual, 149–153; unusual, 149–151, 154; in zoos, 149–154
Bailey, K. G., 79, 87–88, 90
Baldwin effect, 46–47
behavior: abnormal, 12, 182–184 (*see also* psychopathology); adaptive, 154–155, 288 (*see also* psychopathology, as result of adaptation); automatic, 361; biological approach to, 115; breakdowns in, 13–14, 17, 74, 85; cognitive changes in (*see* physiomorphic learning model); conforming, 290; cultural standards of, 13–15; deceptive, 17, 83, 85, 134, 205, 304, 327, 330–331; and emotions, 263, 267; environmental factors affecting, 29–30, 44–46, 48–49, 100–101, 108–110 (*see also* developmental manifold; environmental mismatch; Goodwin; probabilistic epigenesis); genetic factors affecting, 29–30, 49; genetic survival value of, 154–155; gratification value of, 154–155; imitative, 288; learned, 288; maladaptive, 154–155; normal, 195; of primates in zoos, 149–154, 165–167; of primates with lesions, 193–195; during protocultural phases, 119; supportive, 62, 64, 67, 68, 72, 196, 225, 357–358. *See also* baboons, behavior of; chimpanzees, behavior of
behavior, anatomically modern human, 276–277, 278; in children, 107, 123, 328–330; and language, 329–330
behavior, hominid, 219–221, 275; begging, 254–255; psychopathological, 222–225
behavioral adaptability, 102–103
behavioral change, constructed, 105
behavioral diversity, 246–247, 250
behavioral ecology, 44–45; framework of, 214–215; of hominid psychopathology, 71, 242
behavioral embryology, 102, 103
behavioral mechanisms, 27, 60–61, 74, 190, 210
behavioral neophenogenesis, 103–105
behavioral plasticity, 29, 101–103, 323
behavioral routines, 52, 65–66, 68, 209–210, 353, 356, 357
behavior dispositions, 27–31, 54
behavior programs, 54–56
behavior space of hominids, 348, 350
beliefs, 303, 311, 318; among Archaic humans, 69, 71, 333, 335–336, 337, 340, 364
Berkson, Gershon, 193–195
Bickerton, Derek, 259–260, 262, 271, 354–355
biological goals, 241
biological imperatives, 350. *See also* survival and reproduction
biological residuals, 365. *See also* evolutionary residuals
bipolar affective disorders, 81
body, 9, 62, 65, 317, 338, 350, 360
Bourguignon, Erika, 298, 306
Bowlby, J., 27, 53–54
Boyer, P., 364, 365
brain, 79, 82, 266–267; evolution of, 49–50, 274, 300, 307–308, 321–323, 328–330; lesions of, 16, 57–58, 61, 137, 195–196, 263–264, 266–269; size of, 49–50, 103, 217–219, 308, 314–316
Bunney, W. E., 13

Caine, M. G., 198
Call, J., 255–257
cannibalism, 158–160, 162–164, 177, 296, 302
Cant, J. G. H., 252–253
Capitanio, J. P., 183
care: among monkeys, 194–196, 197–202; of invalids, 198–199, 205; of psychopathologic group members, 146, 196–197
cascade hypothesis, 181
Chance, M. R. A., 34
Chase, P., 274, 278
children: behavior of, 107, 123, 328–330; development of, 49–50, 53–54, 107, 236, 253. *See also* infants
chimpanzees: awareness of, 203–208, 306;

changes in food provisioning for, 177–178; communication among, 203, 210; compassion among, 204; and evolutionary relationship to humans, 207–208; Pan Paniscus, 144; premeditated actions of, 205; social organization of, 174–179; and stress, 175, 302
chimpanzees, behavior of, 98, 284, 302–303, 349; aberrant, 157–164, 179–181; cannibalism, 158–160, 162–164, 296, 302; deceptive, 205, 327; dissociation, 162, 164; infanticide, 158–159, 161–163, 169, 172–173, 177, 302; and orphaned mothers, 163; psychological, 201–202; psychopathological, 168–172, 174–179, 207–210, 333; rescuing of injured, 161–162; sexual, 171, 176; during sickness and healing, 202–206; social, 201–202
Chisholm, James S., 52–54, 57
Clarke, A. S., 182–184
clinical psychiatry and evolution, 32–36
coevolution, 96–98
cognition: and culture, 6, 21–24, 30, 31, 85, 119, 317–318; definition of, 316; and dissociation, 301; evolution of, 85–86, 108, 119, 313, 314, 316; hominid, 63, 267; human, 256–258, 279, 334, 355; and language, 317–318, 324–325, 329–330, 335; mimetic, 87, 251–255, 262, 354; primate, 255–258; and psychopathology, 353; social, 256
cognitive changes in behavior during evolution. See physiomorphic learning model
cognitive modules, 23, 265–270; and language, 267–269; and psychopathology, 266–270; social, 265, 266
communication: among hominids, 219; mimetic, 88. See also gesturing; language; protolanguage
community, human, 221–222
compassion, 204, 349
competition, 16–17, 34, 64, 80, 85, 175, 178, 285, 330; sexual, 233–235
conceptual reference, 355
consciousness: altered states of, 294–295 (see also dissociation; possession by external agents; trance); core, 359–360; extended, 354, 359–360
constructed behavior change, 105
cooperation, 274, 285, 287

Cosmides, Leda, 276–277
Crow, Timothy, 8, 32, 82
cultural biases, 4–5
cultural evolution, 211, 227, 302, 316
cultural learning, human, 256–258, 279
cultural relativism, 73, 240
culture, 56, 274; and cognition, 6, 21–24, 30, 31, 85, 119, 317–318; definition of, 225–228, 317–318; and dissociation, 295–298, 300–303, 305–307, 309–311, 339; of early hominids, 62–63, 220–221, 222; and environment, 336–337; and language, 248, 250, 257, 273, 282, 318–319; mature and elaborated, 319; nascent and impoverished, 319 (see also protoculture); partible view of, 277–281; and politics, 278, 279; and psychopathology, 136–138, 279–282, 289–293, 347, 353; qualitative view of, 277–278, 279–281; and sense of reality, 281–282; and social emotions, 284, 286–290. See also art; symbolic culture; Upper Paleolithic Era, culture of

Daly, M., 173
Damasio, A. R., 354, 359
D'Amato, R. F., 165–167
Darwinian and neo-Darwinian theory, 22, 83, 96
Darwinian Psychiatry (McGuire and Troisi), 34–35, 85, 132
Dawkins, M. S., 11, 57, 242, 278
Deacon, T. W., 119, 307–309
deafness, 282
deception, 17, 83, 85, 134, 205, 304, 327, 330–331
decision theory, 245
defects, 191–197, 200; visual, 193–195
defense or response system, 118–123, 129–135, 134
delusions, 281, 282, 283–284
demography, 38, 47, 54, 56, 314–316
denial, 268–269
Dennett, D. C., 354
depression, 59, 79–80, 116–118, 304
developmental evolutionary ecology, 52
developmental manifold, 101–105
developmental plasticity, 46–47, 48–49
development and psychopathology, 100–105, 125, 267
Devereux, George, 13

de Waal, Frans, 60, 146–148, 196–197, 205
disorder of attachment and rank, 34
disorders, treatment of, 167
dispersal of anatomically modern humans, 315–316
displacement, 355
dissociation, 91, 162, 164, 294–312; and culture and language, 295–298, 300–303, 305–307, 309–311, 339, 340; and evolution, 299–304, 309–312; and stress, 302–304, 310–313, 360–361, 365. *See also* psychopathology and dissociation
Donald, Merlin, 87–88, 119, 218, 251–255, 270–271, 334, 354–355
Draper, P., 83, 105, 109–110
Dunbar, R. I. M., 215, 275–276, 314
Dusun Baguk (Malay village), 286

early cultural phase, 363–367
Eaves, L. J., 77
ecological inheritance, 97–99
ecology and environment, 119–120
egalitarianism, 174–175
emotional routines, 285
emotions, 89, 261, 263, 316, 329, 332, 338, 340; primary, 285, 286; and psychopathology, 9, 61, 62–63, 85, 289–293, 361–362; secondary, 285, 286 (*see also* pride; shame); social, 284, 286–290
emotions profile index (EPI), 169
Endler, J. A., 83
environment, 119–120; and behavior, 6, 17, 44–46, 56, 93, 174–179; and culture, 336–337; and psychopathology, 100–105, 199
environmental mismatch, 88, 91, 92, 113, 119, 132
environment of evolutionary adaptedness (EEA), 28, 31, 46, 112–113, 118, 173, 309
epigenesis through selective stabilization, 46–47
ethnocentrism, 13, 14, 16, 90
ethological phase, 350–354
eu-culture, 227
Europe, 273–274, 316, 320, 322
evolution, 83; and clinical psychiatry, 32–36; cognitive changes during (*see* physiomorphic learning model); cultural, 211, 227, 302; human, 214–215, 308
evolutionary bias, 90–91
evolutionary biology, 35–36, 38, 119, 133, 347
evolutionary conception of psychopathology, 13–18, 40–41, 43–44, 50–54, 56–61, 64, 71–75, 124–126, 208–209; advantages of, 129–133; criticisms and limitations of, 133–136
evolutionary functionalism, 54–56, 115
evolutionary hypotheses, 131–133
evolutionary imperatives, 190–191, 366
evolutionary pathway, 103
evolutionary psychiatry, 125, 137, 138
Evolutionary Psychiatry: A New Beginning (Stevens and Price), 32–34, 132
evolutionary psychology, 26–28, 60–61, 125
evolutionary residuals, 31, 44, 55, 112, 137–138, 173–174, 190, 276, 347, 356; and psychopathology, 350, 353, 357, 362
evolution of hominids, 215–222; Australopithecines, 217–218; and social ecology, 216–222
exaptation, 91–92
Eysenck, H. J., 77

feeding behavior, 248–249. *See also* food
female alliances, 275, 288, 300
Fessler, Daniel, 286, 287
fitness, 11–13, 16, 20, 78, 83–84, 86, 89, 92, 94; components of, 47, 53; inclusive, 33–34, 43, 45–46, 49–50, 68, 198; negative, 19, 58–59, 87; and psychopathology, 240–241
Fodor, J. A., 89
Foley, Robert, 214–215, 219, 275–277, 315–316
food, 314; and chimpanzees, 175, 177–178
Fuster, J. M., 180

Gamble, C., 322–323
Gardner, R., 80–81, 92
gender, 347–348
general performance mechanisms, 116
genetic assimilation, 46–47, 108
genetic defects, 196–197
genetic inheritance, 96–97
genetic variation, 90
genotype, 8–9, 43, 47, 48, 52–53, 83

gesturing, 63, 64, 66, 252, 255, 327–328, 334, 357, 361
Gibson, K. R., 322–323
Gilbert, P., 80
Gombe (Africa), 168, 175–176, 178
Goodall, J., 104, 158–163, 168–172, 177
Goodenough, Oliver, 288–289, 290
Goodman, Felicitas, 299, 300, 306
Goodwin, B., 105–107, 109
Goosen, C., 181–182
Gottlieb, G., 101–105, 106, 109
Gould, S. J., 92–93, 118
grammar, 317, 318, 324, 325, 335, 337–338, 363. *See also* language
group history, 67, 71, 119, 336, 353, 363–364
group identity, 62, 67, 69, 71, 209, 219, 292, 326, 349, 364
group life, 279–280, 284
group needs, 66, 247, 250, 290
group selection, 94, 98, 175

Hallowell, Irving, 286
hallucinations, 281–282, 283
handicaps, 191–196, 203–204
harmful dysfunction (HD) analysis, 126–129
Harpending, H., 83, 105, 109–110
Hayden, B., 321–322
Hockett, C. F., 355
Hofer, M. A., 82
hominids, 15, 217–218, 246–247, 252, 275, 277, 280, 282–283, 308–309, 348–350; and cognition, 63, 266, 267; culture of, 62–63, 220–221; and dissociation, 307, 309; evolution of, 215–216; life history of, 314–316; mental traits of, 318, 327; sexuality, reproduction and parenting among, 66, 222–223, 233–236, 249, 316, 351, 354; and sickness and healing, 70, 72, 136, 219–220, 263–264, 305–306; social ecology of, 46, 50, 216, 219–223, 242, 358; stress in, 222, 224, 229, 233, 234, 236, 242, 360; and symbolization, 23, 62, 65–67, 355; and technology, 315, 322. *See also* psychopathology in hominids
Homo, 46, 99, 125–126, 280, 282–283, 313, 323; and dissociation, 307, 308, 309; erectus, 50, 65, 218–220, 247, 251, 313–314, 316; habilis, 62; lineage, early, 90, 217–218; sapiens, 15, 23, 84, 99, 275, 315, 320–321

humans, 49; anatomically modern, 46, 315–316, 320–323, 329; Archaic and early anatomically modern, 220–221, 315; and cognition, 256–258, 279, 334, 355; community of, 221–222; and emotions, 332, 338; evolution of, 214–215, 308; in later phases of evolution, 234–235; mothers and infants, 256, 257, 299, 306; social behavior of, 276–277, 336, 364; social organization of, 221–222. *See also* Archaic humans; children; Neanderthals; Homo; psychopathology in humans
Huntington's disease, 77
Hyland, M. E., 93–94

identity, 19, 66–67, 69, 279, 291, 296. *See also* group identity
ideology, 331
imaginative thought, 316–317, 325–326, 335, 336, 338, 339–340
infanticide, 172–174; among baboons, 153; among chimpanzees, 158–159, 161–163, 169, 172–173, 177, 302
infants: abuse of, 164–168, 197; hominid, 267; human, 256, 299, 306; needs of, 47, 50–51; pongid, 164–168, 193–195, 197, 219. *See also* children; mother-infant relationships
information processing and handling, 276–277, 346–347, 350, 354, 356, 359, 361–363
Integrated Causal Model (ICM), 276
intelligence, 82, 266, 268, 327, 330
intentionality, 354, 359
intergroup affinity, 220
internal model principle, 107–108
Isaac, B., 215
isolation, 181–184

Jung, Carl Gustav, 33

Kasakela (Africa), 158
Kendon, A., 355
kin selection, 45–46
Kirmayer, L. J., 127–128
Knight, Chris, 277–278, 279, 288
knowledge, 70, 322, 338–339, 340, 353, 357, 360, 364
Kraemer, G. W., 181
Kummer, Hans, 148–156

language, 208–209, 215–216, 220–221, 226, 274, 278, 284, 289, 317; and behavior, 329–330; and children, 253, 328–330; and cognition, 317–318, 324–325, 329–330, 335; and cognitive modules, 267–269; and conceptual capacity, 324; and culture, 248, 250, 257, 273, 282, 318–319; and dissociation, 297, 300–303, 305, 307–308, 339, 340; evolution of, 108, 258–259, 300, 308, 314–315, 329–331; and mimetic cognition, 251–253; and physiomorphic learning model, 247–248, 250, 288; and sense of reality, 282–283; and thought, 326–328, 335, 337–338. *See also* protolanguage; psychopathology and language
langurs, 173
last common ancestor (LCA), 12, 98, 214–217, 223, 286, 350
Latour, B., 231–232
learning, 17, 48, 49, 147–148, 226, 227, 277–278, 288, 357
Lee, P. C., 219
lesions: induced, 137, 179–181, 193–196; and psychopathology, 57–58, 268. *See also* brain, lesions of
Lewontin, R. C., 118
lexicon, 324–325, 334–335, 336, 337, 339
life cycle, 49–51
life history, 51, 56, 314–316, 362
life history theory, 47–50, 51–54
life routines, 244–250, 285, 355
linguistic coding, 355
Lower Paleolithic Era, 321
Ludwig, A. M., 302

macaques, behavior of: abusive, 165–168, 198; drug-induced, 180; emotional disorders, 166; and lesions, 193–196; psychopathological, 196–197; response to abuse, 166–167; in zoos, 165–167. *See also* monkeys, rhesus
MacLean, P. D., 88
Maestripieri, Dario, 167–168, 197
Mahale (Africa), 175, 178
maladaptation, 18, 58–59, 61, 74, 88–90, 94, 108, 129, 262–263, 306
male competition, 233–235
male kinships and alliances, 217–220, 315
Marks, I. M., 81–82
Martin, N. G., 77
Maryanski, A. R., 216–217

maternal deprivation syndrome, 57, 184, 185
mating and parenting, 28, 48, 53; in hominids, 66, 233–236, 249, 351, 354
McFarland, D. J., 245
McGuire, M. T., 34–35, 77–78, 85, 112, 198–199
McKinney, W. T., 13
medicine, 126, 364
memes, 278, 289
menstruation, 275, 288, 300
mental capacity, 219. *See also* brain size
mentalese, 326–328, 331–332, 334, 336
mental retardation, 230–231
mentation, 12, 323–326, 328, 357
methamphetamine intoxication, 179–180
Middle Paleolithic Era, 320–321
mimetic cognition, 87, 251–255, 262, 354
mimetic representation, 62, 254, 328, 354
mind, theory and prehistory of, 89–90, 265–270, 266, 353, 355, 358
mind-altering substances, 57, 179–180, 297, 299
Mitchell, G., 182–184
Mithen, Steven J., 89–90, 266, 267–269, 271
modular mind theory, 89–90, 265–270
monkeys: alarm calls of, 16–17; and artificially-produced lesions, 193–196; and care giving, 194–198, 199–202; and distress signals, 194; and learning, 147–148, 277; Old World, 184; and psychopathology, 137, 179–182, 196–197, 199–200; and responses to group mates with defects, 193–197, 200; rhesus, 17, 104, 137, 184, 185, 187; and social deprivation and isolation, 181–182; and stress, 185–186; in the wild, 194. *See also* langurs; macaques
morality, 17, 46, 63, 83, 119, 279, 289, 311, 316, 339; in Archaic humans, 69, 70, 72, 125–126
mother-infant relationships: human, 257, 299, 306; pongid, 57, 104, 163–168, 177, 184–186, 194, 196–197. *See also* attachment
motor programming, 252
multiple personality disorder, 299
myths, 278, 279, 299, 303, 309–311; among Archaic humans, 69–70, 71, 316, 323, 331, 333, 335

natural selection, 26–27, 34, 36, 43–47,

79, 97, 104, 119, 360; and behavioral mechanisms, 27; effects of, 48–49; and psychopathology, 78, 84, 86, 96, 115–116. *See also* group selection; negative selection; neural selection; sexual selection
Neanderthals, 220–221, 315, 319–323
negative selection, 86–88
Nelson, K., 88, 89, 253
nervous system, 55, 79, 88
Nesse, R. M., 117–118
neural selection, 46
neurobiology, 4–5
neuroendocrine responses, 186
neuropsychiatry, 7–8, 332
neurotransmitters, 117, 186
niche construction, 97–100
Norikoshi, K., 163–164
normality *vs.* abnormality, 14–17

Odling-Smee, F. J., 88, 92–93, 96–98, 106
Oubré, Alondra, 299–300, 306

Parker, S. T., 56, 227–228, 300
pathology, responses to, 203–204
personality, 27–31, 30, 77, 80, 92, 98, 364
Peterson, D., 57
phenocopy, 106–107
phenotype, 8, 43, 47, 48, 52–53, 83–84, 91, 93, 97–100
phobias, 78, 81–82
phylogenetic regression, 79, 87–88
phylogeny, 8, 13, 15, 37, 346, 347
physiomorphic learning model, 244–251, 355; and language, 247–248, 250, 288; and psychopathology, 248–251; stages of, 245–250
Piaget, J., 106–107
Plotkin, H. C., 88, 92–93, 96–98, 106
Plotkin-Odling-Smee model, 97–100, 109
pongid/hominid split. *See* last common ancestor
pongids, 15, 17. *See also* baboons; chimpanzees; macaques; monkeys
possession by external agents, 294–295, 297–300, 302, 303, 305–307, 311–312, 365
postreproductive phase of life, 50–51
Povinelli, D. J., 252–253
Power, Margaret, 57, 104, 174–179, 224
preadaptation, 357
precultural phase, 227, 354–359

Price, John, 8, 32–34, 77–78, 80, 92, 112–113, 116
pride, 284–291
primate cognition, 255–258
probabilistic epigenesis, 103
problem solving among hominids, 246–247
protoculture, 68, 119, 227, 306, 319, 360, 362–363; and protolanguage in Archaics, 65–68, 69–71, 332–334, 339–340
protolanguage, 23, 65, 259–264, 319, 323–326, 354–355; and psychopathology, 62–63, 66, 67, 262–265, 269, 307, 308, 327–328, 328. *See also* protoculture, and protolanguage in Archaics
prototype theory of categorization and cognition, 325
psychiatric disorders, Roschian, 126–127
psychiatry, 77–78, 115
psychological adaptations, 60, 68
psychological mechanisms, 27–31, 64, 68
psychopathologic group members, caring for, 146, 196–197
psychopathology: and adaptation, 85–86, 91–94, 113–117, 122–124, 130, 133–135, 172; and behavior dispositions, 30–31; and behavior programs, 55–56; biological conceptions of, 18–19, 241; and body, 9, 62, 65, 317, 338, 350, 360; causal/mechanistic approach to, 115–117, 130, 133–134; causes of, 5, 42, 267, 366; centers of interest of, 348–349; in chimpanzees, 168–172, 174–179, 207–210, 333; classification of, 36; criteria for, 18–20; and culture, 136–138, 279–282, 289–293, 347, 353; date of origin of, 146; definition of, 3–4, 9, 11, 36, 76; developmental factors affecting, 100–105, 125, 267; diagnosis of, 18–20, 38, 78, 193, 283–284, 332; and dissociation, 296–299, 303–307, 310–312, 333–334, 352–354, 358, 363, 365; during early cultural phase, 363–367; and emotions, 9, 61, 62–63, 85, 289–293, 361–362; and environmental change, 100–105; during ethological phase, 350–354; and evolutionary functionalism, 54–56; and evolutionary residuals, 350, 353, 357, 362; explanations for, 30–31, 240–241; as failed strategy, 124; and feeding

psychopathology (*continued*)
behavior, 248–249; and fitness, 240–241; functional approach to, 55–56, 115–117, 130, 133–135, 211; as general defense or threat defense system, 118–123, 129–135; genetic factors in, 77–79, 135, 196; as harmful condition, 126–129; in humans, 7, 329–330, 331–335, 338–340; laboratory manufacture of, 179–181, 345–346; and language, 319, 326–331, 332–335, 338–340; and life routines, 248–250; as long-term *vs.* short-term strategy, 123–124; manifestations of, 5, 13, 240–242, 241–242, 268–269, 283, 311–312, 328–330, 333; and mating, 249; medicalization of, 240; and mimetic cognition, 253–255, 262; in monkeys, 137, 179–182, 196–197, 199–200; and natural selection, 78, 84, 86, 96, 115–116; as negatively selected trait, 86–88; in nonhuman primates, 144–146, 241, 304; origins of, 5–6; as passive reaction, 113; and physiomorphic learning model, 248–251; positive functions of, 99–100, 107–108, 115; and possession by external agents, 298, 303, 305–307, 311–312; during precultural phase, 354–359; during protocultural phase, 359–363; and protolanguage, 262–265, 327–328; quantitative deviations in, 185–187; and rank, 34, 64, 80, 249, 349, 353; response of group mates to, 15–16, 166–167, 189–190, 195–199, 202, 240–241, 349; as result of social deprivation, 181–184; and sense of reality, 281–283; and sexual conflicts and assault, 233–236, 236–240; and sexuality, 234, 235–240; and social ecology, 222–223; social science conceptions of, 19–20; and suffering, 54, 61, 64, 67, 69–70, 72–73, 80, 353, 356–357, 362 (*see also* harmful dysfunction (HD) analysis); temporal profile of, 240–241; and thought, 326–328; and trance, 68, 294, 298, 303–307, 311–312, 339; and trauma, 296, 299, 304–305, 333, 351, 361, 365. *See also* infants, abuse of; infanticide
psychopathology, conceptions of: clinical, 18, 41–43, 51–53, 56–59, 61–62, 64–65, 71–75, 114, 124–126, 282–283; similarities between, 68, 73
psychopathology, evolutionary conception of, 13–18, 40–41, 43–44, 50–54, 56–61, 64, 71–75, 124–126, 129–133, 208–209; criticisms and limitations of, 133–136
psychopathology in Archaic humans: with protoculture and language, 335–340; with protoculture and protolanguage, 332–334, 339–340
psychopathology in hominids, 57, 222, 228–233, 241–242, 257–258, 306; causes of, 223–225; factors affecting, 232–233, 309; hypothetical scenario H1, 59–62; hypothetical scenario H2, 62–65; hypothetical scenario H3, 65–69; hypothetical scenario H4, 69–71; response of group mates to, 61, 63–64, 67–68, 70, 71, 72
psychosis, 80, 82, 282, 283, 298, 340
psychosomatic ailments, 93–94

rank, 33, 81, 178, 196, 285, 362–363; and psychopathology, 34, 64, 80, 249, 349, 353
reality: sense of, 281–283, 318, 335; social, 351, 353
reconciliation behavior, 147
Reite, M., 198
Relational Model of hostility, 147
relations between the sexes in hominids, 233–236, 316
religion, 67, 69, 298, 299–300, 305–306, 310, 364
repression, 303–304
Richards, Graham, 244–250, 270, 288, 355
rituals, 22, 67, 275, 278, 279, 316, 323, 331, 335, 337; among Archaic humans, 69–70, 71, 364; and dissociation, 298, 299, 301, 303, 309–311, 333
Russon, A. E., 56, 227–228

Sadler, J. Z., 128
Savage-Rumbaugh, E. S., 284
scapegoating, 176
schizophrenia, 8–9, 77, 82, 135
security, threats to. *See* threat defense or general defense system
self-awareness, 63, 246–250, 284, 296, 306, 354, 355, 356
self-conception, 63, 88, 252–253, 256, 282, 308

self-other distinction, 286
sex and gender, 347–348
sex differences, 28, 233–234, 351
sexual conflicts, 233–236, 236–240
sexual selection, 115, 119, 121, 135, 174, 234
shame, 122, 284–287, 289–291
sibling relationships, 64, 196
sickness and healing, 42, 47; among chimpanzees, 202–206; among hominids, 70, 72, 136, 219–220, 263–264, 305–306
Smuts, B. B., 57
social behavior, 215, 353; in chimpanzees, 201–202; evolution of, 98–99, 182–183, 316; human, 276–277, 336, 364; species differences in, 147–148
social behavioral pathology, 145–146
social competition hypothesis, 116
social conflict, 307, 312, 333
social contracts, 66, 69–70, 279
social deprivation, 181–184
social ecology, 5–6, 44–45, 214–215, 216–222
social ecology of hominids, 46, 50, 216–222, 223, 242, 358; Archaic and early anatomically modern humans, 71, 220–221; earliest, 217–218; Homo erectus, 218–220
social economy, 71, 93, 322, 362
social emotions, 284–293; and culture, 284, 286–289, 290
socialization, 47, 51
social mutualism, 287
social organization, 221–222, 315–316
social solidarity, 280, 349
social status, 284, 285, 291
social support, 62, 64, 67, 68, 72, 196, 225, 357–358
Sovner, R., 230
spacing disorder, 34
species barrier, 40–42
speciesism, 13, 125
species selectionism, 175
speech, 325–326
Standard Social Science Model (SSSM), 276
Stevens, Anthony, 8, 32–34, 77–78, 80, 112–113
stress, 12, 33, 79, 81, 85–86, 87–88, 113–114, 301; in Archaic humans, 333, 339, 340; and dissociation, 296–299, 302–304, 307, 310–313, 360–361, 365;
in hominids, 222, 224, 229, 233, 234, 236, 242, 360; in nonhuman primates, 104, 167, 168, 175, 185–186, 302
structuralist biology, 105–106
Strum, S. C., 156–157, 231–232
Suomi, S. J., 185
surgical ablation, 180
survival and reproduction, 11, 16, 47, 48, 51–52, 54, 55, 58, 61, 345; among hominids, 348, 349, 351, 353, 356, 357
survival of the fittest, 41
symbolic culture, 274–281, 286–287, 292, 295; and dissociation, 295, 297, 310, 311; Foley's view of, 275–276, 277; Knight's view of, 277–278, 279; Tooby and Cosmides' view of, 276–277
symbolic idiom, 362
symbolic niche, 257, 284, 287, 360
symbolic reference, 274, 309, 355
symbolization, 208–209, 300, 302, 305, 307–309, 317, 327, 334; in hominids, 23, 62, 65–67, 355
syntax, 66, 314–315, 317, 324, 335, 337–338, 363. *See also* grammar

technology, 315, 322
temperament, 27–31, 81, 328–330
territoriality, 175
thought and language, 326–328, 335, 337–338
threat defense or general defense system, 118–123, 129–135
time, notions of, 335, 355, 363, 364
Tomasello, M., 255–257
tonic immobility, 301–302, 306
Tooby, John, 276–277
tools, 217, 221, 320
traditions, 62, 136, 226–227, 254, 255, 278, 311, 349, 353
trance, 68, 294–295, 297–303, 305–307, 311–312, 339
trauma, 296, 299, 301, 302, 304–305, 333, 351, 361, 365
Troisi, A., 34–35, 77–78, 85, 112, 165–167, 197–199

Upper Paleolithic Era, culture of, 136, 273–274, 305–306, 315, 319–322, 334, 337, 339
ur-culture, 227

Valzelli, L., 88

van Praag, H. M., 9
violence, 77, 176–177, 236, 301. *See also* infants, abuse of; infanticide; aggression and hostility; sexual conflicts

Wenegrat, B., 37, 78
Wilson, E. O., 26
Wilson, M., 173

Wrangham, R. W., 57

Young, A., 127–128

Zihlman, A. L., 144
zoos, behavior of primates in, 149–154, 165–167

About the Author

Horacio Fábrega Jr., M.D., is a clinical psychiatrist and professor of psychiatry and anthropology at the University of Pittsburgh. He is the author and editor of several books and numerous articles, most recently *Evolution of Sickness and Healing*. His area of academic emphasis is on biological and cultural aspects of disease and behavior as well as on comparative systems of medicine.